Grasshoppers, Locusts, and Crickets of the World

Grasshoppers

Locusts, and Crickets
of the World

Edited by Martin Husemann
and Oliver Hawlitschek

Princeton University Press

Princeton and Oxford

CONTENTS

PREFACE   9

FOREWORD BY THE AUTHORS   11

CRICKETS, HOPPERS, LUBBERS: WHO IS WHO IN THE GRASSHOPPER WORLD?   13

1   EVOLUTION AND SYSTEMATICS   15
1.1   THE KINFOLK OF ORTHOPTERA   16
1.2   ANCIENT WORLDS: ORTHOPTERA IN ROCKS AND AMBER   21
       TAPHONOMY: HOW ORTHOPTERA BECOME FOSSILS   25
1.3   THE PHYLOGENETIC HISTORY OF ORTHOPTERA: FROM 1 TO 30,000 SPECIES IN 300 MILLION YEARS   26
1.4   CONVERGENT EVOLUTION IN BAND-WINGED GRASSHOPPERS   29
1.5   ISLANDS AS CRADLES OF SPECIES DIVERSITY: THE EXAMPLE OF HAWAIIAN CRICKETS   32
1.6   THE EVOLUTIONARY RADIATION OF CAMEL CRICKETS   36
1.7   PYGMY GRASSHOPPERS: TAXONOMY UNDER CONSTRUCTION   41

2   BIOLOGY AND ECOLOGY   45
       THE STAR PARADE: A PECULIAR FORM OF COURTSHIP IN GRASSHOPPERS   46
2.1   EAT AND BE EATEN: THE ROLE OF ORTHOPTERA IN FOOD WEBS   48
2.2   CRYPSIS, MASQUERADE, AND MIMICRY: ORTHOPTERA HIDING IN PLAIN SIGHT   55
2.3   ANT CRICKETS: A LIFE IN CHEMICAL DISGUISE   60
2.4   SECRETS OF BUSH-CRICKET MATING BEHAVIOR   62
2.5   RASPY CRICKETS AS POLLINATORS OF ORCHIDS   66

3   PESTS   71
3.1   LOCUSTS AS PESTS: PAST, PRESENT, AND FUTURE   72
       LOCUSTS AND HUMANITY THROUGH MILLENNIA   78
       MAIN LOCUST PEST SPECIES AND EXAMPLES OF OUTBREAK FREQUENCIES   80
       FROM UNKNOWN TO OUTBREAKING: THE QUAINT CASE OF *BARBITISTES VICETINUS*   82

4   SONG AND HEARING   85
4.1   MUSICIANS OF THE INSECT REALM: SOUND PRODUCTION BY ORTHOPTERA   86
4.2   A SOPHISTICATED AUDIENCE: THE SENSE OF HEARING IN CRICKETS AND THEIR KIN   90
4.3   MANY DIFFERENT STYLES OF SINGING IN CRICKETS, KATYDIDS, MOLE CRICKETS, AND GRIGS   94
4.4   FIELD GRASSHOPPERS, SOME OF THE MOST VERSATILE SINGERS AMONG ORTHOPTERA   98
4.5   DOES COMPLEX COURTSHIP PREVENT OR PROMOTE HYBRIDIZATION BETWEEN GRASSHOPPER SPECIES?   102

5	**THE DIVERSITY OF ORTHOPTERA AROUND THE WORLD**	107
5.1	AUSTRALIA AND PACIFIC	108
5.1.1	THE ASTOUNDING DIVERSITY OF AUSTRALIAN ORTHOPTERA	108
5.1.2	TWO CONTRASTING ENDEMIC GENERA OF AUSTRALIAN MOUNTAIN GRASSHOPPERS	115
5.1.3	SANDGROPERS: A UNIQUE GROUP OF UNDERGROUND ORTHOPTERA	119
5.1.4	SPUR-THROATED GRASSHOPPERS OF NEW ZEALAND'S MOUNTAINS	121
5.1.5	WETA AOTEAROA: DIVERSITY OF NEW ZEALAND'S ENDEMIC ANOSTOSTOMATIDAE	124
5.2	ASIA	129
5.2.1	GRASSHOPPERS OF THE VAST CENTRAL ASIAN STEPPES	129
5.2.2	SONG AND SIGNALING BEHAVIOR IN AN INDIAN WĒTĀ	133
5.2.3	LITTLE MONSTERS: SPLAY-FOOTED CRICKETS OF PAKISTAN	135
5.2.4	CAVES WITHOUT CAVE CRICKETS IN BHUTAN	137
5.2.5	SINGAPORE: A MICROCOSM OF HIGH ORTHOPTERA DIVERSITY IN A DENSE URBAN ENVIRONMENT	140
5.2.6	WHERE CAN'T YOU FIND THEM? ECOLOGY OF PYGMY GRASSHOPPERS IN SOUTHEAST ASIA	144
5.3	EUROPE	
	THE MEDITERRANEAN AND THE BALKANS: HOW GEOLOGY SHAPED BIOGEOGRAPHY	147
5.3.1	THE MEDITERRANEAN HOT SPOT I: THE BALKANO-ANATOLIAN REGION OF THE EAST	149
5.3.2	THE MEDITERRANEAN HOT SPOT II: MOUNTAIN RANGES OF THE WEST	155
5.3.3	THE MEDITERRANEAN HOT SPOT III: A PLETHORA OF ISLANDS AND ISLETS	160
	NATURALISTIC SERENDIPITY: THE DISCOVERY OF A NEW CRICKET VIA ENVIRONMENTAL BIOACOUSTICS	
	WHILE STUDYING PELAGIC BIRDS	164
5.3.4	UNRAVELING THE LIFE CYCLE OF THE ATLANTIC BEACH CRICKET	166
5.3.5	THE PALMENHAUS CRICKET, A MYSTERY LOST TO SCIENCE?	168
5.3.6	BEI-BIENKO'S PLUMP BUSH CRICKET, ONE OF EUROPE'S RAREST INSECTS	170
5.4	AFRICA	173
5.4.1	ORTHOPTERA OF THE SAHARA AND THEIR ADAPTATION TO DESERT LIFE	173
5.4.2	WEST AFRICAN GRASSHOPPERS: FOLLOWING THE RAINS	177
5.4.3	AFRICAN JUNGLES: A WHOLE WORLD OF UNDISCOVERED GRASSHOPPER DIVERSITY	180
5.4.4	AFRICAN GAUDY GRASSHOPPERS: PRETTY POISONOUS PESTS	185
	THE AFRICAN PAINTED GRASSHOPPER, A BEAUTIFUL PEST AND DANGER	189
5.4.5	THE BALLOON BUSH CRICKETS: MYSTERIOUS DENIZENS OF EAST AFRICAN FORESTS	190
5.4.6	STICKS AND STONES: FLIGHTLESS GRASSHOPPERS OF THE SOUTH AFRICAN VELD AND FYNBOS	194
5.5	AMERICAS	198
5.5.1	SKY ISLANDS: HOT SPOTS OF ENDEMIC GRASSHOPPER DIVERSITY IN THE AMERICAS	198
5.5.2	BIG AND SHOWY, BUT POORLY KNOWN: MEXICAN ORTHOPTERA	202
5.5.3	SOUTH AMERICA, A CRADLE OF ENDEMIC GRASSHOPPER RICHNESS	206
5.5.4	COLORFUL LUBBERS: THE DIVERSITY OF AMERICAN ROMALEIDAE	212
5.5.5	THE ENDEMIC JUMPING STICKS FROM MESO- AND SOUTH AMERICA	218
5.5.6	THE NEOTROPICAL MONKEY GRASSHOPPERS: SOME OF THE MOST COLORFUL INSECTS IN THE WORLD	221

6   RESEARCH AND RESOURCES   227
6.1   DIVERSITY IN BOXES: NATURAL HISTORY MUSEUMS AS ARCHIVES AND RESEARCH PLATFORMS OF ORTHOPTERA   228
       HOW IT ALL STARTED: THE LINNAEAN ORTHOPTERA COLLECTION   234
       A DYNASTY OF ORTHOPTERISTS: THE NATURALIS ORTHOPTERA COLLECTION   236
       ARTIFICIAL INTELLIGENCE, THE FUTURE OF ORTHOPTERA IDENTIFICATION?   238
6.2   ORTHOPTERA SPECIES FILE (OSF): THE TAXONOMIC DATABASE OF THE WORLD'S ORTHOPTERA   240
6.3   DNA BARCODING: A GENETIC TOOL FOR CATALOGING THE DIVERSITY OF ORTHOPTERA   243

7   CONSERVATION   247
7.1   CONSERVING ORTHOPTERA DIVERSITY: RESRACH AND MANAGEMENT   248
7.2   THE RESPONSES OF CENTRAL EUROPEAN ORTHOPTERA TO CLIMATE CHANGE   254
7.3   ALPINE GRASSHOPPERS OF THE MEDITERRANEAN: ISOLATED REFUGES SHRINKING FROM GLOBAL WARMING   258
7.4   THE SPECKLED BUZZING GRASSHOPPER –THREATENED AND EXTINCT DESPITE SUITABLE CLIMATE   263
7.5   THE ROCKY MOUNTAIN LOCUST: FROM MAGNIFICENT PROFUSION TO MYSTERIOUS EXTINCTION   265
7.6   BALANCING BIODIVERSITY: HABITAT MANAGEMENT AND GRASSHOPPER RESILIENCE IN INDIA'S PROTECTED AREAS   267

8   CULTURAL ASPECTS   271
8.1   CRICKET FIGHTING IN CHINA, A TRADITION MORE THAN TWO AND A HALF MILLENNIA OLD   272
8.2   THE GIANT HOODED KATYDID, A PET LIKE NO OTHER   274
       CARE OF *SILIQUOFERA GRANDIS*   275
8.3   ON THE HUNT FOR THE BEST PHOTO: ORTHOPTERA ECOTOURISM IN SOUTHEASTERN EUROPE   276
8.4   GRASSHOPPERS AS TRADITIONAL ROYAL FOOD AND MODERN PROTEIN SOURCE IN MADAGASCAR   280
8.5   SALT, LIME, AND *CHAPULINES:* THE REVIVAL OF AN ANCIENT MEXICAN CULINARY TRADITION   283
       RECIPE: GRILLED *CHAPULINES*   285
       CHAPULINES IN THE FLORENTINE CODEX   286

THE AUTHORS   289

GLOSSARY   293

REFERENCES   295

INDEX   311

PREFACE

It is not often that a treatise as comprehensive as *Grasshoppers, Locusts, and Crickets of the World* becomes available. Experts on Orthoptera from all over the world have provided novel insights into the evolution, systematics, biology, ecology, and bioacoustics of this group and present in-depth accounts of the diversity of Orthoptera adapted to a wide range of habitats throughout the world.

While some Orthoptera are pests, the vast majority are not, and together they play a critical part in the biodiversity of ecosystems. Consequently, it is only fitting that their conservation has been investigated thoroughly in the face of human activities, including climate change, especially since they are easily observable and so provide a ready measure of the effects of human activities.

Humankind has interacted with Orthoptera for millennia, and there are traditions in various countries with Orthoptera as a food source, as entertainment (cricket fighting in China), and even as pets!

I commend this most comprehensive work as being a valuable resource for all those who have an interest in Orthoptera – whether scientists working on members of this group, naturalists who recognize their importance in ecosystems and as indicators of human activities, or those who want a ready reference for this most interesting group of insects.

David Hunter
Orthopterists' Society President 2019–2023
Executive Director 2013–2019

FOREWORD BY THE AUTHORS

by Oliver Hawlitschek and Martin Husemann

A key moment in the career of (we suppose) everyone studying, working with, or interested in Orthoptera is when they realize how surprisingly poorly known they are. Orthoptera are widespread and often abundant; they can be observed in or near almost any place people live, from isolated forest refuges to the centers of sprawling cities. They are often conspicuous, as they sing and because many of them jump away instead of just hiding. From an entomologist's point of view, they are a "not very diverse" group of insects; the roughly 30,000 known species hardly match the hundreds of thousands of beetles, wasps, and flies. Yet, in many ecosystems, they represent a substantial part of the biomass. As such, Orthoptera make perfect study organisms for weekend naturalists and professional researchers alike. Nevertheless, every visit to a library, every search on the web, and every meeting with naturalists makes clear that Orthoptera are understudied and probably also underrated.

Other groups of insects have been studied by generations of specialists and hobbyists, mostly lepidopterists and coleopterists. And it is easy for laypersons to understand what the work of these experts is about: the formal term Coleoptera translates to "beetles," and the taxon Lepidoptera comprises butterflies and moths. These are the animals these experts work with. Orthoptera, on the other hand, consist of what we call grasshoppers, locusts, crickets, bush crickets, katydids, and some other, more obscure names. There is no English term that describes the same composition of animal groups as Orthoptera, and this may be a reason why the group remains so inaccessible to naturalists and scientific study. Then, again, the English names of groups within Orthoptera often raise negative associations. Locusts are known as agricultural pests and biblical plagues, crickets mostly as food for reptilian pets. There are a multitude of books describing the amazing world of butterflies, beetles, and bees, but hardly any on Orthoptera.

This book is intended to fill this gap. Just like other insects, Orthoptera have, over hundreds of millions of years, diversified into thousands of species that have adapted to countless different environments. They can be colorful, they produce amazing songs, and they show a range of remarkably complex behaviors. They are abundant, are comparatively easy to observe, and are good targets for nature photographers, both locally and in ecotourism. They make wonderful pets and outreach animals. They also offer much as research systems in physiology, ecology, and evolutionary biology:

they have a wide variety of sensory functions, they are central parts of food webs, and their giant genomes and hybrid complexes never stop puzzling geneticists.

As Orthoptera inhabit all continents except Antarctica, we invited orthopterists from around the world to fill this project with life. Our authors explain how Orthoptera evolved through geological history, how they live, how they die, and how diverse they are. This book presents stories about monkey hoppers, Cooloola monsters, king crickets, wetas, sandgropers, and their wondrous allies. Our hope is that this book will spark the fascination we have for Orthoptera in other naturalists (or in naturalists-to-be) and show researchers the potential of Orthoptera groups as study systems. Perhaps it will help some people who have never had much interest in the natural world to appreciate some of the huge variety of organisms with whom we share this planet. We will be particularly proud and honored if even researchers who spent their entire careers on the study of Orthoptera will learn a new fact or two from this book. We certainly did along the way.

Writing this book has been a collaborative effort of many people. We thank 72 authors from 25 countries around the world, as well as many photographers, for their invaluable contributions. We also thank the Orthopterists' Society and the Deutsche Gesellschaft für Orthopterologie for supporting this work.

Deutsche Gesellschaft für Orthopterologie

Orthopterists' Society

CRICKETS, HOPPERS, LUBBERS: WHO IS WHO IN THE GRASS-HOPPER WORLD?

by Martin Husemann and Oliver Hawlitschek

Orthoptera are not as diverse as other groups of insects, but still comprise thousands of different species. Despite their species-richness, all Orthoptera have some common features, mainly in the organization of the body. The most distinctive feature that orthopteran species share is the presence of a thickened, strong pair of jumping legs. Evolutionary biologists call this an autapomorphy, a character that evolved specifically within a certain group of organisms. But beyond that, members of Orthoptera express a staggering diversity in "looks," which may be why human observers gave them a variety of vernacular names.

Disregarding the notorious term "bug" as a generalizing and somewhat looked-down-upon synonym for all kinds of insects, there is no English term that really applies to all members of Orthoptera. Most people likely have never even heard the term Orthoptera. The most general term that is sometimes used in an attempt to include all Orthoptera is "hoppers." This can, however, be very misleading, as there are a variety of different "hoppers" among insects. "Leaf hoppers" is a name commonly used for cicadas, and "plant hoppers" describes a subgroup of cicadas, the Fulgoromorpha, neither of which are related to Orthoptera.

Within Orthoptera, the situation becomes more complicated. The main division recognized by biologists within the group is the one into the suborders Caelifera, the short-horned grasshoppers, and Ensifera. Ensifera has been translated as "long-horned grasshoppers," but in reality, these insects often do not match the common picture of a grasshopper. These two groups are so distinct that, while many studies indicate a common evolutionary origin, the theory that Caelifera and Ensifera are not closely related is still popular among scientists. The length and shape of the antennae is the most distinctive character of the two groups: short and thickened in Caelifera, long and threadlike in Ensifera. They differ further in their sound production and hearing.

Caelifera have an "ear," called a tympanum, on each side of the body, and the most common mode of sound production is rubbing the hind legs against the wing. In Ensifera, the hearing organs are located on the fore legs, and sounds are produced most often by rubbing the wings against each other. Finally, the two differ in the shape and size of the female egg-laying organ, the ovipositor, which is indistinct in Caelifera, but typically long and sword- or needle-shaped in Ensifera.

The English terms assigned to members of Caelifera and Ensifera are hardly congruent with these two major groups, and the situation is similar in other languages. Caelifera is the somewhat easier group of the two to describe and comprises most of the aforementioned "hoppers," as most of its species come close to what most people would picture as a grasshopper. However, depending on the definition, "grasshopper" may specifically apply only to the family of Acrididae, or even to its subfamily Gomphocerinae. Ground hoppers represent the family Tetrigidae, also called pygmy hoppers. Caelifera also comprise all "locusts," but locusts are not an evolutionary group: the name is used for an assemblage of migratory species belonging to different subfamilies of Acrididae. "Lubber" is the name of a single species of Romaleidae, *Romalea microptera*.

The variety in naming is even higher within Ensifera. Many people use the term "crickets" (or sometimes "cricket-like" insects). Yet, systematically, the term describes a single large superfamily, the Grylloidea. The "tree crickets" are a subfamily within the true crickets, the Oecanthinae, whereas the "mole crickets" are a closely related superfamily (Gryllotalpidae). "Pygmy mole crickets," or Tridactylidae, while similar to mole crickets in shape and lifestyle, are completely unrelated to all other "crickets"; instead, they are members of Caelifera. The name "bush crickets" is often applied to the family Tettigoniidae and comprises the majority of non-grylloid Ensifera, but may conversely also describe a specific subfamily within the Gryllidae, the Eneopterinae. Especially in North America, Tettigoniidae are also called "katydids," a name that is supposed to mimic the song of the common true katydid *Pterophylla camellifolia*.

This summary is far from conclusive. There are a plethora of other common names for which specific meanings often differ locally or regionally. Common names may be misleading, which is why it is important to also refer to the scientific names of organisms. In this book, we try to do both and provide common names as well as scientific ones to provide the background for the stories we want to tell.

Figure > Orthoptera have evolved to be among the dominant insects of some ecosystems. This photo shows males of *Euchorthippus declivus* (above) and *Omocestus rufipes* (below) sharing a perch on a meadow in eastern Central Europe. Photo: Oliver Hawlitschek.

1 EVOLUTION AND SYSTEMATICS

1.1   THE KINFOLK OF ORTHOPTERA

by Roberto Battiston, Paolo Fontana, and Bruno Massa

Within insects, Orthoptera belong to the larger group called Polyneoptera, also known as orthopteroid insects. It is one of the three primary subdivisions of neopteran insects whose emergence dates back to the Carboniferous Period. Though far less species-rich than the large holometabolous orders (those that undergo complete metamorphosis), Polyneoptera represent a diverse and fascinating group encompassing various orders with unique characteristics and remarkable adaptations. The exact phylogenetic relationships within Polyneoptera remain poorly resolved, and their branching patterns remain ambiguous.

At the current state of research, Polyneoptera comprise 10 orders of insects, one of which is Orthoptera. The others are described below.

Zoraptera are a tiny and poorly known insect order with 30 known living species that have been found in tropical regions of South Asia, West Africa, Oceania, and South America. All known species exhibit a unique dimorphism of winged and wingless forms. The winged individuals have pigmented bodies, compound eyes, and ocelli, while the more common wingless forms have pale bodies and no compound eyes or ocelli; hence, they are completely eyeless and have a distinctive, Y-shaped groove on their heads. Zoraptera have an elongated and flattened body, hind legs with thickened femora, and an abdomen consisting of 10 segments. They are primarily found underground, under decaying logs, in decomposing wood, and even in termite galleries. Their primary foods are fungi, but occasionally they consume nematodes, springtails, and mites.

Dermaptera, commonly known as earwigs, have a worldwide distribution, but the largest part of the 2,000 species are found in tropical regions. Earwigs have an elongated, flattened body, are typically dark brown or orange-brown, and have characteristic pincers or forceps-like cerci at the end of the abdomen; they are primarily nocturnal. Their legs are short and adapted for running, and in winged individuals, the wings are modified, with short tegmina on the fore wings and broader hind wings that fold under the fore wings. They exhibit a range of ecological roles and may feed on live plant tissues, damaging flowers and fruits, or participate in the decomposition of organic matter. Some species are solitary predators, while others have gregarious behavior.

Plecoptera, commonly known as stoneflies, comprise approximately 3,500 known species worldwide. Their nymphs primarily inhabit cool freshwater habitats, with a notable preference for mountainous regions and clean, oxygen-rich streams and rivers. They are sensitive to pollution and are often used as bioindicators of water quality. The distinctive features of stoneflies are two long tail filaments (cerci) and tough, flattened bodies. Adult stoneflies are poor fliers, and some species are even wingless, relying on their strong legs for movement. They play a crucial ecological role in freshwater ecosystems, serving as both predator and prey, and are an important food source for fish and other aquatic organisms.

Mantophasmatodea were newly described in 2002, which makes them by far the most recently discovered of all insect orders. They are known as "gladiators" because of their predatory habits, capturing prey between their fore and mid legs. The roughly 20 living species are wingless, resembling nymphs of mantises or stick insects, and are mainly found in arid savanna regions of southwestern and eastern Africa. Fossil evidence dates this order back to the Middle Jurassic, when it was much more species-rich than today. Females produce saclike egg cases containing 10 to 16 eggs, which can withstand several months of drought. Cannibalism has been observed among both nymphs and adults in captivity. They have specialized mechanoreceptors for detecting ground vibrations,

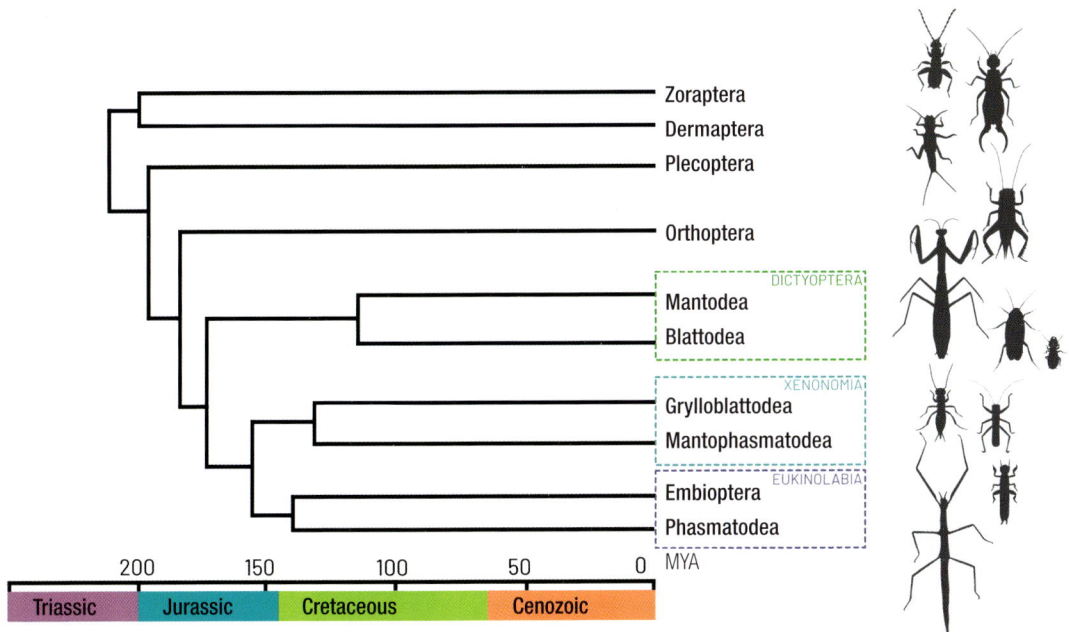

Figure 1 Phylogeny and divergence times based on large-scale transcriptomic data generated in the 1Kite Project. Orthoptera are recovered as sister taxon to a large clade comprising Xenonomia (= Grylloblattodea + Mantophasmatodea), Eukinolabia (= Embioptera + Phasmatodea), and the superorder Dictyoptera (= Mantodea + Blattodea, incl. termites), while Plecoptera and Zoraptera + Dermaptera diverged from them, likely in the late Triassic Period.

Figure 2 Some representatives of the polyne-opteran insect orders: (A) Zoraptera, *Zorotypus asymmetricus*, photo: Petr Kočárek; (B) Dermaptera, *Forficula apennina*, photo: Paolo Fontana; (C) Plecoptera, *Dinocras cepalotes,* photo: Roberto Battiston; (D) Mantophasmatodea, *Minutophasma richtersveldense*, photo: Benjamin Wipfler, Tobias Theska, Reinhard Predel (Wikimedia); (E) Grylloblattodea, *Grylloblatta,* photo: Alex Wild (Wikimedia); (F) Blattodea, *Periplaneta americana*, photo: Roberto Battiston; (G) Mantodea, *Geomantis larvoides*, photo: Roberto Battiston; (H) Embioptera, *Embia thyrrenica*, photo: Toni Puma; (I) Phasmida, *Clonopsis gallica*, photo: Roberto Battiston.

which they use to detect their prey, but also for communication by drumming on the substrate.

Grylloblattodea are an order of 33 species of wingless insects with elongated bodies and legs, somewhat resembling cockroaches but with a more slender form. They are found sporadically in northwestern North America and northeastern Asia. Their legs are adapted for moving on rocky, often icy surfaces, which is why they are sometimes called "rock/ice-crawlers". Grylloblattodea are typically active at night and have a diet that includes detritus and small invertebrates. They are considered a very primitive group of insects, possibly dating back to the Permian Period.

Blattodea are among the larger orders of Polyneoptera, comprising 7,600 species of cockroaches and termites ranging in size from 2 mm to 12 cm. From a phylogenetic point of view, termites are not their own insect order (called Isoptera until 2007), but just very unusual cockroaches. The order has an ancient history dating back to the Carboniferous Period and can today be found on all continents except Antarctica. The social organization varies from solitary in cockroaches to highly eusocial, with complex caste systems in termites. Reproductive castes include the royal couple, substitute royals, supplementary royals, and reproductive neotenics; but the majority of individuals belong to the sterile castes of soldiers, workers, and so-called pseudoergates. The abdomen of termite queens is usually extremely enlarged (physogastry) due to egg production.

Mantodea, known as mantises or praying mantises, are specialized predatory insects. Their 2,500 species, ranging in size from 1 to 17 cm, are found primarily in warm regions, displaying diverse forms, extraordinary morphotypes, and cryptic behaviors. Their most characteristic feature is the raptorial fore legs, which they use for grasping prey. Femora and tibiae are armed with spines and hooks and close like tongs. Further characters include the triangular, highly mobile head with large compound eyes able to offer stereoscopic vision, and an elongated prothorax, often with serrated margins or lateral expansions to mimic leaves. While some species are brachypterous or apterous, the wings are usually well-developed, are often used in mimicry or displays, and have conspicuous coloration. Mantises adapt to environments with striking camouflage that mimics flowers, barks, leaves, lichens, or mosses. Sexual cannibalism is common; males may copulate even if partially eaten.

Embioptera, commonly known as webspinners, are gregarious insects capable of constructing tubular galleries using silk secreted from their front tarsi. They are found in tropical zones and other warm-climate regions. There are approximately 250 known species worldwide, with another 750 identified but not yet described. Fossils date back to the Upper Permian, with many discoveries from Baltic amber in the Tertiary Period. Adult webspinners are small, rarely exceeding 2 cm in length. Webspinners exhibit remarkable sexual cannibalism despite otherwise being herbivorous. After fertilization, females lay eggs within the silk galleries, often attaching them to debris and silk, and exhibit parental care by guarding them.

Phasmida or Phasmatodea, commonly known as stick insects (or leaf insects in the case of species with broad and flattened bodies), are a group of 3,000 described, mainly tropical species of insects. They are characterized by elongated bodies and remarkable cryptic mimicry resembling twigs, leaves, or lichens. Stick insects display notable sexual dimorphism; females are typically larger and flightless, while males are often smaller and capable of flight. They reproduce primarily through sexual reproduction but also exhibit widespread parthenogenesis (asexual reproduction in which an egg can develop into an embryo without being fertilized by a sperm). Stick insects lay eggs that resemble seeds or excrements, aiding in their dispersion by ants. They are herbivorous, feeding on specific plant species, which they often imitate through their cryptic mimicry. These mainly nocturnal insects are masters of cryptic camouflage, remaining still during the day and using various behaviors to deter predators, such as simulating twigs or leaves and producing stridulating sounds or repellent substances. Phasmids also hold the record for being the longest among all insects (the genus *Phryganistria*, 62 cm).

1.2 ANCIENT WORLDS: ORTHOPTERA IN ROCKS AND AMBER

by Ulrich Kotthoff, Ole-Kristian Odin Schall, and Martin Husemann

The fossil record of Orthoptera (see the box titled Taphonomy) dates back around 300 million years into the late Carboniferous, the geological period in which the first forests developed and when several continents formed a large landmass around the South Pole. In the following Permian Period, most of Earth's continents drifted northward and joined to form one huge landmass, the so-called Pangea ("all earth"). This continental situation resulted in vast arid and semi-arid areas and the spread of plant taxa better adapted to dry conditions, such as conifers and other seed plants. Flowering plants (angiosperms), including grasses, had not yet developed at this time.

Genetic data, however, suggest that the common ancestor of all Orthoptera and closely related groups (see chapter 2.1, The kinfolk of Orthoptera) lived even earlier, around 350 MYA. One of these related groups, the Titanoptera (the name refers to the enormous size of some titanopterans), is sometimes even regarded as a subgroup of orthopterans and, indeed, shows many similarities. Among others, they had stridulation organs similar to those of Orthoptera. Findings of possible fossil insects from earlier than the Carboniferous are still under scientific debate; in any case, orthopterans or their earlier relatives, together with the relatives of dragonflies, damselflies, and cockroaches, belong to the more ancient insect groups. They are called hemimetabolous (characterized by incomplete metamorphosis) insects, compared with the so-called holometabolous insect orders that show a complete transformation from larva to imago, which appeared later during the dry conditions of the Permian and the following Triassic. But also the hemimetabolous orthopterans adapted well to these conditions. The transition from the Permian to the Triassic Period ended with the greatest mass extinction of all time, dubbed "the Great Dying." The following Mesozoic, comprising the Triassic, Jurassic, and Cretaceous Periods and starting at ca. 252 MYA, witnessed the separation of the main continents. During the Jurassic, the Atlantic Ocean opened, and geological processes led to global greenhouse conditions that became even more extreme during the Cretaceous. The Mesozoic not only witnessed the rise and spread of famous vertebrate groups, such as the dinosaurs, ichthyosaurs, pterosaurs, and mammals, but also the development and/or diversification of many highly successful insect groups, including numerous families of the Orthoptera.

The varying geological and climatic conditions of different eras also resulted in changes of fossilization. Fossil-lagerstätten (see box titled Taphonomy) yielding fossil insects do not occur along large time scales. Instead, there are certain short time intervals for which there are very important lagerstätten featuring a high diversity of insects, alternating with long intervals in Earth's history for which there are no fossil insects at all. It is therefore no surprise that the fossil record of Orthoptera is relatively scarce; only about 900 species from more than 500 genera are currently known. Even if specimens are found during excavation campaigns, they frequently remain unstudied. Often only wings are preserved, making the systematic position of many fossils difficult to determine. Nevertheless, the diversity of fossil Orthoptera is high, and most extant groups are represented. Also, several families are only known from the fossil record and are today extinct. Some of the more species-rich extinct families are the Locustopsidae of Caelifera, but also the Elcanidae, Haglidae, and Oedischiidae of Ensifera.

Three of the most remarkable lagerstätten with an abundance of insects are the Burmese amber biota of northern Myanmar (ca. 100 MYA), the Crato Formation of northeastern Brazil (ca. 122–113 MYA), and the Yixian Formation of northeastern China (ca. 130–125 MYA). All three lagerstätten are of a relatively similar Cretaceous age, yet the diversity of Orthoptera found in each is very different. Several families are known only from one of the locations (e.g., Locustopsidae from Crato, Haglidae from Yixian, and Burmecaelidae from Burmese amber). Burmese amber especially has revealed a stunning selection of Orthoptera diversity. The Burmecaelidae resemble Tridactyloidea in many ways, but some characters are clearly more similar to Tetrigoidea. Also, the diversity of the family Ripipterygidae, or mud crickets, is much higher in amber than today, with extant species only occurring in South and Central America. But the fossilized resin is not only an extraordinary sanctuary for phylogenetic diversity. The often outstanding state of preservation of the insects trapped inside it allows for ecological reconstructions of the long-gone inhabitants of the tropical forest. In Elcanidae, for example, a remarkable morphological diversity has been found in the large characteristic spurs located on the hind legs of these animals. Some are elongated and slightly curved (ensiform), others are leaf-shaped, and some feature rows of minute teeth along their margins. Completely straight spines occurred as well. The spurs of Elcanidae are thought to have been used to maneuver on the water's surface, similar to some modern tridactylids. However, why there were so many different spur types present in Burmese amber Elcanidae is a question yet to be answered and one sure to come up alongside many others in future studies that try to reveal the mysteries of the world of the dinosaur-age grasshoppers from the Mesozoic.

After the end of the Cretaceous, ca. 66 MYA, during the still-ongoing Cenozoic Era, the Elcanidae and other Mesozoic lineages became extinct, but members of the still successful Ensifera and Caelifera survived and remained relatively diverse compared with other hemimetabolous insects. The fossil record of the Danish Fur Formation (ca. 50 MYA) revealed a variety of Ensifera and Caelifera in different sizes. Numerous orthopteran inclusions are also known from the slightly younger Baltic amber and other European amber deposits, as well as from lake sediments. During the later Cenozoic, the worldwide trend toward a cooler and dryer climate, due to the isolation of Antarctica and silicate weathering, led to the expansion of grasslands. Among the Orthoptera, several groups have adapted particularly to such ecosystems. In several fossil-lagerstätten – for example, Öhningen – members of groups preferring open habitats, such as the Oedipodinae, are widespread. In other lake deposits from the middle to late Cenozoic, orthopterans are surprisingly rare. Of course, not all orthopterans adapted to cooler conditions, since substantial areas of tropical ecosystems, though declining in size during the later Ceno-

zoic, remained close to the equator. Dominican Amber (around 15 million years old) and Malagasy Copal (a few thousand years old) are among the lagerstätten reflecting tropical or subtropical orthopterans. Findings from Dominican Amber comprise Caelifera, including pygmy grasshoppers, and Ensifera, including crickets. During the past million years, the fossil record of insects, including orthopterans, becomes particularly patchy in the mid to high latitudes due to further climatic cooling and the regularly occurring glacial periods since around 2.5 MYA (the Quaternary). Yet there are also examples of fossil orthopterans from younger lagerstätten – for example, members of the Oedipodinae from the fossil Willershausen lake in Germany, members of Melanoplinae from the La Brea Tar Pits in California, and various findings from South America.

IMPACT OF MASS EXTINCTIONS ON ORTHOPTERANS AND OUTLOOK

The impact of mass extinctions on orthopterans and insects in general appears to be weaker than on other animal groups. The most significant extinction at the end of the Permian, caused by huge volcanic activity in the area of present-day Siberia, probably also affected insects. Several ancient orthopteran lineages, such as the Permelcanidae and orthopteran sister groups, went extinct, but it is difficult to determine whether these extinctions were directly related to Siberian volcanism or happened gradually over a longer period of time. The Cretaceous-Paleogene extinction, which killed off the dinosaurs, seems to have had an impact on certain previously successful orthopteran groups, such as the Elcanidae and the Locustopsidae. Considering that the overall climate conditions during the early Paleogene were similar to those of the Cretaceous, it is possible that the meteorite impact at the Yucatán Peninsula (and perhaps the volcanism of the Deccan Traps in South Asia) had direct influence on these groups.

Today orthopterans are threatened by the very different danger of anthropogenic influence (see chapter 7.1, Conserving Orthoptera diversity). While they have shown a significant adaptability, the questions remain if modern orthopteran species and families can adapt to the multiple hazards caused by humans, such as alteration, isolation, and destruction of ecosystems, use of herbicides and insecticides, and climate change, as even previously very common species, such as the Rocky Mountain locust have already gone extinct (see chapter 7.5, The Rocky Mountain locust).

Figure 1 Orthoptera in Cretaceous Burmese amber. A Burmecaelidae, B Elcanidae, C Tridactyloidea. Photos: A Zhendong Lian, B, C Ole-Kristian O. Schall.

Figure 2 The fore wing of a Titanoptera from the Triassic of Australia. Image of Titanoptera kindly provided for this publication by Dr. Patrick M. Smith (Australian Museum).

Figure 2b Fossil wing of an orthopteran, around 3 MYA, from the Willershausen site in Germany. Photo: Ulrich Kotthoff.

TAPHONOMY: HOW ORTHOPTERA BECOME FOSSILS

by Ulrich Kotthoff, Ole-Kristian Odin Schall, and Martin Husemann

While they also occur in sinter, bitumen, and glacial deposits, fossil orthopterans and insects in general are particularly often preserved in perimarine deposits, lacustrine shales, and "fossilized" resin, generally called amber. A site with an extraordinary concentration of very well preserved fossils is called a lagerstätte (after the German word for "storage site") or fossil-lagerstätte. While most often only wings are preserved, even behavior can be assessed from the fossil record in some cases – for example, via feeding traces, morphology, and the insect's situation at the time of embedding. Recently, even million-years-old orthopteran eggs have been found. Of course, the chance of an insect being fossilized also depends on its behavior. For example, few insects falling into a lake become fossils: they would try to reach the shore by swimming or – in the case of small insects – by walking on the water surface. On the other hand, surface water tension can trap winged insects. The surface tension often causes insects to remain at the water surface for so long that they partly degrade before sinking to the bottom. Of course, predators may also damage or eat insects; interestingly, fish excrements may thus also be a way to find insects in the fossil record. For all these reasons, specimens in aquatic deposits are often only partially conserved.

Insect inclusions in amber, on the other hand, are generally conserved very quickly. They are mostly complete and three-dimensionally preserved, and they allow us to deduce what a specimen did before embedding. However, small specimens have a higher chance of becoming trapped than others and are therefore overrepresented. In addition, amber deposits are related to the presence of certain tree types producing substantial amounts of suitable resin; hence, the surrounding fauna is adapted to this environment. These examples show that fossil-lagerstätten cannot reflect general diversity, but are linked to certain ecosystems and many other factors. In the case of orthopterans, this could mean that the fossil record particularly reflects species living in lake-related or forest-related ecosystems.

1.3 THE PHYLOGENETIC HISTORY OF ORTHOPTERA: FROM 1 TO 30,000 SPECIES IN 300 MILLION YEARS

by Oliver Hawlitschek, Hojun Song, and Martin Husemann

The currently known 30,000 species of Orthoptera are the result of a very long evolutionary time period: the oldest known fossils of Orthoptera date back to about 300 MYA (see chapter 1.2, Ancient worlds). Understanding their relationships and major evolutionary steps is the aim of phylogenetic research. For a long time, phylogenetic research on Orthoptera relied only on morphological and bioacoustics traits, but more recently, the study of genetic and genomic data has helped illuminate the origin and evolution of this group. While the monophyly – that is, the common evolutionary origin – of most families has been relatively well established by now, the relationships within many families remain largely unresolved. Today, sampling of the highly diverse clades is a major limiting factor to phylogenetic studies, and many changes may still be expected in this dynamic field.

The first well-supported split in the evolution of Orthoptera is the division into the short-horned grasshoppers (Caelifera suborder) and the long-horned bush crickets and crickets (Ensifera). This is confirmed by all genetic and morphological analyses: good apomorphic (or diagnostic) traits – such as the structure and shape of the antennae, the stridulation and hearing organs, and the sexual organs – define the groups. Within the Ensifera, the crickets, mole crickets, and ant crickets form a well-supported group. This group is sister to another large group of the remaining Ensifera. The cave or camel crickets (Rhaphidophoridae) (see chapter 1.6, The evolutionary radiation of camel crickets) are basal here. The position of the Schizodactylidae, or splay-footed crickets, is not entirely clear; they have been grouped either within the aforementioned

clade, with crickets, or outside as close relatives of the cave crickets and bush crickets (Tettigoniidae). The Gryllacrididae, Anostostomatidae, and Stenopelmatidae form another closely related group within this cluster.

Study of the relationships within Caelifera are complicated by conflicts between different types of genetic and genomic data; they are in many ways still not fully resolved and remain a matter of debate. It seems certain, though, that a group consisting of the Tridactylidae (the pygmy mole crickets), the Cylindrachetidae (the sandgropers; see chapter 5.1.3, Sandgropers), the Ripipterygidae (the mud crickets), and the Tetrigidae (the pygmy grasshoppers; see chapter 1.7, Pygmy grasshoppers) take positions as sister lineages to all other Caelifera. Yet the position of the strange-looking Australian endemic Cylindrachetidae, as well as the South African Pneumoridae, remain questionable. Similarly, the positions of most of the Caeliferan families have not been entirely clarified yet, and even superfamily composition remains a matter of debate.

Within the Caelifera, the Acrididae are the most diverse group. Phlyogenetic studies of Acrididae have been considered notoriously difficult. Many analyses found established subfamilies to be likely not monophyletic. Similar problems have been found at the levels of genera and species, which is why some scientists have preferred to speak of "fuzzy sets" rather than actual taxa. On one hand, this may be a result of the relatively young evolutionary age and genetic similarity of many clades. On the other hand, large-scale convergent evolution of traditionally used morphological characters, like wing shape,

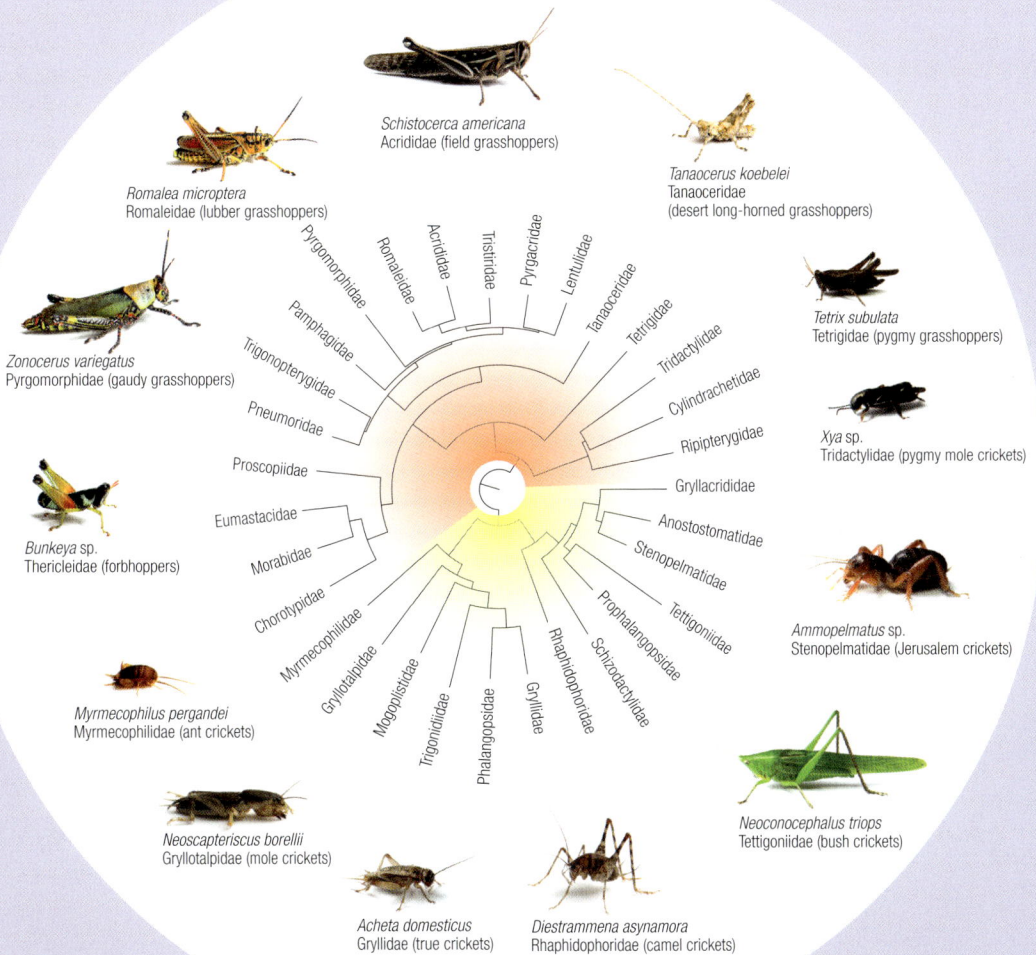

Figure 1 A phylogenetic tree of Caelifera (red) and Ensifera (yellow), the two suborders of Orthoptera, based on genomic data. The figure is modified from a tree published in Shin et al. (2024). For simplicity, only the most species-rich families are shown. The photographs surrounding the tree (by Brandon Woo, with permission) exemplify the diversity of Orthoptera. Despite their positions in this tree, the true relationships between these groups are not yet fully clarified, and future studies may change details of our phylogenetic hypotheses.

can be observed. Similar phenotypes are repeated in multiple groups across the Americas and the Palearctic. This pattern is further complicated by biogeography: many recent studies indicate multiple events of lineages crossing the Atlantic Ocean, resulting in frequent faunal exchange between the Americas and Eurasia plus Africa.

Genomics have been able to solve similar problems in the phylogenetic study of other insect groups, but these studies are impeded by the extremely large genomes of Orthoptera. Acrididae, and specifically the Oedipodinae, are known to have giant genomes with sizes almost an order of magnitude larger than that of humans. While all Orthoptera studied so far have larger genomes than the typical insect size of a few hundred thousand base pairs, the overall range is also enormous, spanning from fewer than 1 million base pairs in some crickets to more than 23 million in the speckled buzzing grasshopper *Bryodemella tuberculata*. The reasons for the existence of such large genomes remain unknown. They have not derived from chromosome duplication or polyploidization, as the number of chromosomes is relatively stable at 11 pairs of chromosomes plus sex chromosomes in most lineages. European Gomphocerinae even have a reduced set of 8 pairs of chromosomes, but no reduction in genome size. Future projects sequencing full genomes will help to elucidate the phylogeny of Orthopterans and help us to understand why genomes became so large.

The timing of the evolution is another field of study in which many new results can still be expected. The current estimates of the origin of Orthoptera at about 300 MYA have been derived from studying a framework of all insects and arthropod groups and are likely rather solid. The origins of clades within Orthoptera are less well known. Dating based on molecular data relies on many assumptions and is often conducted with a calibration by fossils. However, the fossil record is limited, and many fossil taxa are difficult to assign to nodes in the tree (see chapter 1.2, Ancient worlds). A better understanding and more studies on these fossils, as well as new robust phylogenies with a broad taxon sampling and based on large genomic datasets, will help to further resolve the relationships of the different orthopteran taxa and help us to better understand their evolution.

Figure 2 (A) The three-pulsed song of *Pterophylla camellifolia*, the 'true katydid' from Southern North America, is the source of the English term 'katydid' that applies to all bush-crickets. (B) Not all grasshoppers live on grass. Band-winged grasshoppers (see opposite page), such as this member of the genus *Thalpomena*, are perfectly adapted to rocky surfaces. Photos: (A) Delise and Matthew Priebe, (B) Martin Husemann.

1.4 CONVERGENT EVOLUTION IN BAND-WINGED GRASSHOPPERS

by Lara-Sophie Dey and Martin Husemann

Band-winged grasshoppers can be found in most arid and semi-arid areas around the world. They comprise more than 700 species whose outward appearance is very similar. Their diversity is hidden: the hind wings, covered when the animal is resting and only shown in flight or, in some species, during courtship, display a wide variety of patterns and colors all over the rainbow. This hidden diversity inspired us to learn more about how these species evolved.

Typically, we expect that species that are closely related with each other also live in geographical proximity. In band-winged grasshoppers, interestingly, we find species with similar hind-wing patterns in very different regions of the world (fig. 1). For example, *Trachyrhachys kiowa* from the USA and *Mioscirtus wagneri* from southwest Asia share a very similar morphology; their hind wings are bright yellow with a broad dark band, and their body shape is very similar.

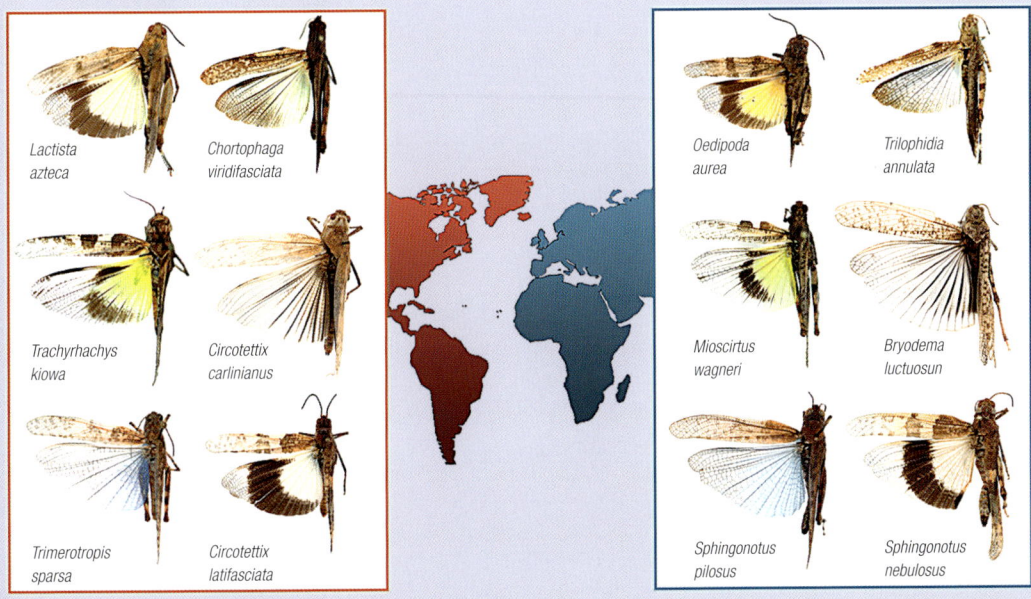

Lactista azteca

Chortophaga viridifasciata

Trachyrhachys kiowa

Circotettix carlinianus

Trimerotropis sparsa

Circotettix latifasciata

Oedipoda aurea

Trilophidia annulata

Mioscirtus wagneri

Bryodema luctuosun

Sphingonotus pilosus

Sphingonotus nebulosus

Figure 1 Similar morphology in unrelated species from the Americas and Eurasia and Africa. Figure: Lara-Sophie Dey.

But if *T. kiowa* and *M. wagneri* are closely related because they are so similar, how is it possible for them to live so far from each other? One possible answer: at some point some millions of years ago, the ancestor of both species traveled to a faraway land, crossing the Pacific Ocean either from Asia to the Americas or in the other direction. The two isolated populations then, over time, evolved into two separate species. This would have been a long journey for a small insect, and therefore unlikely; but it would not have been impossible, as band-winged grasshoppers are capable fliers.

However, there is an alternative explanation based on the concept of convergent evolution. Convergent evolution in two species means that the two are similar in morphology, behavior, or ecology, but are not closely related; the traits that make them similar to a human observer have evolved independently instead of being inherited from a common ancestor. This is often the result of similar ecological and selective pressures. Bats and birds are a common example: the fore limbs of both groups function as wings for active flight, but the two groups are not closely related, and both derived from ancestors that did not have wings. If the wing patterns of species pairs of band-winged grasshoppers had evolved convergently in different parts of the planet, this would not require the assumption of many unlikely transoceanic journeys in the past.

We found the answer to this question in the genes. The genes of any organism hold the information about the evolution of its species and also on its phylogenetic position; that is, its evolutionary relationships with other species. If the similar hindwing patterns are the result of common ancestry and close phylogenetic relatedness, we expect species with similar patterns to group together in the phylogenetic tree. However, if species group by their geographic origin, we can infer convergent evolution of these patterns. But some more analyses are needed to quantify the similarities, such as fine measurements of their morphology (shape) beyond superficial perception. These can be compared with the environmental conditions they live in and may provide some insights on which selective pressures have determined the evolution of these species.

Our reconstruction of the phylogenetic relationships in figure 2 provides a clear answer: we see two large groups according to geographic regions, Eurasia and Africa versus the Americas. Species with similar wing color and body shape were not found to be closely related but occurred in both of these groups.

Summing up, we asked ourselves, what are the factors that caused this convergent evolution? When extrapolating the distributions of species pairs with mathematical models, we found in many cases that both species within a pair prefer very similar climatic conditions. But how might climate relate to wing color? The truth is that we don't know. What we do know – though not in detail for most species – is that the conspicuous hind wings play a role in avoiding predators and signaling to potential mates. We may speculate that certain patterns are particularly efficient under the specific visual conditions of certain habitat types relating to climate zones, but we do not have any data or evidence to support this claim.

Much remains to be learned about band-winged grasshoppers. Their evolutionary complexity goes way beyond the convergence of morphology, as they have the largest genomes of all insects, some of the most complex patterns of biogeography, and a diverse range of reproductive isolation mechanisms. We expect many more interesting insights into speciation from them in the future.

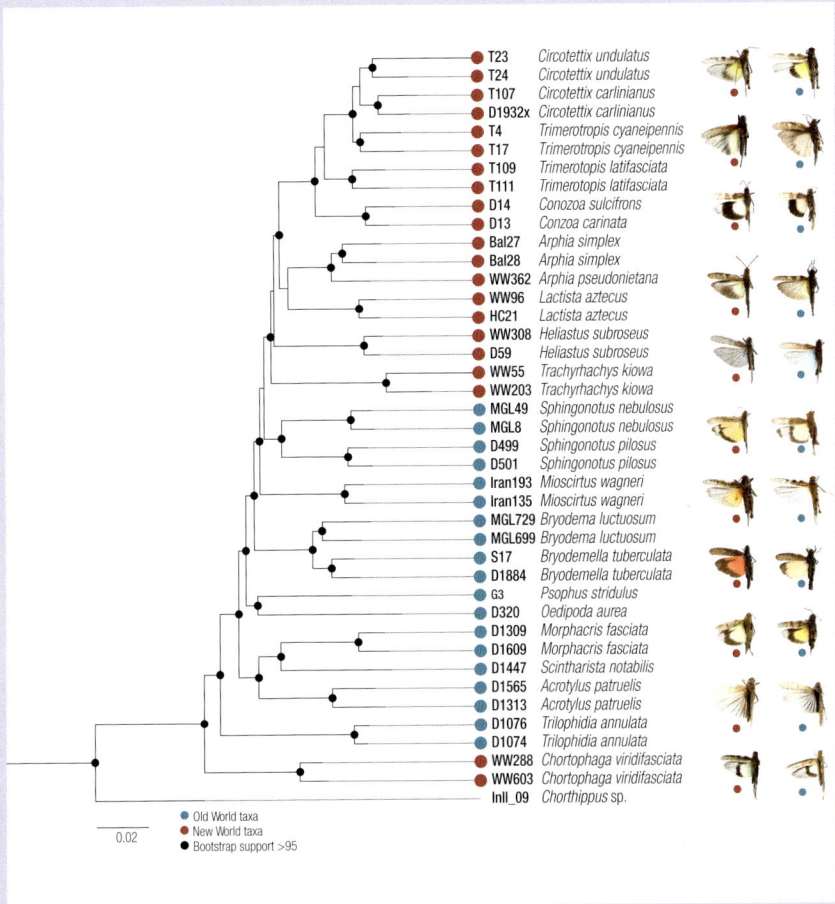

Figure 2 Phylogenetic tree based on species from the American and Eurasian African continents. A clear separation by geography is visible, while the species show high similarity in their morphology. Figure: Lara-Sophie Dey.

1.5   ISLANDS AS CRADLES OF SPECIES DIVERSITY: THE EXAMPLE OF HAWAIIAN CRICKETS

by Kerry L. Shaw

Hawaiian swordtail crickets (subfamily Trigonidiinae) are the most diverse of the native Hawaiian crickets, with over 170 species currently described. They are currently grouped into three genera (*Trigonidium*, *Prolaupala*, and *Laupala*). However, it was not until Otte's (1994) extraordinary taxonomic efforts in Hawaii that the extent of this diversity became known.

Among the Hawaiian swordtail crickets are many consummate examples of cryptic species (fig.1) – that is, pairs or clusters of species within closely related groups that look very similar. The most well-known example of cryptic species are members of the genus *Laupala* (fig.1), within which currently 38 species are recognized. Species of *Laupala* are small (~ 1 cm in length), range in color from nearly black to gold, sport a "smile" on the face (fig.1), and present a distinctive banding pattern on the hind femur (fig.2). Their small size and inconspicuous color keep them well-hidden in the foliage near to the ground. Prior to Otte's work, this group was formally recognized as only a single species. Otte (1994) recognized the extensive acoustic diversity within these cricket communities, indicating the presence of multiple species, despite the fact that the insects were nearly identical in appearance. A nature enthusiast can easily hear this acoustic diversity by visiting mid-elevation rain forests in Hawaii.

Although little work has been done on most of the Hawaiian swordtail diversity, studies of *Laupala* have revealed that speciation (the process by which new species evolve) has been extremely rapid in this group. Using DNA sampling of the different named species in the genus, an evolutionary tree was constructed that revealed three important features. First, the different named species were found to be genetically unique. This discovery means that, despite overall similarity in appearance, species that sing differently are in fact distinct groups of organisms that ordinarily do not interbreed. Second, these acoustically and genetically distinct species were found to be restricted to single islands of the archipelago. And third, species within a single island were found to be most closely related to other species restricted to that same island. This means that an ancestral species that arrived on a given island in the geologic past evolved into multiple species within that island. Together with knowledge of the well-studied geologic history of Hawaii, these patterns allow us to put an upper age limit on the time it took for species to evolve. It turned out that the speciation rate of *Laupala* is the fastest known among all invertebrates. For example, six species evolved from a common ancestor on the Big Island of Hawaii in the last 400,000 years. This represents a speciation rate that is four times faster than the average rate of speciation in other insects.

One naturally wonders if the evolution of the acoustic behaviors of *Laupala* crickets influences their rate of speciation. As with most crickets, males produce songs with the fore wings, using a specialized wing vein that, when struck, causes the rest of the wing to resonate (see chapter 4.1, Musicians of the insect realm). In Hawaii, *Laupala* crickets make such sounds in regular, slow rhythms. The song structure is a simple "tink...tink...tink...tink...", etc. In some places, males number in the thousands, and a pulsating tinkling can be heard for most of the day, giving the impression of sleigh bells. The rate of the tinkling varies little within a species, but there are recognizable differences between species (e.g., two pulses per second versus four pulses per second). One can usually hear two species of *Laupala* singing in the

Figure 1 A comparison of Big Island species *Laupala kona* (above) and *L. cerasina* (left) exemplifies the exceptional morphological similarity among different species. Both photos display the long, fusiform antennae of these species. A red face band can be seen in the *L. cerasina* image. Photo: Kerry Shaw.

same location; however, some localities on the Big Island of Hawaii and in the Koolau Range of Oahu have three or possibly even four species singing within the same forest.

Males produce these sounds to announce their location and interest in mating. Females do not make sounds, but display an eagerness in locating the source of these sounds and walk toward them. The result, upon contact, is the meeting of potential mates, with courtship and mating to follow. Songs of different species of *Laupala* living together always differ in pulse rate, and tests of female acoustic preferences show that females prefer the songs of their own species. These songs, and song preferences, function to bring these small crickets together within the expansive forest in order to reproduce. It may be that different species living together prosper by filling out the acoustic space, thereby reducing acoustic interference with one another. Thus, the acoustic system functions to bring together individuals that are reproductively compatible.

Acoustic behavior and its evolution may not, however, explain everything. The above scenario begs the question, if two species with similar songs evolve away from each other to reduce acoustic interference, what makes them different species in the first place? Other aspects of Hawaiian cricket reproductive behavior may contribute to the evolution of different lineages before new species come together. Different species of *Laupala* have characteristic differences in the surface chemicals on their bodies (called cuticular hydrocarbons) that we hypothesize are involved in chemical communication, mediated through the use of their long, threadlike antennae (fig. 1). In addition, strong reproductive competition exists among males to fertilize a female's eggs, as females are known to mate with many different males during their reproductive lifetimes. When mating, *Laupala* crickets spend all day in courtship. During courtship, males repeatedly provide small, spermless nuptial gifts over the course of the day to their female mating partner before transmitting a spermatophore, and evidence suggests that this behavior enhances the male's chances of paternity (see chapter 2.4, Secrets of bush-cricket mating behavior). These reproductive features may be under intense selective pressure to change according to what brings about the best success in mating. If this competition plays out differently in different populations, it could form the initial basis of population change.

Another factor to consider is the landscape in which this evolution occurs. Like all native crickets, *Laupala* are flightless and likely do not disperse very far during their lifetimes. *Laupala* are found in both very young parts of Hawaii (forests less than 1,000 years old) and very old parts (millions of years old). In this varied landscape, young habitat is often naturally fragmented by lava flows, and old habitat is divided by erosion and valley formation. The flightless lifestyle, coupled with the dissected landscape, can explain the deep genetic breaks observed both within species and between species. Thus, there are many independent geographic "pockets" where the above-mentioned reproductive competition might play out and in which new successful strategies could evolve. When formerly separated populations meet again in the same geographic space, new interactions between them may arise, causing differences in the acoustic system. Further research is necessary to test this idea.

The study of these little creatures has much to teach us about how and why life on Earth is so diverse. Studying evolutionary history alongside animal behavior has revealed deep insights into how new species form. Many questions are still unanswered, and opportunities remain for further insights, as many species have never been studied in detail.

Figure 2 A male (above) and female (left) cricket belonging to the species *Laupala pruna*. Note the banded pattern on the hind leg, characteristic of all *Laupala* species. Photo: Kerry Shaw.

1.6 THE EVOLUTIONARY RADIATION OF CAMEL CRICKETS

by Mary Morgan-Richards, Danilo Hegg, Cheten Dorji, and Steven A. Trewick

Living around the world are 863 described species of camel cricket (Rhaphidophoridae) divided among nine subfamilies. Many of these camel cricket species are specialized to living in caves, which explains one of their common names: cave crickets (fig. 1a). They all have long legs, long antennae, and long, compressed tarsi without pads. They all lack wings and auditory organs. Adult females lay eggs into soil or soft wood using a scimitar-like ovipositor (fig. 1b,c). During the day, they can be found in groups, hiding in large, dark, cool spaces created by underground caves. These crickets scavenge on debris that is washed into caves, or they come out at night to nibble on fungi, plants, and dead animals (fig. 1d–g). Species within the subfamilies Dolichopodinae and Troglophilinae are found only in the Mediterranean region; Rhaphidophorinae representatives live in southeast Asia, and members of Ceuthophilinae, Gammarotettiginae, and Tropidischiinae are restricted to North America (fig. 2).

All 52 species within European Dolichopodinae are found exclusively in caves, but other members of the family are more versatile. One Aemodogryllinae species has traveled with humans from Asia to greenhouses in Europe and is now common in homes in eastern North America (the greenhouse camel cricket *Tachycines asynamorus*). This invader is an omnivorous scavenger and has been reported foraging on living and dead plant tissue and dead insects. Other camel cricket species are familar to humans because they use cool damp sheds, cellars, and outside toilets; Japan, Australia, and New Zealand all have endemic toilet Rhaphidophoridae. In Australia, the species *Parvotettix domesticus* was initially recorded only from houses, and in southern New Zealand, *Pleioplectron simplex* is common under buildings and inside cellars.

The subfamily Macropathinae is found only in the Southern Hemisphere: in South Africa, South America, Australia, and Aotearoa/New Zealand, and on many oceanic islands. In total, 108 species circle the south. For example, *Heteromallus spina* inhabits Chile (fig. 3c), *Parudenus falklandicus* is found on the Falkland Islands in the South Atlantic Ocean, and *Notoplectron campbellense* lives on Campbell Island in the Southern Ocean. On subantarctic islands, these orthoptera are scavengers that share burrows with seabirds and use empty tuatara burrows for shelter. At least seven geologically relatively young oceanic islands that have never had direct contact with other land are home to one or more endemic camel cricket species (cave wētā). This distribution pattern on far-flung islands suggests that the Macropathinae are good at transoceanic dispersal despite all being without wings. Crossing the ocean is probably accomplished by rafting on forest flotsam. Eggs laid into wood might survive weeks of ocean travel. Many of the smaller Macropathinae species hide during the day in hollow branches and inside decaying wood – homes that are easily washed downstream and out to sea.

Figure 1 The majority of northern Rhaphidophoridae (camel cricket) species are specialized to living in caves, as illustrated by this group of *Miotopus richardsae* on a cave wall in New Zealand (A). Females lay eggs into soil or soft wood using a scimitar-like ovipositor, as demonstrated by *Talitropsis sedilotti* (B) and *Neonetus variegatus* (C). Rhaphidophoridae scavenge on debris that is washed into caves, and in forests, it can be found at night feeding on almost anything, as seen here: *Pleioplectron simplex* eating a snail (D), *Pleioplectron hudsoni* eating lichen (E), *Neonetus* n. sp1 eating a crane fly (F), and *Isoplectron parallum* eating a leaf (G). Photos: (A, B, D) Danilo Hegg; (C, E, F, G) Steven A. Trewick.

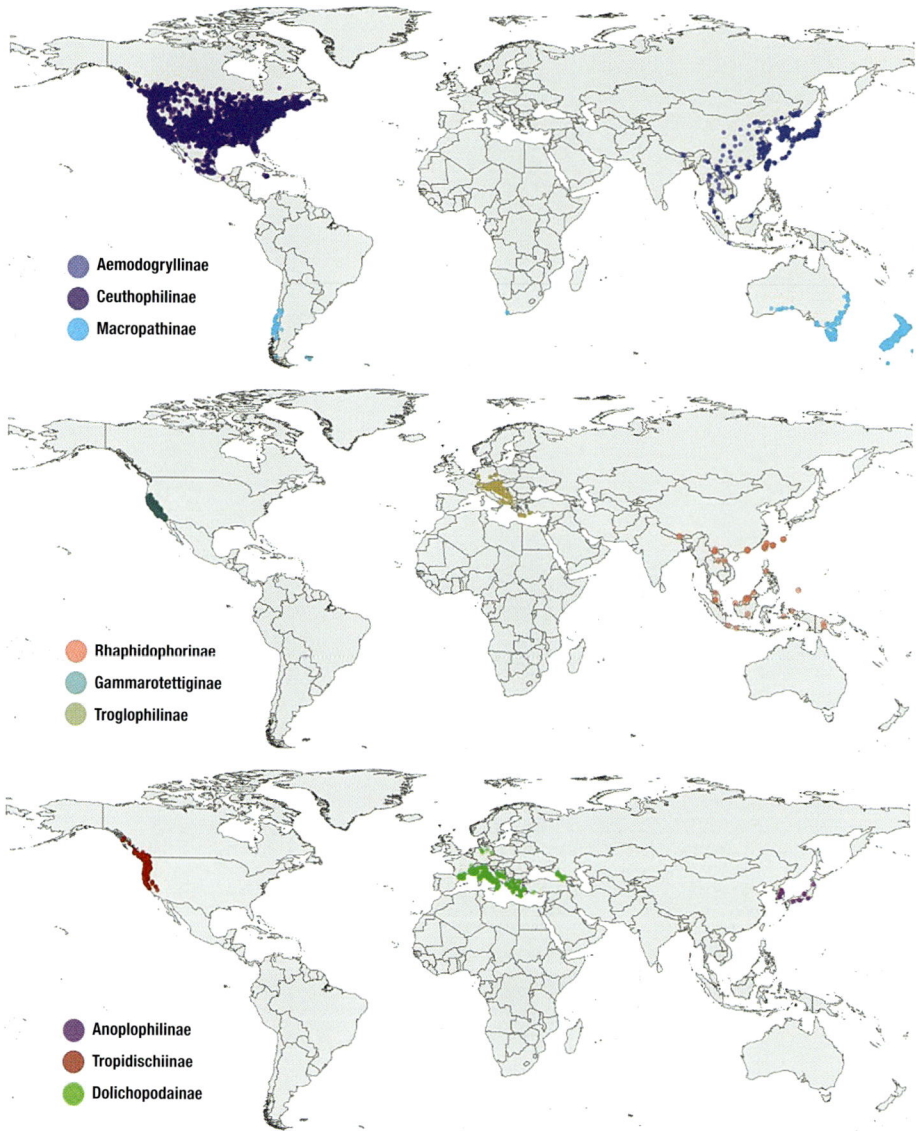

Figure 2 Global distribution of rhaphidophorid subfamilies based on observations from iNaturalist, the Global Biodiversity Information Facility, and Barcode of Life Systems. Note that Tropidischiinae comprise a single species (*Tropidischia xanthostoma*) of uncertain affinity. The invasive range of the greenhouse camel cricket *Tachycines asynamorus* is excluded. Map: Mary Morgan-Richards.

> **Figure 3** Macropathinae species use a range of habitats: daytime concealment of large *Pachyrhamma longipes* in tree-trunk cavity (a), aggregation of *Pachyrhamma edwardsii* in cave (b), *Heteromallus spina* cryptic among leaf litter (c), and *Maotoweta virescene* among moss in temperate rain forest (d). *Pharmacus cochleatus* is part of an alpine radiation (e). Common species in North Island, New Zealand, lowland forest are *Neonetus variegatus* (f), *Neonetus* n. sp2 (mating pair; g), and *Neonetus* n. sp3 (mating pair; h). Photos: a, b, c, g, h: Steven A. Trewick; d, e, f: Danilo Hegg.

In New Zealand, a radiation of at least 63 species of Macropathinae has spread across the country from mountain to rocky shore. Within this single subfamily are a huge diversity of species placed within 18 genera. They all share the same flightless, nocturnal habit but range in size from 10 mm to 190 mm. None are known to make any sound, and all are without a tympanum. Some species are cave dwellers and tolerant of each other, forming clusters on the ceilings and walls of caves and tunnels (e.g., *Pachyrhamma*, fig. 3a,b). But many other species are forest specialists hiding among arboreal moss and lichen (*Maotoweta virescene*, fig. 3d), in holes in tree trunks (*Talitropsis sedelloti*, fig. 1b) or under bark (*Isoplectron armatum*). Other species have found shelter in damp sheds or cool basements (*Pleioplectron simplex*). A radiation of alpine species has resulted in eight camel cricket species living at high elevation. One species known as the Mount Cook flea (*Pharmacus montanus*) is almost completely black and is renowned for leaping out like showers of rain from rock crevices onto unsuspecting climbers (fig. 3e). All the *Pharmacus* species live above the tree line, where snow can cover the ground for half the year. The highest elevation where they have been recorded is 3,400 m above sea level. Due to the short alpine summers with cold nights, *Pharmacus* are often active during the day. They eat lichen and can be active even when the air temperature is below zero degrees. Peculiarly, New Zealand also has some species of camel cricket that feed by the sea, using coastal caves and rock crevices during the day and at night finding scraps at the high-tide line. However, the majority of New Zealand Rhaphidophoridae are small, secretive species of the forest (fig. 3f–h).

Figure 1 A glimpse into the morphological diversity of pygmy grasshoppers, Tetrigidae, showing members of 21 genera from all over the world. Specimens are not to scale. Source of images: Orthoptera species file, Museo Nacional de Ciencias Naturales Madrid, Josip Skejo and Josef Tumbrinck.

1.7 PYGMY GRASSHOPPERS: TAXONOMY UNDER CONSTRUCTION

by Josip Skejo and Niko Kasalo

Pygmy grasshoppers are a morphologically and ecologically unique group of tiny grasshoppers forming the large family Tetrigidae. With 2,000 described species assigned to almost 300 genera, they are the second-largest among 36 caeliferan families, second only to Acrididae. New taxa at all levels are being described regularly, but there is still very little order. Although most of what we have are questions, some fascinating information has crystallized over this group's centuries-long but discontinuous history. In this chapter, we

Subfamily or tribe placement

Batrachideinae **n, r**	Scelimeninae **h**	a *Hymenotes*	g *Trypophyllum*	m *Cladoramus*	s *Truncotettix*
Cladonotinae **a, i, j, k, t**	Tetriginae **e, l**	b *Rhynchotettix*	h *Discotettix*	n *Paraselina*	t *Deltonotus*
Fijitettigini **u**	Tripetalocerina **e-q**	c *Potua*	i *Cladonofus*	o *Cota*	u *Fijitettix*
Metrodorinae **b, f, o, p, s, v**	Xerophyllini **c, d, g, m**	d *Royitettix*	j *Nesotettix*	p *Xistrella*	v *Amorphopus*
		e *Paratettix*	k *Diotarus*	q *Tripetalocera*	
		f *Systolederus*	l *Dinotettix*	r *Scara*	

will provide a short summary of the general diversity, distribution, and evolution of these insects.

For such small animals, an incredible morphological diversity has been observed among tetrigids. Their body length varies from the barely perceptible 6 mm of the New Caledonian *Nesotettix cheesmanae* to the comparatively gigantic 25 mm of the Jamaican *Phylotettix rhombeus*. The most important feature of any tetrigid is the pronotum, the shield that drapes over the insect's back like a stiff cape. It may be covered with dents, grooves, tubercles, and spines, but the most impressive feature is the median carina that extends over the entire length of the pronotum and can form impressive structures such as spines, wavy shapes, or dome-, fin-, and leaf-like elevations.

The head is often neglected in descriptions but has recently been shown to bear crucial information about the supposed deep divergences within the family due to the very different arrangements of facial elements. The vertex – the forehead – is especially interesting as it sometimes forms sizable horns or rostrums. On the other hand, legs seem to correspond more to ecology than to phylogeny and can be long and slender, robust and bumpy, or even oar-like.

Other grasshopper families have long fore wings, called tegmina, which extend over the hind wings and protect them. In Tetrigidae, this function is carried out by the extended pronotum, while tegmina are reduced to two small, scaly structures. Although the pronotum is not movable, the species with elongated hind wings can use them for flight, some with impressive maneuverability. A good example is the Oceanian *Paratettix nigrescens*, which is present on islands such as New Caledonia, Vanuatu, and Palau, while its supposed close relatives are present in Fiji and Australia. This suggests that this tiny animal is able to fly across hundreds of kilometers of open ocean. On the other end of the spectrum, the above-mentioned islands house many local endemic species, which are usually completely wingless.

Pygmy grasshoppers are found on all continents except Antarctica. Curiously, despite extensive research efforts, no tetrigids have ever been found in New Zealand. Their diversity is much larger in the tropical zones than in the temperate ones. Many genera of Tetrigidae are constrained to relatively small areas, with only two having a global distribution: *Tetrix* and *Paratettix*. It is, however, abundantly clear that the wide ranges of some taxa may in large part be due to taxonomic errors. This is not to say that there are no widely distributed genera or species, only that many things remain uncertain while revisions are underway.

Molecular phylogenies place the origin of Tetrigidae at the end of the Paleozoic (Permian) or the beginning of the Mesozoic (Triassic) some 250 MYA, but more work needs to be done to refine this estimate. Few fossil specimens have been found, mostly from different ambers formed in the Cenozoic. Most of the fossil representatives belong to Batrachideinae, the oldest tetrigid subfamily, but some Cretaceous fossils elude interpretation – they may represent a new family related to Tetrigidae.

Barring a few molecular studies, the current system of Tetrigidae classification is based on morphology and is a direct continuation of the system established in the 19th century. This system is plagued with groups that are composed of superficially similar morphologies but do not form true evolutionary groups. The modern approach to taxonomy places a big emphasis on examining type specimens, studying the wider context of the region being worked on, and clearly identifying evolutionary significant characters. Although future molecular phylogenies may bring further insights, we are now finally at a point where we can offer some broad hypotheses on the evolution of the tetrigid subfamilies.

The subfamily Batrachideinae is well-supported by molecular phylogenies, which reconstruct it as a sister group to all the other tetrigids. The members of this subfamily have some shared morphological peculiarities – most notably, a spine at the front edge of the pronotum, and antennae with more than 20 segments. This cosmopolitan and ancient group may even be proven to represent a separate family. The distribution of Batrachideinae will be important for elucidating

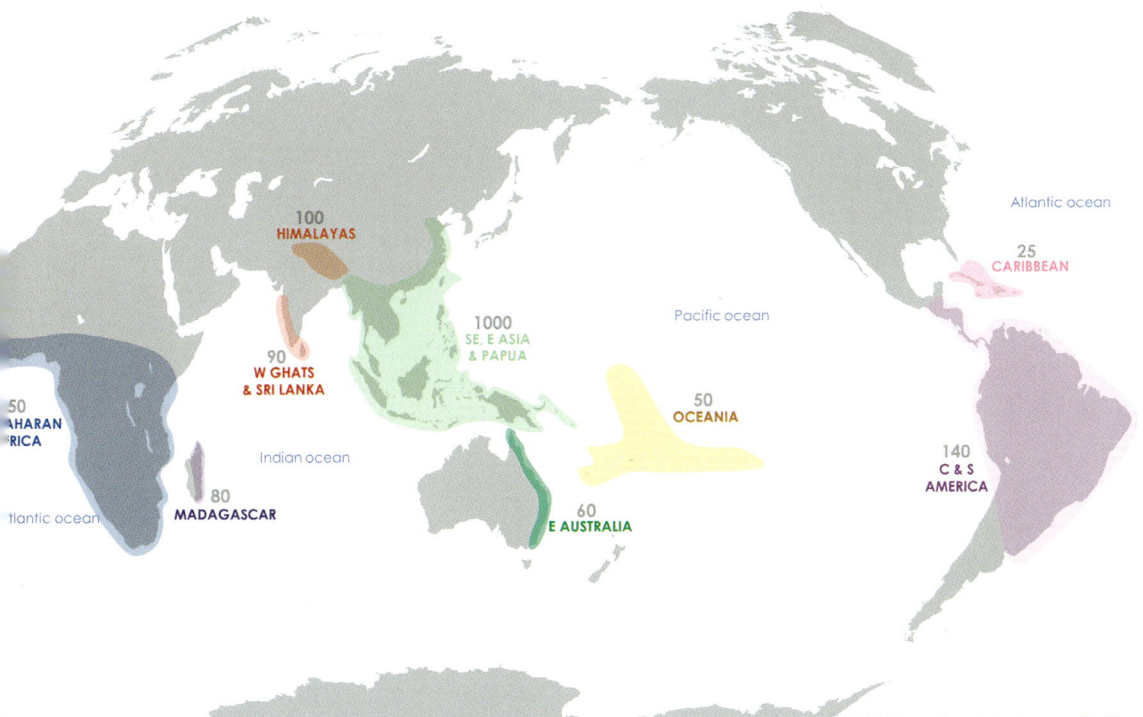

Figure 2 Tetrigidae world biodiversity hot spots with an approximate number of species living in the region, according to the Orthoptera Species File. Altogether, 80% of known pygmy grasshopper species live in the marked regions. Map modified from Wikimedia Commons (adapted from https://commons.wikimedia.org/wiki/File:World_map_Pacific_Center.svg).

the biogeographic patterns of Gondwanan insects, since the current distribution of its members in South America, Africa, southeast Asia, and Australia implies an origin on that ancient continent. The subfamily Cladonotinae resembles Batrachideinae in its distribution and general morphology. Unfortunately, the taxonomy of this group is in urgent need of revision. Lately, the prevailing hypothesis has been that Cladonotinae and Batrachideinae are both ancient and are in some way related.

The subfamilies Metrodorinae, Scelimeninae, and Tetriginae are thought to be younger than Batrachideinae and Cladonotinae, and may have a Laurasian origin. Most Scelimeninae are clearly identifiable by their pronotal projections, and they likely stem from a common ancestor. On the other hand, Tetriginae and especially Metrodorinae are commonly considered to be the most difficult to work with due to the many disparate morphologies that are assigned to them. The work on these groups is underway, and many new, smaller subfamilies may be found hiding within them.

Lophotettiginae (South America) and Tripetalocerinae (southeast Asia) are examples of small subfamilies, each encompassing only two genera, but are nonetheless problematic. They resemble each other at first glance, due to their leaf- or fin-like crests and expanded antennal segments, but they are not closely related. While Lophotettiginae resem-

43

ble Metrodorinae and likely form an evolutionary unit with them, there is no clear candidate for the sister taxon of Tripetalocerinae.

Throughout the long history of Tetrigidae research, only a few people have worked on this group at any given time. Most of them did not have access to all the previous material and publications, and new material kept pouring in from around the world. The modern epoch in tetrigidology has been mostly focused on resolving inherited problems and producing a practical framework for future research. There is still a tremendous amount of work to be done on all fronts: from making the type specimens more accessible and establishing more local research teams to diversifying the topics of research to include molecular and ecological data. Even southeast Asia, in which nearly half of all described species live and which could represent the ancient homeland of Tetrigidae, is severely understudied. Even there, many more species await discovery.

There has never been a better time for studying Tetrigidae. The myriad of problems are awaiting young and eager researchers with open arms, and any of the possible answers promises to impart to us a greater understanding not only of this extraordinary family, but also of the ecosystems these creatures have been witnessing for hundreds of millions of years.

Figure 3 Tetrigidae also includes species with very unusual shapes, such as this *Ophiotettix rohwedderi* from Papua New Guinea. Photo: Philipp Hoenle, CC0 1.0 Universal.

Figure > Like in all insects, molting represents an essential part of the life cycle of Orthoptera. This raspy cricket (Gryllacrididae) is just completing its final molt to the adult stage. Photo: Chien C. Lee.

2 BIOLOGY
AND ECOLOGY

THE STAR PARADE: A PECULIAR FORM OF COURTSHIP IN GRASSHOPPERS

by Paolo Fontana and Roberto Scherini

Many species of Orthoptera are known to exhibit varying degrees of gregariousness. More than a century ago, Australian entomologist Walter Wilson Froggatt (1858–1937) reported a strange behavior for the plague locusts *Chortoicetes terminifera* (not *C. australis*, as reported by Chopard); he observed many males (up to 30 to 50) of the mentioned species, living in Australia, assisting the egg-laying of a conspecific female. Two males in particular remained in antennal contact with the egg-laying female. A similar behavior was observed in Italy for the Oedipodinae *Acrotylus patruelis*, a non-gregarious species never cited as harmful. The first time we observed this phenomenon was in Tuscany (central Italy) in 2002. Many males were arranged in a star shape around an ovipositing female. All males were in antennal contact with each other and with the female, and from time to time, they moved their posterior femora up and down all together for a period of one or two seconds. The female occasionally also moved the posterior femora in the same way and at the same time as the males. These movements did not appear to produce any sounds: a highly sensitive microphone was placed in close proximity to the specimens involved. The phenomenon was observed for over 40 minutes, during which time some males abandoned the choreography and others

joined in, always remaining in number from six to eight. The final result of this behavior was not observed by these authors, who called it a "star parade".

We recently observed the star parade for *A. patruelis*, also in Lombardy (northern Italy). On a warm day (20°C) at the end of October 2022, four star parades were observed a few meters apart in the fields on the edge of a town. In each of these star parades, a laying female was surrounded by eight to ten males. The males rhythmically moved their hind legs and palps. In turns, some males would emerge from the formation to attempt mating with the female. On the same site, the behavior was observed until the beginning of November and again, in 2023, at the end of October. A similar behavior, although involving a small number of males (two to four), was observed in *Oedipoda caerulescens* by Buzzetti (personal communication).

Other pseudo-gregarious behaviors are known in the literature, all linked to the oviposition of Caelifera. Yuri Alexandrovich Popov observed groups of females of *Acrotylus longipes*, a non-gregarious species, laying eggs together, assisted by some males. A similar behavior is reported by D. P. Clark for the Australian species *Austroicetes pusilla*. This behavior is frequent for gregarious spe-

cies, such as *Locusta migratoria migratoria* and *Schistocerca* species, as reported by Boris Uvarov. In other Acridids – for example, in *Dociostaurus maroccanus*, as observed in Libya by Bruno Massa – the males often mate with females who have just completed oviposition; this may explain the gathering of males around a female. Even if the case studies for this behavior do not allow us to advance solid interpretations, the star parade could be a phenomenon linked to sexual competition between males, perhaps in the presence of an unbalanced sex ratio for some reason, such as greater predation of the larger and less mobile females, especially during oviposition.

Figure 1 Males of *Acrotylus patruelis* surrounding a female and forming a star shape. Photo: Paolo Fontana.

2.1 EAT AND BE EATEN: THE ROLE OF ORTHOPTERA IN FOOD WEBS

by Sebastian König and Jens Schirmel

FEEDING IN ORTHOPTERA

Not a single green leaf remains when voracious locusts clear entire stretches of land of any plant, including agricultural crop plantations (see chapter 7.1, Conserving Orthoptera diversity). As we think about Orthoptera feeding habits, such phenomena bring back memories for those concerned; however, most will regard grasshoppers in the meadows as harmless, polyphagous herbivores that mainly chew on a range of grasses and herbs. While both perceptions are true, the feeding strategies of Orthoptera are far more multifaceted (fig. 1); yet much remains to be understood. Since observing feeding interactions in nature can be challenging, collecting gut content or fecal samples are alternative approaches to unravel the mysteries of resource use and dietary preferences through morphological determination or by means of molecular genetic methods.

There are marked differences between Ensifera and Caelifera in terms of feeding habits. On one hand, most caeliferans are indeed vegetarians and feed mainly on grasses, herbs, and other plant material with a certain degree of selectivity (fig. 2). Compared to the high number of specialized phytophagous insects of other groups, however, the average dietary niche of a grasshopper is quite broad. While some can utilize an astonishing spectrum of different host plants, others prefer a narrow range of closely related food plants, such as grasshoppers (Gomphocerinae) feeding on grasses and sedges (order Poales), the cactus-feeding species in North America (e.g., *Chloroplus cactocaetes*) or the euphorb-feeders (e.g., *Acrostira euphorbiae*) on the Canary Islands. A special case are the tetrigids,

which consume detritus, mosses, lichens, and algae, which they may even graze on underwater.

The feeding behavior of ensiferans is even more diverse. The diet of many bush crickets and katydids contains plant material. Which part of the plant is consumed is specific to certain feeding groups; some feed on flowers and seeds; others prefer the leaves. Thus, some Orthoptera may even be important seed dispersers and pollinators, as has been demonstrated in gryllacridids (e.g., genus *Glomeremus*) visiting orchid flowers (see chapter 2.5, Raspy crickets as pollinators of orchids), while some Australian species are even specialized feeders of pollen and nectar (e.g., *Anthophiloptera dryas*). Crickets are often detritivores, and many ensiferans supplement their diet with other arthropods, especially soft-bodied ones such as caterpillars. Even opportunistic cannibalism is not a rare phenomenon. This goes as far as some groups, such as the oak bush crickets (e.g., genus *Meconema*) and the predatory bush crickets (Saginae), being strict carnivores, preying exclusively on other insects, including grasshoppers. The proportion of plants or arthropods in the diet can vary during their development. In some extreme environments, such as harsh deserts, species may only feed on detritus blown by the wind. As for the species with a more hidden lifestyle, mole crickets (Gryllotalpidae) can be pests in gardens under certain circumstances, as they feed on the roots of vegetables and other plants, in addition to ground-dwelling arthropods that they find in their underground tunnels. In similar tunnels, pygmy mole crickets (Tridactylidae) mainly consume detritus. Finally, cave crickets (Rhaphidophoridae) scavenge on dead arthropods, but also eat fungi, plants, or live insects in caves.

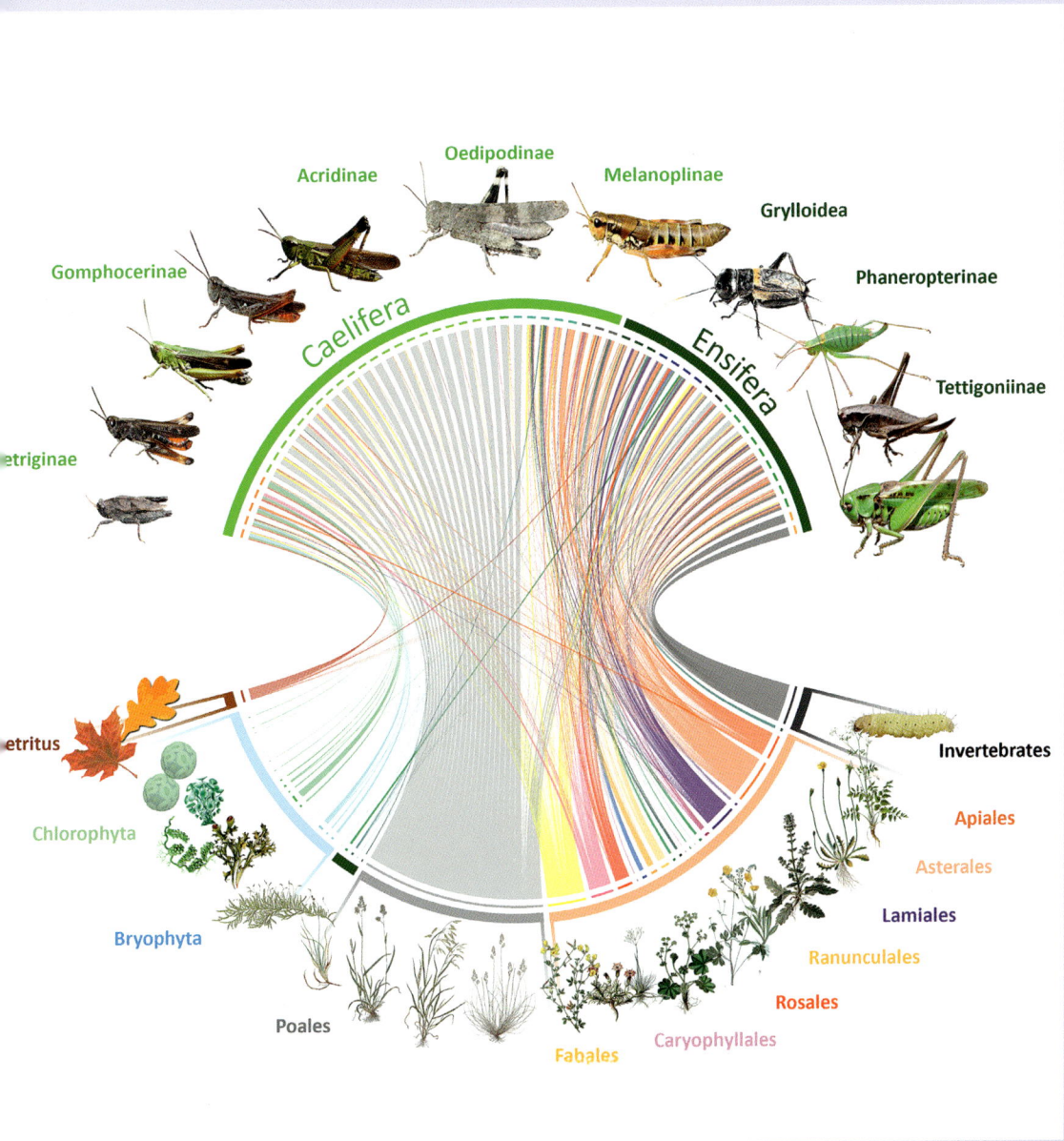

Figure 1 Schematic overview of feeding interactions from central European grassland sites. Dietary interactions were reconstructed based on observations and metabarcoding of the DNA content in fecal pellets. Modified figure from König et al. (2022).

Since most feeding links between Orthoptera and plants are rather loose, they are anticipated to respond directly to changes in micro- and macroclimatic conditions in their habitats by range shifts.

NATURAL ENEMIES OF ORTHOPTERA

The majority of Orthoptera are relatively large and soft-bodied insects, making them an important source of protein for a diverse range of predators, including humans (see chapters 8.4, Grasshoppers as traditional royal food and modern protein source in Madagascar, and 8.5, Salt, lime, and chapulines). Insectivorous birds, reptiles, and amphibians are among the predators of Orthoptera. The white stork (*Ciconia ciconia*) is perhaps the best-known example of a grasshopper-eating bird. In Europe, white storks exhibit a selective feeding behavior, feeding primarily on larger grasshoppers. They have been observed to

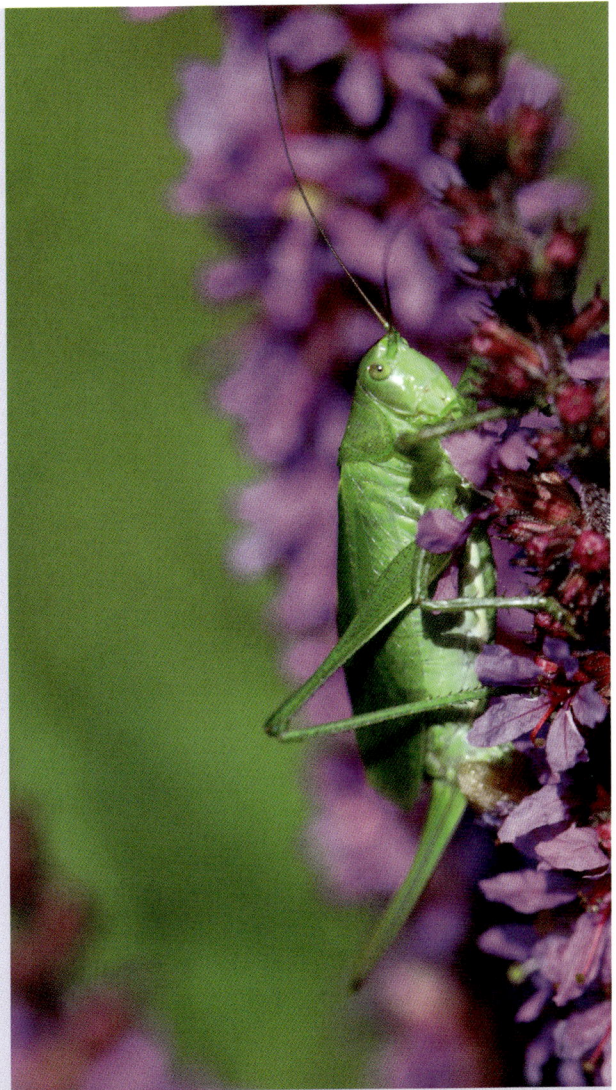

Figure 2 Most Caelifera are herbivores, while Ensifera can be omnivores or carnivores. (left side above) A male of the Siberian grasshopper *Gomphocerus sibiricus* feeding on moss, (left side below) a black color morph of the green mountain grasshopper *Miramella alpina* feeding on herbs, (right side above) an upland green bush-cricket *Tettigonia cantans* eating flowers and (right side below) a predatory bush-cricket *Saga pedo* devouring, in turn, a *T. cantans*. Photos: (left side) Sebastian König, (right side above) Martin Husemann, (right side below) Christian Roesti.

consume several hundred individuals within minutes. Gregarious locusts frequently constitute the principal food source for migrating white storks returning to Africa. In North America, the grasshopper sparrow (*Ammodramus savannarum*) is an example of a specialized grasshopper predator. Other birds that frequently feed on Orthoptera include shrikes (Laniidae), bee-eaters (Meropidae), and several birds of prey, such as the common kestrel (*Falco tinnunculus*). Furthermore, mammals are also known to consume Orthoptera. New World monkeys belonging to the Cebidae family, such as capuchin and squirrel monkeys, are known to prey on Orthoptera. In some cases, they have even learned to avoid toxic grasshoppers, such as stick grasshoppers (Proscopiidae), which employ a rare form of chemical defense as a predator-avoidance strategy (see chapter 5.5.5, The endemic jumping sticks from Meso- and South America). In addition to insectivorous vertebrates, Orthoptera are prey to a variety of invertebrates, with spiders, especially orb-weavers, representing significant predators. The wasp spider (*Argiope bruennichi*), for instance, is highly adapted for capturing grasshoppers with its nets, which are strategically placed at the optimal height for jumping grasshoppers. Other insects – such as mantids, wasps, ants, and even other Orthoptera – may also feed on Orthoptera. In particular, predatory bush-crickets (genus *Saga*) can be important predators.

However, not only crawling, jumping, and flying Orthoptera are eaten, but also their immobile eggs. The larvae of dipterans, such as bee flies (Bombyliidae), are known to prey on eggs, as are the larvae of beetles, particularly those of blister beetles (Meloidae, genera *Mylabris* and *Epicauta*). Additionally, some generalist predatory insects, such as ground beetles (Carabidae) and rove beetles (Staphylinidae), are capable of attacking grasshopper eggs.

Parasites, parasitoids, and pathogens represent further natural enemies of Orthoptera and can attack all life stages. Ectoparasites of Orthoptera adults or nymphs include larvae of some wasps (e.g., Rhopalosomatidae) and red velvet mites (Trombidiidae). Typical endoparasites include some dipterans,

Figure 3 (left) A female sphecid wasp *Isodontia mexicana* carrying a paralyzed oak bush cricket *Meconema meridionale* individual to her nest. (right) The breeding chambers may be full of paralyzed crickets stored as larval food. In the middle, a larva of *I. mexicana* can also be seen (arrow); it has already consumed all bush crickets in the chamber. Cocoons after pupation of the larvae are shown in (below). In most cases, 6 to 8 Orthoptera individuals (here mostly nymphs of *Meconema* spp.), but sometimes up to 16, were deposited into a breeding chamber. Photos: (left) PJT56/Wikimedia Commons/ CC BY-SA 4.0; (right) modified figure from Schirmel et al. 2020.

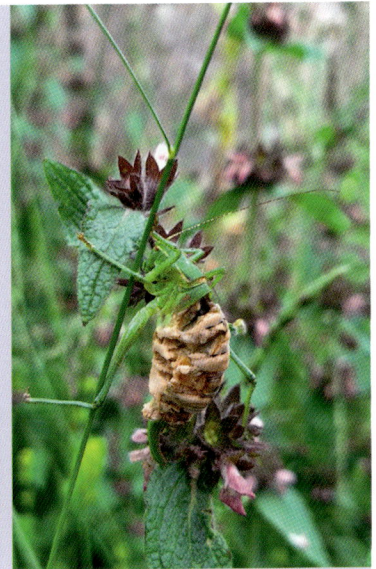

such as tachinid flies (Tachinidae) and flesh flies (Sarcophagidae). Flies deposit their eggs or larvae on Orthoptera, which penetrate the cuticle and develop within the body. Some of these flies (e.g., *Ormia ochracea*) exhibit phonotaxis – that is, the attraction to Orthoptera songs. Some sphecid wasps (Sphecidae) are well-known and specialized predators of Orthoptera. For instance, the Mexican grass-carrying wasp (*Isodontia mexicana*) preys on Ensifera, transporting the paralyzed, yet still alive, individuals to its nest in hollow branches, where the larvae feed on them. *I. mexicana* is native to Central and North America but has become established in large parts of Europe, where it frequently feeds on tree crickets (Oecanthinae) and oak bush crickets (genus *Meconema*, fig. 3). Further-

more, different entomopathogenic fungi infest Orthoptera. The globally distributed complex *Entomophaga grylli* has a broad host spectrum of mostly short-horned grasshoppers (Acrididae). The infestation of the fungus alters the behavior of grasshoppers, causing them to climb up and remain in higher vegetation. The fruiting bodies of the fungus emerge from the body of the grasshopper, and spores are released and can be distributed from the exposed location of the grasshopper. Finally, some nematomorphs (horsehair worms) are known to be specialized endoparasites of Orthoptera. As these cause infected grasshoppers to enter water bodies, they can influence the food web and energy flow in riparian systems.

GRASSHOPPERS OF KEY IMPORTANCE IN (GRASSLAND) FOOD WEBS

Although Orthoptera do not rank among the most diverse groups of insect herbivores, they play a substantial role in food webs of many ecosystems worldwide, with a particular importance in grassland ecosystems. Here, grasshoppers often reach high densities and constitute more than 50% of the total arthropod biomass. Studies have shown that grasshoppers can consume about one-third of their body mass in food per day. Thus, they often represent the dominant insect herbivore group in many grassland ecosystems and can consume up to 20% of the net primary

Figure 4 (left) The red bulbs visibly attached under the pronotum of this Alpine dark bush cricket *Pholidoptera aptera* are parasitic mites. (right) The ornate bright bush cricket *Poecilimon ornatus* is normally green. The abdomen of this unfortunate and still living individual is completely infested with the fungus *Entomophaga grylli*. Photos: (left) Sebastian König; (right) Roberto Battiston.

production and up to 30% of the standing phanerogam biomass – for example, in central European Alpine meadows. Of course, these rates are highly variable and depend on grasshopper densities. These numbers can fluctuate year by year, depending on the weather conditions, ranging from fewer than one to more than 10 grasshoppers per square meter; under certain circumstances, grasshoppers may have even larger impacts and can be severe pests (see chapter 7.1, Conserving Orthoptera diversity).

Through their feeding activity, grasshoppers have a top-down impact on the vegetation structure and plant composition of grasslands. Although most grasshoppers are generalist herbivores, they often select specific plants (see above). Species preferentially feeding on fast-decomposing plants can reduce annual overall plant production, whereas those preferentially feeding on slow-decomposing plants, such as grasses, can increase nitrogen availability and productivity. In both cases, nutrient cycling and, therefore, ecosystem processes and functioning are affected. By experimentally increasing grasshopper density in a high-elevation system to simulate conditions under a warming climate, researchers were able to show that the enhanced herbivore density may even lead to a promotion of plant coexistence by reducing the dominant high-stature grasses. While Orthoptera influence the plant community composition, they are, in turn, affected by the nutritional and structural characteristics of these communities.

Through their consumption – but also through the waste of uneaten plant material (grasshoppers can be wasteful feeders) and their feces – grasshoppers play an important role in the energy budget of grasslands. Their high biomass makes them an important source of protein for many predators. Therefore, grasshoppers are key organisms in grassland food webs worldwide.

Figure 5 Many animals feed on Orthoptera. The red-backed shrike *Lanius collurio* even hoards them by impaling its prey, such as this great green bush cricket *Tettigonia viridissima*, to thorns of plants. Photo: Wolfgang Brandmeier.

2.2 CRYPSIS, MASQUERADE, AND MIMICRY: ORTHOPTERA HIDING IN PLAIN SIGHT

by Chien C. Lee and Oliver Hawlitschek

Stick and leaf insects of the insect order Phasmatodea are widely known for mimicking parts of the plants they live on. Orthopterans are less famous for their mastery of this art, but the camouflage of some species, especially in the tropics, rivals that of the phasmids – or goes even further.

All orthopterans, whether they are herbivores or predators, face a whole army of natural enemies. Most are able to escape via jumping or flight, but many species of grasshoppers and katydids have evolved other means of defense, such as spines or chemical substances (see chapter 5.4.4, African gaudy grasshoppers: Pretty poisonous pests). Others rely on stealth to avoid being caught and eaten. The biological term for any strategy of an animal to avoid detection by another animal is called crypsis. Visual crypsis is also called camouflage.

The most common type of camouflage is background matching, which means that the animal's color pattern makes it resemble its surroundings. This can be comparatively simple and intuitive: a green katydid is generally difficult to spot when sitting among the leaves of a tree or bush, as is a gray grasshopper resting on a gravel bed or rock (chapter 1.3, fig. 2b). A very high number of orthopteran species have evolved this form of camouflage. Some, however, have taken it further. Species living on the bark of trees or on lichens are often colored in a way that dissolves their outline against the background of their habitat. Many species have evolved body structures that complement the camouflage and make them even harder to spot (and, at the same time, harder to ingest, as often these body structures are spines; fig. 1).

Camouflage may also be supported by behavior. In addition to staying on a substrate that matches the cryptic coloration, some species are able to flatten their bodies and hide their legs to further dissolve their shapes. Others completely look like a living or dead leaf, in some cases even a half-eaten one. This form of crypsis, in which the animal resembles a specific structure or item of its surroundings, is called masquerade or mimesis (figs. 2, 3). Other than leaves, some orthopteran species masquerade as a pebble, a patch of lichen, or even a twig.

Some orthopterans even pose as different kinds of animals. The biological term for an animal masquerading as a different species of animal is mimicry (figs. 4, 5). Model species for mimicry (i.e., species that are mimicked by another species) are typically not very attractive to a potential predator because they have some kind of defense mechanism. Stinging wasps and ants or beetles with a tough exoskeleton are common models for mimicry by other insects, such as Orthoptera. As with camouflage, mimicry is often supported by behavior: the mimic not only looks like its model, but also moves like it.

Most, but not all of these spectacular forms of camouflage occur among tropical species, often tree-dwelling katydids, possibly because the population densities of the orthopterans in these habitats are lower than in temperate zones and grassland or bushland areas, and a single individual is subject to a particularly strong predation pressure. Species from different tropical regions of the world show remarkably similar forms of camouflage, even though they are not closely related to each other. These highly specialized forms are striking examples of convergent evolution (see chapter 1.4, Convergent evolution in band-winged grasshoppers) and make us aware of the enormous selective pressures driving the evolution of Orthoptera.

< **Figure 1** The peacock katydid *Pterochroza ocellata* occurs widely throughout the Amazonian rainforest region. Like many other katydids, this species masquerades as foliage, but it also has a secondary defense if this crypsis fails. When threatened, the insect raises its wings in dramatic fashion, exposing two large eyespots that resemble the eyes of a large animal. This startling sight has certainly scared more than one would-be predator away. Photos: Chien C. Lee.

Figure 2 (above left) Central American *Championica*, such as this *C. montana* from Costa Rica, are masters of disguising themselves as the moss growing on the bark of trees and leaves in humid forests. Not only their color, but also the array of spines on their body and legs matches the structure of their microhabitat. (above right) In the southeast Asian tropics, *Sathrophylliopsis longepilosa* and related species have evolved similar camouflage that is also effective when remaining flat against tree bark or other surfaces. Photos: Chien C. Lee.

Figure 4 (top) Some species of bush crickets have evolved mimicry of other animals, such as this *Aganacris velutina* from Ecuador (left), which resembles a wasp of the genus *Pepsis* (right). Potential predators that avoid wasps will likely also leave a wasplike katydid alone. Photos: Chien C. Lee.

< Figure 3 The genus name *Mimetica* speaks for itself. The various species of this tropical Central and South American genus, such as this *M. incisa*, masquerade as leaves of all sizes, colors, or states of decay or consumption. Members of the southeast Asian genus *Systella* follow a similar strategy, also resulting in an effective form of camouflage. Unlike *Mimetica*, which are katydids and belong to Ensifera, *Systella* are grasshoppers that belong to Caelifera. Photos: Chien C. Lee.

Figure 5 (bottom) Some katydids follow different strategies throughout their lives, mimicking other animals as nymphs and later masquerading as leaves as adults. This nymph of a member of Amblycoryphini resembles a stinging ant (*Diacamma* sp.) (right), but as the katydid matures, it sheds this appearance in favor of camouflage. Photos: Chien C. Lee.

2.3   ANT CRICKETS: A LIFE IN CHEMICAL DISGUISE

by Thomas Stalling

Ant crickets (Myrmecophilidae, Myrmecophilinae) often elude the eye of the grasshopper expert because of their small size and their secretive, specialized way of life. With a size of 1.5 mm to 8 mm, they are the smallest known species in the order Orthoptera. The first species of ant cricket was discovered as early as 1799 in Germany (fig. 1), but was thought to be and was described as a cockroach at the time. This is not surprising, as these inconspicuous brown, wingless crickets, with their oval body shape, their conspicuous cerci, and their darting movements, resemble the nymphs of cockroaches.

Ant crickets occur in North and Central America, the northwesstern part of South America, the central and southern parts of Europe, North Africa, middle, southern and southeast Asia, and Australia, and on various tropical and subtropical islands, but they are absent in sub-Saharan Africa, the main part of South America, and the northern regions of the Holarctic. Globally, there are currently 67 described species in 3 genera, but it seems there are still many species that remain undescribed. After an initial revision of the European species in the 1960s, the subfamily Myrmecophilinae has recently been studied more intensively, and numerous new species have been described in Europe and Asia. All 11 species in Europe belong to the genus *Myrmecophilus*, including *M. quadrispinus*, which usually occurs in tropical and subtropical areas and was presumably introduced to Malta together with invasive ants. The different species can be identified by their coloration, type of setae, and shape of the ovipositor, for which a stereomicroscope is usually required due to their small size.

Ant crickets occur in various habitats, from moist alluvial forest habitats to semi-open landscapes with evergreen scrub and woodland to semi-deserts. They can also be found in the centers of large cities, such as Berlin, Copenhagen, and Lima, where they live in fallow areas at railroad tracks or in gardens. All known species live as kleptoparasites in ant nests, which they rarely leave, taking advantage of the stable conditions in the nests. They share this life strategy with other myrmecophilous – that is, ant-parasitizing – species of various arthropod groups, such as silverfishes, beetles, (larvae of) hoverflies, and butterflies. Ant crickets avoid attacks from their hosts through their great physical agility and by using chemical camouflage. Ants recognize their nest mates by their specific cuticular hydrocarbons, which the crickets glean from the ants to coat their own bodies, preventing recognition as intruders. Their diet comprises discarded matter – the "waste" of the ant nest, ant eggs, ant larvae, and food obtained from ants by begging them to feed them directly through trophallaxis.

Studies by Japanese scientists have shown that there are ant cricket species that specialize in one or a few host species and others that use a wide range of host ants. Perhaps the most extreme example is the species *Myrmecophilus albicinctus*. This species is found in southeast Asia and Japan and lives exclusively on the ant species *Anoplolepis gracilipes*. Due to its specialized mouthparts, it cannot feed itself, but is exclusively fed by the ants. Other species, such as the common ant cricket *Myrmecophilus acervorum*; fig. 2), are host-generalists and live in nests of many different ant species from various genera and subfamilies (fig. 3).

Most ant crickets reproduce sexually, but the common ant cricket, which is widespread throughout Europe and large parts of Asia, reproduces parthenogenetically, as is also known from other Orthoptera species, such as the common predatory bush cricket *Saga pedo*. Recently, males of the common ant cricket were found for the first time in a small part of the distribution area. The occurrence of males in southeastern Europe is a typical case of geographical parthenogenesis, which means that asexual individuals have a larger distribution area than their sexual relatives.

There are still many unanswered questions regarding the systematics, ecology, and biology of this interesting orthopteran group, which can be researched in the coming years.

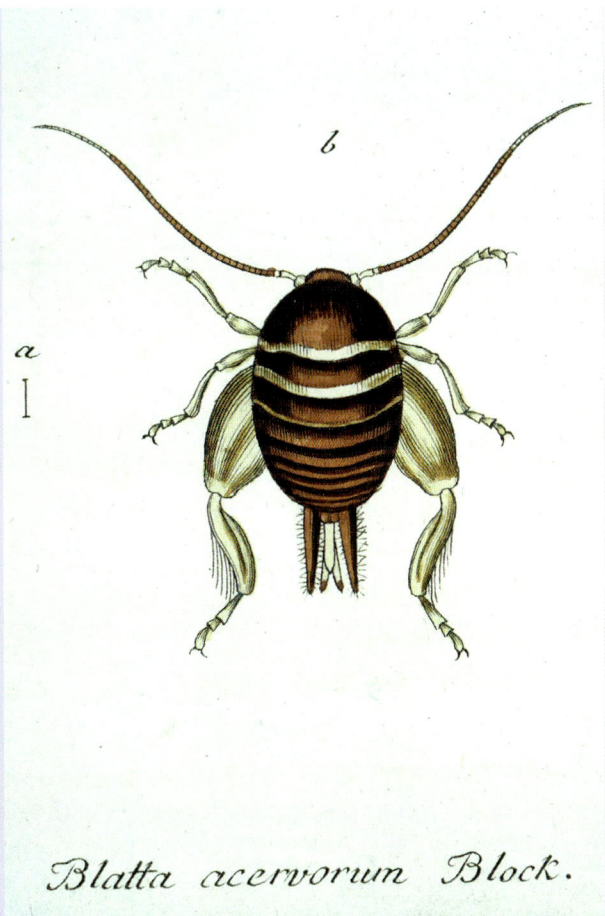

Blatta acervorum Block.

Figure 1 First-ever illustration of an ant cricket from Panzer 1799, originally described as a cockroach, *Blatta acervorum*. Illustration: Public domain.

Figure 2 Adult female of *Myrmecophilus acervorum*. November 2019, La Roque-sur-Pernes, France. Photo: Thomas Stalling.

Figure 3 Nymph (left) and adult female (right) of *Myrmecophilus acervorum* in a nest of *Lasius* sp. November 2019, La Roque-sur-Pernes, France. Photo : Thomas Stalling.

2.4 SECRETS OF BUSH-CRICKET MATING BEHAVIOR

by Gerlind U. C. Lehmann and Arne W. Lehmann

When it comes to reproduction, Orthoptera exhibit some fascinating behaviors. Bush crickets (Ensifera, Tettigoniidae), also known as long-horned grasshoppers or katydids, in particular, are famous for the huge physiological investment males make in their mating strategy. The French researcher and writer Jean-Henry Fabre, who described the mating in the wartbiter species *Decticus albifrons,* observed the occurrence of a huge, white, sticky mass transferred from the male to the female during mating, which he called the spermatophore (Greek for "sperm carrier"). Closer study of the spermatophore revealed that it consisted of two parts: the smaller ampulla, which contains the sperm, and a large white mass surrounding the ampulla (figs. 1 and 2). The latter is called the spermatophylax (Greek for "sperm guard"), produced by special glands inside the male's body.

THE MALE NUPTIAL GIFT

The spermatophylax is recognized as a "nuptial gift" (or "wedding present"). Males from a variety of insect orders provide such gifts prior to or during copulation. These gifts include prey items captured by the male, specifically adapted parts of the male's body, or, in sexually cannibalistic species like mantids, the male itself. In bush crickets, the male spermatophylax represents a substantial investment: bush-cricket males sacrifice up to one-third of their body mass. The spermatophylax contains, aside from 80% water, which itself can be a valuable resource for egg production, hundreds of different proteins. This massive effort involved in producing the spermatophylax can only be understood in the light of sexual selection; therefore, the evolution of courtship feeding has been extensively studied by biologists. These studies have yielded evidence supporting a role both for increased male fertilization success and for male parental investment.

FERTILIZATION SUCCESS

Supporting male fertilization success, the spermatophylax is believed to have evolved to act as a sperm-protection device. After mating, females immediately start to feed on the spermatophore. The position and mass of the spermatophylax prevents females from consuming the ampulla, or at least until a substantial amount of the male's sperm has entered the female's body and been stored in the receptive organ, the spermatheca. Females of most bush-cricket species mate with several males, and the spermatheca can store the sperm of all these potential fathers. As in most insects, eggs are not fertilized directly after mating but only shortly before laying, so sperm from different males may compete for fertilization in the spermatheca. The larger the spermatophylax, the longer the female feeds on it before removing the ampulla, and this prolonged time span increases the number of sperm transferred. This, in turn, increases the chances that the male will fertilize more eggs. Furthermore, the spermatophylax is believed to transmit substances that make females temporarily unreceptive to other males. The length of this post-mating refractory period is variable, both within and between species, but typically stretches over several days. Within a species, its duration is positively associated with the attachment period of

Figure 1 Bush-cricket mating pair separating. The female, on top of the male, has received the "nuptial gift." *Poecilimon gracilis,* collected from Slovenia and mated in a garden at the Humboldt University of Berlin. Photo: Gerlind U. C. Lehmann and Arne W. Lehmann.

Figure 2 Freshly mated female with the spermatophore, consisting of the yellowish ampullae and the white spermatophylax. *Poecilimon gerlindae* found in Domokos, Greece. Photo: Gerlind U .C. Lehmann and Arne W. Lehmann.

the spermatophore and thus the amount of sperm transferred. During the refractory period, females lay eggs fertilized only by the sperm already stored in their spermatheca. Thus, spermatophylax size is under positive selection to increase the chance of fertilization of eggs by a specific male.

PARENTAL INVESTMENT

At the same time, larger spermatophylaces are beneficial to females since they form an easily digestible and nutritious meal (fig. 3). Ingredients in the spermatophore are incorporated into the somatic tissue by females and can be quickly metabolized. As a re-

sult, consuming the spermatophylax increases a female's survival probability and her reproductive output, in terms of both the number and size of eggs. Larger eggs are more resistant to desiccation and increase the survival of offspring. The size of the spermatophylax is positively correlated with male size and body mass. As a result, when given a choice,

Figure 3 Mated female feeding on the spermatophylax. *Poecilimon gracilis* from Slovenia. Photo: Gerlind U. C. Lehmann and Arne W. Lehmann.

females prefer heavier males over lighter ones. The males' potential for providing larger gifts seems to be signaled via their courtship song; when males have been artificially muted, females mate at random in regard to male body mass.

MALE MATING INVESTMENT
As a consequence, male bush crickets are under selection for increased body size due to female choice. However, body size is constrained by the amount and quality of available food and by the requirement of spending energy on courtship songs. The huge investment into the spermatophore also comes at a cost to the male's lifetime reproductive success. A large spermatophore takes longer to produce, so the periods between matings are greater. The remating period in male bush crickets is exceptionally long, lasting several days – a post-mating refractory period similar to that of females. Due to the high costs of reproduction, low-weight males are restricted in their investment capacities. This is well exemplified in bush crickets infected by a parasitoid fly. Such parasitized males produce smaller spermatophylaces and songs of lower quality. Females select against the songs of parasitized males. After mating with a parasitized male, females exhibit a shorter refractory period until remating with the next male. Consequently, parasitized males are less reproductively successful.

FEMALE MATING INVESTMENT
Compared with other crickets, bush-cricket eggs relative to female size are large, so females are comparatively large too because of the metabolic investment they make in reproduction. In insects and other animal groups, there can often be a significant size difference between males and females, which is called sexual size dimorphism. However, due to the selection for large spermatophores, bush-cricket males can be close in size to or even heavier than females. Comparison of a large number of different species demonstrates that males and females become more similar in body size if the male-derived spermatophore is larger.

FEMALE BUSH CRICKETS TYPICALLY TAKE LONGER TO REACH THE ADULT STAGE
As the body size in insects is fixed at adulthood, size differences between males and females are achieved during the nymphal development: female bush crickets have a longer nymphal period than their conspecific smaller males. The phenomenon that males mature earlier than females is named protandry. In the field, this effect is obvious, with males dominating at the start of the adult season and more females being alive at the end. The developmental difference and hence the level of protandry varies greatly between species, ranging from a few days to four weeks.

SUMMARY
Body size, developmental patterns, and spermatophore investment provide a complex picture of how selection has shaped the life histories of animals. Bush crickets became textbook examples for evolutionary biology as their reproduction has been extensively studied and is comparatively well understood. While the mating behaviors and strategies of bush crickets can differ from other animal groups, they provide an excellent subject to understand some of the selective forces shaping reproduction across the rest of the animal kingdom.

2.5 RASPY CRICKETS AS POLLINATORS OF ORCHIDS

by Sylvain Hugel

ORCHIDS OF THE GENUS ANGRAECUM AND DARWIN'S PREDICTION

Charles Darwin, one of the founders of the theory of evolution, famously had a fascination for orchids of the genus *Angraecum*. What intrigued him most was a Malagasy species of the genus, *Angraecum sesquipedale*, whose flowers have spurs about 25 cm long. Since only the tip of the spur contains nectar, Darwin hypothesized that there must be a (then unknown) species of pollinator in Madagascar with a proboscis of the same length. Alfred Russel Wallace, the "other" father of evolutionary theory, later speculated that this pollinator must be a hawk moth (Sphingidae). The species in question was not discovered until 1903 and was then named *Xanthopan morgani praedicta* to commemorate this prediction. This Malagasy subspecies was recently elevated to the species rank: *Xanthopan praedicta*.

The Mascarenes are a volcanic archipelago located to the east of Madagascar. This archipelago includes Réunion Island, Mauritius, and Rodrigues and harbors several species of orchids belonging to the genus *Angraecum*. In Réunion Island, three species of *Angraecum* have a spur not suitable for pollination by a hawk moth: it is very short and very wide (see fig. 1). The pollinators of these three orchid species have long remained a mystery. But what does this have to do with grasshoppers?

THE DISCOVERY OF THE POLLINATORS OF SHORT-SPURRED ANGRAECUM

These short-spurred orchids surprised Claire Micheneau, Jacques Fournel, and Thierry Pailler from the University and the Herbarium of Réunion. who sought to find out who could ensure the pollination of these strange plants. Using cameras during the day, they showed that two of these three species were regularly visited by birds endemic to the island. These birds carried the pollinia on their beaks and deposited them while visiting other flowers. The third orchid, *Angraecum cadetii*, was not pollinated during the day. After ensuring that this orchid was not self-fertilizing, the scientists used infrared cameras to monitor the orchids at night. To their surprise, these orchids were regularly visited by a small species of Orthoptera. These visitors came almost every night, systematically visiting the open flowers to consume the nectar. At that time, I was preparing an article describing new micropterous raspy crickets (Gryllacrididae) that I had just discovered in Réunion and Mauritius. One of these species corresponded to the unknown pollinator. For this reason, we named it *Glomeremus orchidophilus* (figs. 2 and 3).

AN ORCHID ADAPTED TO ITS POLLINATOR

As is often the case with orchids associated with a single pollinator, the dimensions of the flower opening correspond exactly to the width of the head of *G. orchidophilus*. Furthermore, the flower is tough and cannot easily be eaten by insects of this size. Interestingly, the consumption of a viscous liquid such as nectar involves a passage of the liquid through a particular space in the mouthparts of *G. orchidophilus*.

Figure 1 *Angraecum cadetii* in bloom in the forest of Sainte Rose, south of the island of Réunion. Note the very short spur (yellow arrow) and the short petals. Photo: Sylvain Hugel.

This space appears to have another function: through it, raspy crickets expel silk that they use to glue leaves together to form a hiding place for the day. The twofold use of this anatomical structure could therefore be an example of exaptation or pre-adaptation, in which the function of an adaptation is diverted to ensure a distinct function.

TWIN SPECIES IN MAURITIUS?

A raspy cricket very close to *G. orchidophilus* lives in Mauritius's neighboring Réunion Island. Because of this morphological proximity, we named it *G. para-orchidophilus* and described both species in the same work. This Mauritian species has very distinct genitalia compared with its Réunionese sister species and a significantly smaller size. Interestingly, an orchid close to *A. cadetii* has just been discovered in Mauritius: *A. jeannineanum*. This orchid has a slightly smaller flower opening and could speculatively be pollinated by *G. paraorchidophilus*, which has a head dimension corresponding to that of the flower aperture.

A SPECULATIVE SCENARIO LEADING TO THE DESPECIALIZATION OF ANGRAECUM FOR POLLINATION BY HAWK MOTHS

Molecular phylogenetic data of *Angraecum* and their current pollinators suggest that pollination by hawk moths is ancestral and that the species from the Mascarene Islands derived from Malagasy ancestors. The ancestor of the short-spurred *Angraecum* from Réunion could therefore have reached the archipelago bearing a long and narrow spur. Moreover, data on the *Angraecum* phylogeny and their current pollinators also suggest that pollination of short-spurred orchids by gryllacridid crickets is ancestral with respect to pollination by birds. Speculatively, the orchid-Gryllacrididae association could have selected for flowers with wide openings and short floral tubes; this de-specialization in regard to hawk-moths opens the way to other pollinators, such as birds. Such a scenario would be compatible with field observations. Indeed, the Gryllacrididae of the Malagasy region are often observed eating parts of flowers at night, in particular the stamens and pollen. The current association between orchids and Gryllacrididae could therefore have started with an initial phase of florivory.

Figure 2 An adult female of *Glomeremus orchidophilus* is clinging to an *Angraecum cadetii* in bloom in the forest of Plaine des Palmistes, on the island of Réunion. Note that the orchid is fruiting, indicating that it has been previously pollinated. Photo: Sylvain Hugel.

Figure 3 An adult male *Glomeremus orchidophilus* has recently visited a flower of *Angraecum cadetii* and has two pollinaria still attached on the head. Photo: Sylvain Hugel.

Figure > A flying swarm of the migratory locust on the Horombe Plateau, Madagascar. The density of locusts is such that the landscape is completely hidden. The farmers in the foreground seem understandably frightened by this mass of insects. Photo: Michel Lecoq.

3 PESTS

3.1 LOCUSTS AS PESTS: PAST, PRESENT, AND FUTURE

by Arianne Cease, Michel Lecoq, Mira W. Ries, and Clara Therville

Locusts have been intertwined with people for thousands of years due to their unique biology (see box in chapter 3.1:, Locusts and humanity through millennia). Locusts are grasshoppers that, when exposed to specific environmental cues, shift from shy and solitarious to gregarious phenotypes (fig. 1). High population density is the primary trigger and can lead to massive swarms (figs. 2 and 3) that fly hundreds of kilometers and lay waste to crops, causing food insecurity among human populations (see box in chapter 3.1: Main locust pest species and examples of outbreak frequencies). However, this goes both ways, as human activities also greatly influence locusts. Climate change, land-use competition, and land-cover change can all affect the distribution and probability of outbreaks and migration patterns. Long-term monitoring and response programs are necessary for preventive management to avoid plagues, but the sustainable governance of these systems is challenging because outbreaks are generally infrequent in any given region but can be massive when they occur.

LIVELIHOODS

Locust outbreaks can have disastrous impacts on livelihoods, especially for small-holder farmers in regions with simultaneous hazards and hardships. Because swarming locusts have the potential to decimate all vegetation in their path, crops and pastures may be completely wiped out. This results in severe food shortages for people and livestock, followed by wide fluctuations in market prices, leaving farmers no choice but to sell their livestock at very low prices to meet household subsistence needs. Land-use competition may cause displacement and tension between pastoralists and farmers. In addition to the

acute and immediate effects of food shortage, the impacts of locust outbreaks also reverberate into the future. Reports on the amounts of crops saved through control campaigns are a simplified portrayal of success and omit many externalities. For example, children born in affected communities during plague years are less likely to begin school or perform well academically. The increased use of chemical pesticides in attempts to control the outbreaks (figs. 4 and 5) may have serious environmental and human health consequences that can disrupt other industries. It is also important to acknowledge the impact on culture and traditions, and the cost of physical and mental human suffering. Ultimately, most externalities of locust outbreaks and their management are very difficult to track and quantify, so the extent of the true cost remains largely unknown.

MANAGEMENT

Due to these high costs, people have always tried to control locust pests, but for centuries they were unable to contain the invasions and limit the damage. It was not until the discovery of locust phase polyphenism by Uvarov in 1921 that the determinism of locust invasions began to be understood and more effective control strategies started to emerge. Uvarov and his research team revealed that solitarious and gregarious phenotypes developed from the same species and that there were certain environments that fa-

Figure 1 Solitarious (right) and gregarious (left) nymphs of the desert locust. Individuals from laboratory colony at Texas A&M University. Photo: Brandon Woo, 2022.

vored the shift from solitarious to gregarious pheno-types – termed gregarization. The focus then became identifying gregarizing zones to detect outbreaks early. This paved the way for preventive manage-ment, with regular monitoring of these areas and treatment of small locust outbreaks before they formed large swarms and had an impact on agricul-ture. Over the last century, ever greater resources have been deployed on every continent to improve both surveillance and control operations and combat these pests more effectively. Since invasions may affect large parts of entire continents, concerted in-ternational cooperation has been established. At the same time, powerful synthetic chemical pesticides and more effective treatment methods were imple-mented and, combined with preventive management, have been fairly effective in reducing the severity,

duration, and frequency of plagues (see box in chap-ter 3.1: Main locust pest species). However, in recent decades, there has been an increasing recognition of the environmental impacts of pesticide use. The fun-gal biopesticide *Metarhizium acridum*, which is spe-cific to locusts and grasshoppers and thus spares their natural enemies, may be a viable alternative treatment. Although challenges in their use remain, the effectiveness of biopesticides in locust-control programs is increasingly recognized.

GOVERNANCE AND CHALLENGES

Such concerted efforts to control these extreme pests spurred the historical development of plant protection organizations. The origins of the focus of the US Department of Agriculture (USDA) on insect pests can be traced to the now-extinct Rocky

Mountain locust *Melanoplus spretus* (see chapter 7.5, The Rocky Mountain locust). South American locust-control campaigns in the early 1800s led to the first government agency for plant protection, which eventually was integrated into the present-day National Plant and Animal Health and Quality Service of Argentina (SENASA). The desert locust *Schistocerca gregaria* (fig.1) remains among the most devastating; its migratory range includes up to 60 countries across northern Africa and southwest Asia. A cooperative international framework for managing this species was provided by the formation of the Food and Agriculture Organization (FAO) of the United Nations in 1945. Since 1955, the FAO has been mandated by its member states to ensure the coordination of desert-locust monitoring, an early-warning system, and control activities. Furthermore, the FAO provides training and prepares publications on various aspects of locusts, undertakes field-assessment missions worldwide, and coordinates survey and control operations, as well as assistance during locust plagues.

Collaborative governance, such as that led by the FAO for the desert locust, is crucial for managing locusts but poses significant challenges due to the vast scales involved and the relative infrequency of plague events. The high mobility of locusts requires international coordination of responses among multiple stakeholders. Challenges include deteriorated access due to armed conflicts and political tensions between neighboring countries. Cultural differences, language barriers, and varying capacities for investing in robust preventive systems further complicate coordination. Major locust plagues may only occur once every few decades in a given region, particularly if preventive systems are effective (see box in chapter 3.1: Main locust pest species). This pattern often leads to memory loss about the threat, resulting in collapsed funding and management systems. This infrequency also complicates the integration of gradual changes, such as evolving land uses, rising environmental concerns, and calls for decentralized, integrative locust management. Crises often expose these changes, creating both tensions and learning opportunities. To address these challenges, locust managers have developed adaptive and embedded governance structures. In low-risk situations, the responsibility falls to local land managers or protection agencies. Plague situations, on the other hand, are managed by top-down structures at the national and international levels. FAO mobilizes worldwide expertise and financial resources; nations can mobilize national armed and civilian forces. To mitigate problems associated with the low frequency of outbreak events, locust managers have established permanent communication channels, set aside reserve funds, organized annual international meetings of field officers and simulation exercises, and engaged in research projects during recession periods. These strategies help ensure continued vigilance and readiness.

GLOBAL CHANGE

Climate change will make locust governance increasingly difficult due to the heightened unpredictability of weather patterns. As locusts mostly prefer hot environments, climate change is expected to expand the ranges of most locust species to higher latitudes and elevations, though some species or population ranges may contract. Most locust species are adapted to arid environments, where they persist in low numbers by moving among areas with available green vegetation and/or via dormancy periods as eggs or adults. Higher moisture levels leave locusts more susceptible to pathogens, and mesic areas harbor more predators and competitors. However, in arid systems, outbreaks do require well-timed precipitation pulses that provide

Figure 2 Flying swarm of the migratory locust *Locusta migratoria* on the Horombe Plateau, in the extreme south of Madagascar, not far from the city of Ranohira and the Isalo massif (aerial view taken from a helicopter). Photo: Michel Lecoq, 1999.

Figure 3 Desert Locust hopper band in Northern Chad (Mao region). Most hoppers are on the ground, but the bushes are also covered with them, and a few young adults are flying around. Photo: Michel Lecoq, 1988.

enough green vegetation to support juveniles to fully develop into adults that can migrate long distances by flight. These precipitation pulses, such as from cyclones in the breeding zones of the desert locust, will become more difficult to predict.

Beyond climate change, people have more directly impacted locust populations through changes in land use and land cover. Most locusts avoid forests and prefer open grasslands or deserts. Deforestation thus expanded habitat and likely promoted outbreaks of the migratory locust *Locusta migratoria* in Australia and Indonesia and of the Central American locust *Schistocerca piceifrons* in Mexico. Gregarious locusts have high activity levels, forming marching bands as juveniles prior to developing wings, then as adults, aggregating in swarms that can eventually cross continents. These active, migrating locusts thrive on plants with lower protein and higher energy (carbohydrates and lipid) contents, which allow them to eat larger plant quantities to meet both their protein and energy demands. Land-management practices that degrade soils and lower soil nitrogen generally produce plants with lower protein contents (since most nitrogen in plants is found in the form of proteins and amino acids) and higher carbohydrate concentrations. These environments promote locust outbreaks. The positive effect of low-nitrogen environments on locust popula-

tions has been found by a variety of studies in China, West Africa, Australia, and South America. Understanding these relationships opens additional sustainable management options, including rangeland management, soil amendments to suppress populations emerging in cropping fields, and including soil nutrients to enhance distribution and forecasting models.

A CASE FOR TRANSDISCIPLINARY APPROACHES

Finally, despite huge improvements in locust management, upsurges still occur, albeit at a more modest level than in the past. The problem is no longer simply a question of improving biological and technical knowledge, but also of better understanding various forgotten components: people, human behavior, decision-making processes, and the potential consequences of any human action. Better integration of the social sciences and humanities in the process of developing sustainable solutions is increasingly necessary. Transdisciplinary approaches and better continuity of research and surveys between invasions are therefore essential for sustainable locust management. Our appreciation is due to the many researchers, managers, agriculturists, and others who have contributed to our collective understanding and sustainable management of locusts.

Figure 4 Ground treatment with a mycopesticide against the locust *Rhammatocerus schistocercoides* in the cerrado areas of the Chapada dos Parecis, Mato Grosso, Brazil. Photo: Michel Lecoq, 2002.

Figure 5 Campaign to control an invasion of the migratory locust in Madagascar. The helicopter that carries out the treatments is being resupplied with insecticide. Photo: Michel Lecoq, 1999.

Figure 6 South American locust late-instar nymphs marching across the road in northern Argentina. After 60 years with no major outbreaks, in 2015, swarms of South American locusts began to emerge, leading to plague years in 2016 through 2020 in Argentina, Bolivia, and Paraguay. As of 2024, swarms have re-emerged, and plant protection groups remain vigilant. Photo: Rick Overson and Laura Steger, 2020.

LOCUSTS AND HUMANITY THROUGH MILLENNIA

by Arianne Cease, Michel Lecoq, Mira W. Ries, and Clara Therville

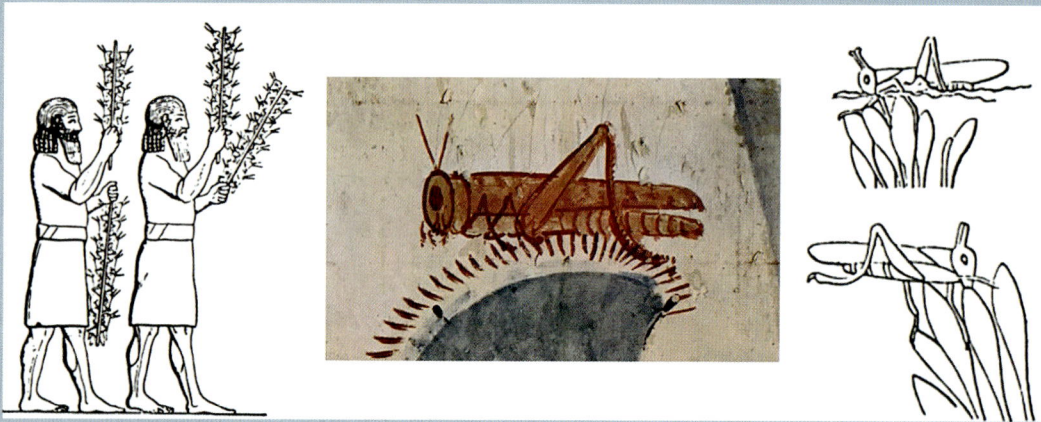

Locusts are featured in a variety of cultural and religious heritages, including Indigenous traditions, the Bible, the Quran, the Sanskrit epic Mahabharata, and the ancient Iranian Zoroastrian text, the Vendidad. In some cultures, locusts were considered a culinary delicacy enjoyed by the social elite. Art from an eighth-century BC Neo-Assyrian palace depicts skewered locusts being presented at a royal banquet. Some of Brazil's Indigenous people attribute to locusts the creation of the world and the mastery of fire. These few illustrations show just how important locusts have been to humanity since ancient times, as both a serious threat to harvests and an abundant source of food.

< **Figure 1** From left to right: Servants carrying locusts to a banquet (Nineveh, Iraq, ca. 705–681 BC). Painting of a locust, probably a desert locust (Tomb of Horemheb, West Thebes, Egypt, ca. 1323–1295 BC). Bas-relief depicting a locust (Saqqarah, Egypt, ca. 2500–2350 BC). Images: Public domain.

Figure 2 Moses invoking the eighth plague of Egypt (wood engraving illustrating the Nuremberg Bible of 1483, the Book of Exodus). The locusts depicted in this engraving probably correspond to the desert locust. The Egyptians who fell victim to this plague are shown dressed in Germanic-style costumes from the engraver's time. Image: Public domain.

MAIN LOCUST PEST SPECIES AND EXAMPLES OF OUTBREAK FREQUENCIES

by Arianne Cease, Michel Lecoq, Mira W. Ries, and Clara Therville

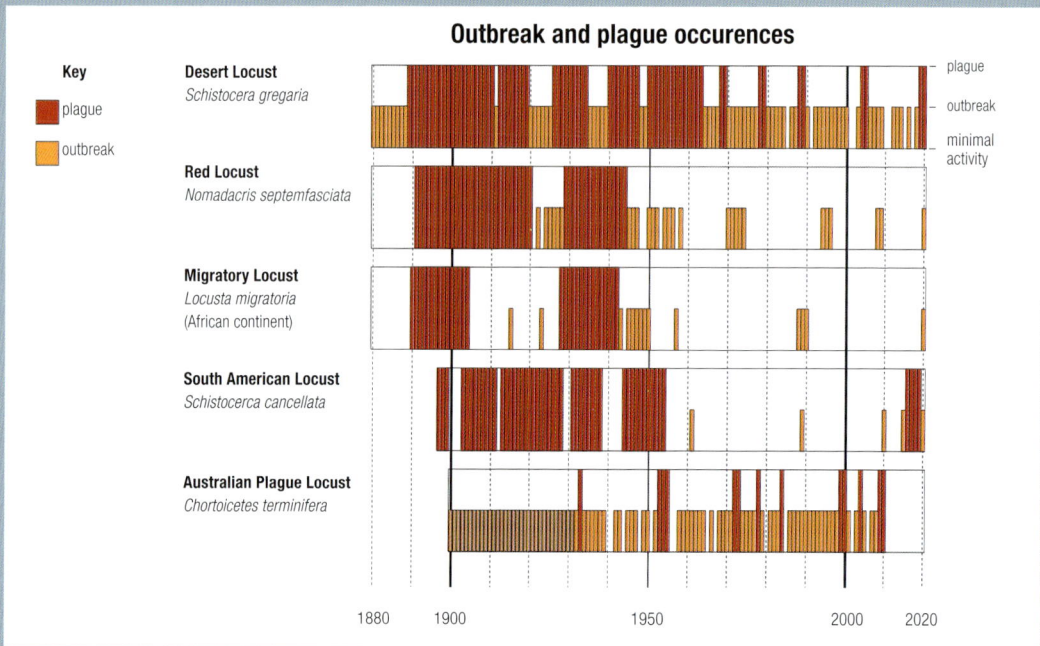

Outbreak and plague occurences

Although the desert locust *Schistocerca gregaria* is considered to be the most important species of locust due to its ability to migrate over large distances and rapidly increase its numbers, there are several other important species of locusts throughout the world. These include, in alphabetical order: Australian plague locust *Chortoicetes terminifera*, Australia; Moroccan locust *Dociostaurus maroccanus*, northwest Africa and southern Europe to Asia; migratory locust *Locusta migratoria*, worldwide excluding the Americas; brown locust *Locustana pardalina*, southern Africa; red locust *Nomadacris septemfasciata*, eastern Africa; South American locust *Schistocerca cancellata*, South America; and Central American locust *Schistocerca piceifrons*, Central and South America. The following locust species are of lesser economic importance: tree locust *Anacridium melanorhodon*, Africa, Near East; Italian locust *Calliptamus italicus*, western Europe to central Asia; Bombay locust *Patanga succincta*, southwest to southeast Asia; Mato Grosso locust *Rhammatocerus schistocercoides*, South America (Brazil, Colombia, Peru, Venezuela); and Peruvian locust *Schistocerca interrita*, Ecuador, Peru, Chile.

< Figure 1 The chart shows temporal variability of outbreaks and plagues for several locust species. The general decrease in plague frequency and duration since the 1960s can be attributed, in large part, to implementation of more effective management programs. It is difficult to apply precise definitions for plagues and outbreaks across different species, but a plague generally involves multiple regions and has substantial impact on agriculture. An outbreak involves smaller areas, with locusts crossing only a few, if any, political boundaries. Where possible, we used definitions of plague and outbreak years as determined by experts on each species and published in earlier papers and/or locust bulletins. Data for the migratory locust is for continental Africa only and does not include locust activity in Madagascar. Data for the Australian plague locust is not precise before 1933, but anecdotal records indicate there were consistent outbreaks between about 1900 and 1932. Likewise, data for the desert, red, and migratory locusts are less reliable before 1900. Sources of data: desert locust (Waloff 1976, Magor et al. 2008, Sword et al. 2010; the Food and Agriculture Organization's Desert Locust Information Service monthly Desert Locust Bulletins), red locust (Steedman 1990, Lecoq and Bazelet 2019), migratory locust (Waloff 1960, Lecoq 2023), South American locust (Gastón 1969, Hunter and Cosenzo 1990, Barrientos-Lozano 2011, Medina et al. 2017, Trumper et al. 2022), and Australian plague locust (Wright 1987, Deveson 2011, Adriaansen et al. 2016; Australian Plague Locust Commission Locust Bulletins, M. Khan pers. comm.

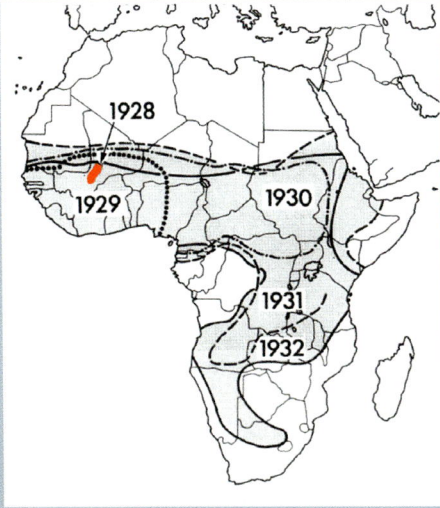

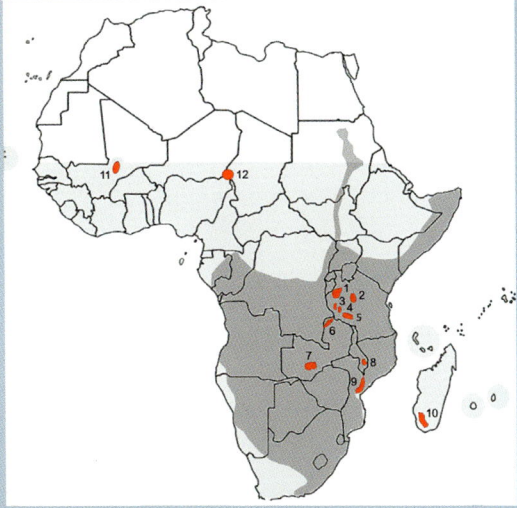

Figure 2 These two maps of the last major invasions of the migratory locust (left) and the red locust (right) show the continental scale of the invasions, which start from very small outbreak areas for each species. The migratory locust plague had a small original outbreak area in 1928 in the central delta region of the Niger River in Mali, as indicated by the red area and arrow (after Batten 1966, Duranton et al. 1982). By 1932, the plague had expanded to eastern and southern Africa; it did not end until 1941. In the red locust map, light gray indicates the maximum distribution area (the northern limit is approximate), and dark gray indicates the maximum area invaded by swarms of the red locust during the 1929–1944 plague (adapted from Steedman 1990). Red indicates the main outbreak areas where swarms originate (Centre for Overseas Pest Research 1982, Duranton et al. 1982, Steedman 1990).

FROM UNKNOWN TO OUTBREAKING: THE QUAINT CASE OF *BARBITISTES VICETINUS*

by Paolo Fontana

Not only locusts, but also other members of Orthoptera may cause plagues. Within the genus *Barbitistes*, some species are known to cause outbreaks, resulting in severe damage to woods or cultivated plants; specifically, *B. constrictus*, *B. ocskayi*, and *B. serricauda* have been recognized as occasional pests.

   Barbitistes vicetinus (Tettigoniidae), is an endemic northeast Italian flightless bush cricket (fig. 1). The species was described based on a single male that was morphologically clearly distinct from any other specimens found in collections. *B. vicetinus* remained a very rare species confined to small hilly areas for the following 15 years. Thanks to the identification of its bioacoustics pattern, *B. vicetinus* has been more intensively researched, and it was found in other localities of a very small area of the Veneto region, between 200 to about 600 m above sea level. In 2004, it was for the first time found outside the Veneto region, in the Trentino district, along the Adige River Valley. During these first years of studies, the species was demonstrated to be associated with trees, especially European hop hornbeam (*Ostrya carpinifolia*), various elm (*Ulmus*) species, and flowering ash (*Fraxinus ornus*). At the

only intense defoliations of broadleaf forests but also severe damage to the neighboring crops (mainly vineyards, olive groves, and fruit orchards). After a few years, the outbreaks were restricted to the Euganean Hills, with limited attacks in Trentino vineyards. In addition to the crop damage, the outbreaks were a source of annoyance to people living nearby, as the bush crickets invaded streets and gardens.

Outbreaks of *B. vicetinus* are still occurring. The causes that may have determined the sudden demographic explosion of this species are probably linked to global warming and have been hypothesized to be at least in part due to the increase in both summer and winter temperatures. High summer temperatures seem to shorten the diapause of the eggs (which in *B. vicetinus* can last up to three years), and mild temperatures during the winter seem to reduce their mortality. The insights from this peculiar case are twofold: it shows, on one side, that rare and localized species may have the potential for unexpected outbreaks, and, on the other side, that even species that we consider pests may break down to very small and isolated populations that may easily evade detection. A further consideration that could derive from the case of *B. vicetinus* is that some animal species react much more sensitively to environmental changes, such as global warming, than humans, even though humans largely caused these changes.

beginning of the third millennium, *B. vicetinus* suddenly left its rarity and cryptic habits behind, and its populations exploded. The first outbreak was recorded in 2006 on the Berici Hills (Vicenza district). In this first event, the species was represented by predominantly green individuals, like all the specimens known until then except for one individual obtained in captive breeding. In 2008, *B. vicetinus* started to produce further severe outbreaks, in both the Berici and the Euganean Hills (Padua district), during which the individuals were mostly black and started becoming highly polyphagous, causing not

4 SONG AND HEARING

4.1 MUSICIANS OF THE INSECT REALM: SOUND PRODUCTION BY ORTHOPTERA

by Laure Desutter-Grandcolas

Orthoptera mostly communicate through sounds – that is, mechanical vibrations that travel from the emitter through the air or the substrate (fig. 1). These signals are complex, and we can distinguish three elements of different nature and intensity: (1) a pressure variation of the air (= far-field sound) that may travel over a long distance and can be perceived by a tympanal membrane (see chapter 4.2, A sophisticated audience); (2) a local movement of air molecules (= near-field medium motion) close to the emitter that can be detected by hairlike structures (setae); and (3) a vibration of the substrate that can be detected by specific vibration sensors, such as the subgenual organs of insect tibiae, of a receiver standing close to the emitter and on the same substrate. These physical components are combined in "acoustic signals" in diverse ways and contribute to the complex act of communication. As an example, a stridulating cricket generates all three types of signals when singing, and all three can be detected and evaluated by potential mates (and also by potential predators). Human observers normally only perceive the far-field component of the signal, which we will treat exclusively as "acoustic signal" in this account.

In Orthoptera, acoustic communication is very ancient: the earliest fossil evidence dates from the middle Permian, ca. 260 MYA. It is also very common, involving a wide diversity of sound-production organs, several types of hearing organs, a large array of signals, and several behavioral contexts. Acoustic communication may have been one of the main drivers of Orthoptera diversification and evolution, similar to sociality for Hymenoptera (ants, bees, wasps, and their allies); all of its components may have been, and still are, under strongest natural and sexual selection.

WHY ARE ORTHOPTERA PRODUCING SOUNDS? BEHAVIORAL CONTEXTS

Most Orthoptera sing in the context of reproduction, one sex calling to attract the other sex from a distance (intraspecific communication). In each species, the role of each sex is usually constant (although role reversals are documented in some katydids, like the Mormon cricket *Anabrus simplex*). These roles allow sharing of the risks between males and females. In crickets (Ensifera, Grylloidea), the male calls, taking the risk of being located by predators that find their prey using acoustic signals, such as parasitic flies or bats. The female walks toward the male she has detected through his song (a behavior called phonotaxis), taking the risk of alerting a visual/vibration-sensitive predator while moving. Some species have circumvented this situation. In some Lebinthini crickets (Gryllidae, Eneopterinae), females answer the calling males by tremulation (i.e., rapidly shaking their body without walking), so that the males are guided toward the females through the vibration of the substrate, taking most of the risks associated with communication. In Gomphocerinae grasshoppers (Caelifera, Acrididae), males call, and females walk to the males, but males and females may engage in complex nuptial parades, associating courting sounds with particular movements of the different parts of their body (see chapter 4.4, Field grasshoppers, some of the most versatile singers among Orthoptera). Calling and courting sounds can be complemented by guarding sounds, emitted by a male to prevent the female from moving away, especially in species that perform multiple matings, a situation often observed in crickets.

Orthoptera may also produce sounds to escape a predator (interspecific communication), a

situation commonly observed in katydids and grasshoppers, but also documented in some crickets, such as the large African genus *Brachytrupes* (Gryllidae, Gryllinae). The insect then emits a loud protest sound to deter a predator that has seized it. Finally, acoustic signals may be emitted during territorial behavior, male-male aggressive interactions, subsocial relations, and even to attract prey (see chapter 5.1.1, The astounding diversity of Australian Orthoptera).

HOW DO ORTHOPTERA PRODUCE SOUNDS? THE SOUND-PRODUCING ORGANS

The most common way of producing sounds in Orthoptera is stridulation – that is, the rubbing of a file located on one part of their body against a scraper located on another part. Some species stridulate by rubbing their palpi (mouthparts), others by scrubbing their pronotum over the tergites or by rubbing their genitalic structures. Most species stridulate either with the hind legs against the lateral vein of the fore wing (fig. 2) or the abdominal tergites, or with the fore wings raised above the body and rubbed together in a to-and-fro lateral movement. The first type is observed mainly in Caelifera and in some species of various groups within Ensifera; the second is exhibited only in crickets and bush crickets. Both modalities may involve different structures in different species and species groups, suggesting that both types of stridulation have evolved several times independently (see chapter 4.3, Many different styles of singing in crickets, katydids, mole crickets, and grigs).

In crickets and bush crickets, the system generating the sound consists of two structures on opposite fore wings: the stridulatory file and the plectrum. In many species, stridulation also involves a system that amplifies the sounds, mostly enlarged areas of the fore wings (see chapter 5.4.5, The balloon bush crickets, mysterious denizens of East African forests). Mole crickets (Gryllotalpidae) additionally use their complex burrow, which has a species-specific shape, as a resonator.

Apart from stridulation, Orthoptera may produce sounds by crepitation (i.e., the production of sounds by a thin wing membrane between veins that suddenly expands and vibrates due to a fast airflow, as performed by some Acridinae, Gomphocerinae, and Oedipodinae grasshoppers when they fly away) or by "playing the castanet" (i.e., making sounds by beating one wing or clashing two wings against each other, as in the grasshopper *Stenobothrus rubicundulus*). The African Pneumoridae, commonly called bladder grasshoppers, stridulate by rubbing their hind femora over a series of ridges located on the side of the abdomen and use their entire body cavity as a resonator to generate loud, low-frequency sounds that can be heard at a distance of up to 2 km.

Figure 1 (left) A singing male of the Guyanese cricket *Lerneca fuscipennis* singing in the leaf litter. (right) A female Zubowski's grasshopper *Stenobothrus eurasius* listening to the courtship song of the male next to her. Photos: (left) Jérémy Anso; (right) Oliver Hawlitschek.

THE SIGNALS?

The shape and the acoustic properties of the signals are determined by the functioning of the stridulatory apparatus of the species. Thus, a grasshopper that uses its hind legs can emit a continuous sound, both legs playing either in the same time or alternating – that is, generating either the same signal or different signals more or less mixed into the song of the species (see chapter 4.4, Field grasshoppers, some of the most versatile singers among Orthoptera). In Ensifera, the file can be hit by the plectrum either during wing closure only (most crickets, mole crickets) or during wing closure and wing opening (katydids, some crickets). The descriptions of the songs of Ensifera take this process into account.

Within Orthoptera, crickets have often been considered "perfect musicians" because their song has, to human ears, a musical quality never heard in Caelifera and rarely heard in katydids: this is due to the functioning mechanism of their stridulum, which is well-tuned, and where the speed of the file hitting corresponds to the main resonant vibration of the main resonators of the fore wings. Having crickets on the summit of acoustic evolution for their songs and for the functioning of their stridulum is a pleasant tale, especially for a cricket taxonomist, but it is not completely true. Some crickets emit non-musical songs, while some katydids emit very nice-sounding musical sounds (see chapter 4.3, Many different styles of singing in crickets, katydids, mole crickets, and grigs).

THE HEARING?

To be used in intraspecific communication, sounds must be detected before being recognized and evaluated at the brain level. This is performed by hearing organs called tympana (see chapter 4.2, A sophisticated audience). Just like the modes of sound production, the tympanal organs are different in Caelifera and Ensifera. Caeliferan tympana are, more or less uniformly, made of a large membrane stretched over a sclerotized frame on the first abdominal segment. Ensiferan tympana are present only in crickets, katydids, and Anostostomatidae (and some Mesozoic Hagloidea): these oval or slit-like openings are located on the inner and/or outer sides of each fore leg, but their inner structures can vary, especially in the arrangement of the sensory cells (fig. 3).

Most insects possess so-called chordotonal organs on their tibiae, which allow the detection of vibrations. These organs are made of sensory cells connected to a trachea, one of the tubes that allow the insects to breath. The ensiferan tympanum can be seen as a chordotonal organ covered by a membrane that is able to vibrate under sound pressure (for details, see chapter 4.2). Thus, many mute Ensifera can perceive the vibrations generated by the sounds produced by others, while many atympanate Ensifera can still stridulate for interspecific communication. Tympana are necessary for an efficient intraspecific communication, but as mentioned above, sound signals can be perceived by their multiple components.

Tympana of Ensifera alone cannot be used to locate a sound source because the wave length of the sounds is much longer than the distance between

< **Figure 2** (top row) Dorsal views of left and right fore wings of (left) *Paragryllus* sp. (Gryllidea, Phalangopsidae); (middle) *Gryllotalpa* sp. (Gryllidea, Gryllotalpidae); and (right) *Nesonotus vulneratus* (Tettigoniidea, Phaneropteridae), showing the diversity of the song-producing structures (stridula) in Ensifera. (bottom row) The corresponding stridulatory files located on the ventral sides of the fore wings (see red arrows in the top row). Photos: Sylvain Hugel.

the tympana. They are complemented by a pair of openings located on the second thoracic segment, the first respiratory spiracles, which are connected to the tympana by the trachea and may be separated from one another by a median septum. The whole structure, made of the two tympana and the two spiracles, is a two- or four-entry structure enabling directional hearing to locate a sound emitter.

SOUNDS AS CHARACTERS TO IDENTIFY ORTHOPTERA

The repertoire of a species includes all the sounds a species can emit in different situations. Among these, the calling song is often the most species-specific, as it allows the recognition of the calling individual by potential mates as a member of their own species. It is also the song that human observers use to identify singing Orthoptera. Thereby, songs are used now in eco-acoustic studies to survey biodiversity in a non-invasive way, allowing observers to detect which singing species are present and when, but also allowing them to detect invasive species or the loss of singing species previously recorded in a biotope (see chapter 7.1, Conserving Orthoptera diversity).

Figure 3 (main photo) Auditory tympana of Ensifera are located on the tibia of the first pair of legs, as shown here on a male of *Homoeogryllus reticulatus* (Gryllidea, Phalangopsidae). (inset) (A) Tympana in a cricket (*Pentacentrus* sp., Gryllidea, Gryllidae) and two katydids: (B) *Holochlora biloba* and (C) *Nesonotus vulneratus* (Tettigoniidea, Phaneropteridae). Photos: (main) Philippe Grandcolas, (A–C) Sylvain Hugel.

A B C

4.2 A SOPHISTICATED AUDIENCE: THE SENSE OF HEARING IN CRICKETS AND THEIR KIN

by Fernando Montealegre-Zapata

The ability to detect airborne sound is expressed in five Ensifera families: Prophalangopsidae, Tettigoniidae, Anostostomatidae, Gryllidae, and Gryllotalpidae, with most species displaying tympana (eardrums) in the proximal part of the fore tibia. While hearing physiology and biomechanics have been intensely studied in model species, there are still many gaps in our understanding of the evolution of hearing in Orthoptera as a whole.

Ensifera produce sound across a wide variety of acoustic channels (1–150 kHz; fig.1). Their hearing sense is equally sophisticated, showing outer-, middle-, and inner-ear components, as in vertebrates. The main purpose of louder sounds produced by ensiferans is intraspecific communication, mostly to attract potential mates (see chapter 4.1, Musicians of the insect realm). At the same time, these loud sounds bear a high risk of giving away the sound emitter's position to potential predators, often nocturnal mammals with a keen sense of hearing. Therefore, Ensifera have presumably been under strong adaptive pressure to produce sounds that are difficult to locate by predators but can still be detected by conspecifics, resulting in the evolution of some of the most elaborate of all hearing systems in the animal kingdom.

ENSIFERA HEARING SYSTEMS

Some ensiferans, such as the Gryllacrididae, are, as far as we know, largely unable to perceive sounds. Most, however, have highly sophisticated hearing systems involving an outer, middle, and inner ear. The outer ear often includes a so-called acoustic trachea (AT; fig.2), whose function has derived from the original tracheal function of gas exchange. In "relic hearing" Ensifera, such as the Prophalangopsidae, the prothoracic tracheae have retained their original function and structure: they are bifurcated at the femora and preserve the tracheal fine ramifications supplying the leg. Prophalangopsidae ears, therefore, are basic pressure receivers, capturing sound on the external tympanal surface only. In most Tettigoniidae and Gryllidae, the AT, along with two active tympana, receive sound externally and internally (via the AT), thus creating a pressure-difference-receiving system. A tympanal plate on the eardrum serves as a middle ear, converting airborne sound into fluid vibrations. Tettigoniidae often have so-called acoustic bullae, or exponential horns, at the proximal ends of their AT, which amplify the sound by 10 to 30 dB. Many Tettigoniidae furthermore have an inner ear with a *crista acustica* (CA) enclosed in a fluid-filled cavity, the auditory vesicle, which enhances traveling waves for frequency mapping. This seems to occur in species that show auditory pinnae; the pinnae function as bat detectors, capturing only bat echolocating sweeps and not conspecific calls (conspecific calls are captured by the AT).

The inner ear of Gryllidae is often considered the most sophisticated of all ensiferans. In many crickets, the left and right AT are connected by a middle septum, which allows sound to travel between the ears. As a result, each ear receives sound in three ways: externally, internally from the same-side trachea, and internally from the opposite ear. This condition provides excellent directional hearing abilities.

Species lacking a functional AT may have other ways of enhanced hearing capabilities. The CA of the Prophalangopsidae (e.g., *Cyphoderris* spp.) is very simple when compared with that of Tettigoniidae. However, it has been recently shown that *C. monstrosa* achieves frequency mapping via traveling waves on the tympanic membrane. The mechan-

Figure 1 (top) *Supersonus aequoreus*; (middle) *Cyphoderris monstrosa*: (bottom) *Typophyllum spurioculis*. Some Ensifera use pure-tone calls for intraspecific communication. Pure-tone calls are difficult to detect by mammalian predators. The purity or sharpness of a tone is measured using the quality factor Q, which is defined as the ratio of center frequency to bandwidth ratio. As shown in the diagrams, all calls displayed here have narrow bandwidths. Sounds with high Q values sound pure and "musical" to human ears; musical instruments often have values of Q >100. The calls of ensiferans using pure tones usually have Q values of >15. Q>10 tones are more difficult to localize by mammals than less pure tones. Figure: Fernando Montealegre-Zapata.

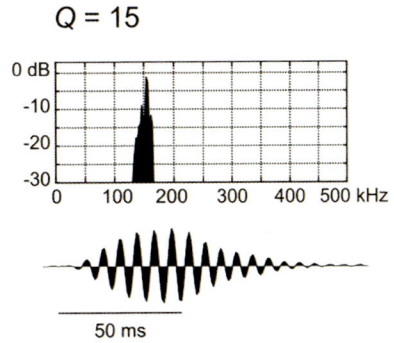

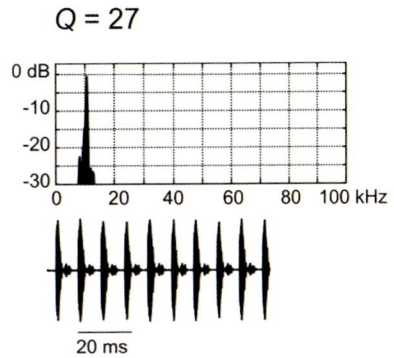

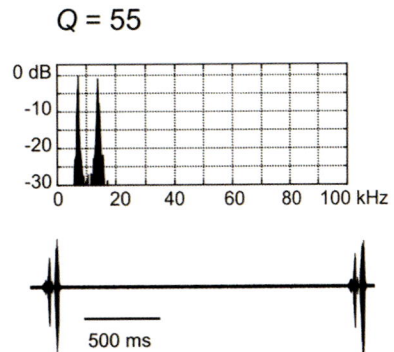

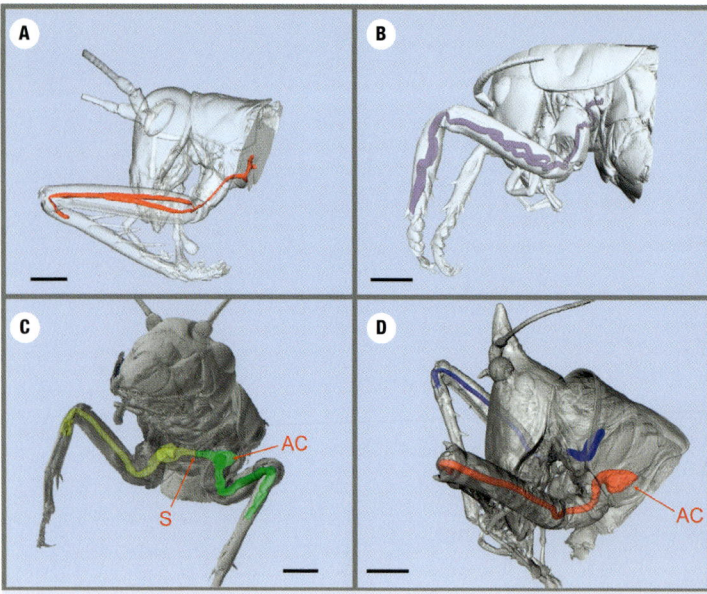

Figure 2 Hearing systems in Ensifera. AC = acoustic spiracle; S = median septum (A). The bifurcated acoustic trachea of atympanate ("deaf") Ensifera (Gryllacrydidae). (B) Bifurcated and non-functional trachea in *Cyphoderis monstrosa* (Tettigoniidae); sound is captured at the external surface only. (C) The acoustic trachea of field crickets (*Gryllus bimaculatus,* Gryllidae) enables cross talk facilitated by a median septum (*Gryllus bimaculatus,* Gryllidae), which enhances directional hearing. (D) The most common form of acoustic trachea in katydids, the exponential horn (in this case *Copiphora gorgonensis,* Tettigoniidae). These ears also work as pressure-difference receivers capturing sound externally and internally via the acoustic spiracle (S) and trachea. Scale bar = 2 mm. Figure: Fernando Montealegre-Zapata.

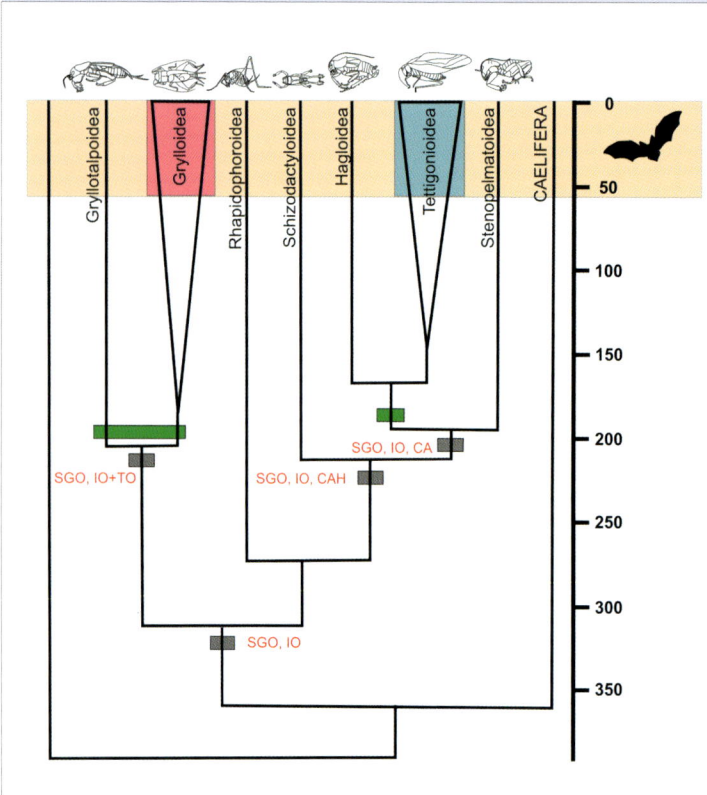

Figure 3 Proposed scenario for the evolution of hearing in Ensifera. (Phylogeny modified from Woodrow et al. 2022.) Tips of the tree represent ensiferan superfamilies. A time scale is given on the right. Gray boxes on the branches represent the evolution of major hearing systems, as inferred from homologies (anatomical structures in red; see text for abbreviations). Green boxes highlight the appearance of low-frequency pure tones. The pink/blue boxes on top of the tree indicate the appearance of the 'insect cochlea' as most notable ear adaptation in the evolutionary history of Ensifera. Both coincide with the emergence of bats (brown background). Figure: Fernando Montealegre-Zapata with data from Gorochov 1995 and Gu et al. 2010.

oreceptors of a precursor CA in living Prophalangopsidae are neither completely isolated from the tympanic membrane nor enclosed in a fluid-filled auditory vesicle, as in Tettigoniidae.

THE ORIGIN OF INSECT EARS

Insect ears derived from sensory organs receptive to vibrations. Vibro-receptors have very different anatomical structures and occur in various body parts. In the insect legs, such organs can be found in the femur and in the tibia. Organs located in the latter take the form of the subgenual organ (SGO) and the complex tibial organ (CTO).

The process of evolution from vibro-sensitive organs to functional ears involved structural changes that affect the types of stimuli that the organ can detect. A tympanum may evolve by thinning the cuticle near key precursor auditory receptors and requires a respiratory trachea backing the vibrating surface. Early hearing-organ precursors could only detect very loud sounds; later, many Orthoptera evolved tympana. Some sensory systems, such as the SGO in cockroaches and the hind-wing chordotonal organ in locusts, can detect high-amplitude sounds without a tympanum. This also happens in the sensory organs of the atympanate middle and hind legs of Tettigoniidae, so the forelegs have gained auditory sensitivity over time. Analogous forms of atympanate sound detection have even been shown in spiders.

Research has found that, in non-hearing Ensifera species (Stenopelmatidae, Gryllacrididae, and Schizodactylidae), the sensory structures that make up the tibial organ are homologous to the auditory receptors found in Tettigonioidea. These taxa have a three-part sensory organ, homologous to the tettigoniid CTO: the proximal SGO, the intermediate organ (IO), and an additional third part with a linear array of sensory cells, resembling the CA of Tettigoniidae. It has been termed *crista acustica* homolog (CAH).

The CAH in atympanate taxa is seen as the precursor organ of the CA in Tettigonioidea. Since in atympanate Tettigonioidea, the entire organ is deemed to function as a vibration sensor, the CAH evolved into the sound receptors known as CA in the hearing Tettigoniidae, Anostostomatidae, and Prophalangopsidae. In contrast, the auditory sensors of crickets (Grylloidea) are only bipartite and include only the SGO and the tympanal organ (TO), which contains the auditory receptors. In Gryllidae, the TO seems to have evolved from the IO and not from a CAH, suggesting that the hearing organs in the gryllid and tettigoniid lineages evolved independently and from different sets of sensilla (fig. 3).

Figure 3 includes a phylogenetic tree of Ensifera mapping tympanal ears and sensory organs, and proposes notable events potentially triggered by the emergence of bats as predators of flying insects, including the "insect cochlea." The tympana of living Prophalangopsidae are like those of their Jurassic relatives, indicating that pressure receivers may be the original method of hearing in Ensifera. The most common type of AT in modern katydids is the exponential horn, which first appeared in the Eocene Epoch (~44 MYA). Numerical models on accurate 3-D geometries suggested that this ancient AT was sensitive to from approximately 10 kHz to 89 kHz, which includes the frequency range of specific calls (~31 kHz) and early bat-echolocation calls. A similar analysis shows the earliest evidence of auditory pinnae in the same period, but also suggests that pinnae appear first as a tympanal protective structure and that then the pinnae cavities were modified to resonate at bat-echolocation frequencies.

Unlike tetrapods, insects independently evolved tympanal organs at least 19 times, with ears found in at least 10 different body parts, across several taxa. Insect ears evolved from proprioceptors scattered throughout their bodies; tympana evolved by the thinning of their exoskeleton in relevant body regions supported by such vibro-receptors. In Ensifera, they evolved in the fore legs and not in the head (unlike tetrapods), which was likely to maximize the separation of the ears to hone directional hearing. Would the molecular mechanisms involved in hearing (mechanoreception, mechanotransduction, and amplification) in katydids and tetrapods be conserved or did they evolve separately, given their similar foundational processes?

4.3 MANY DIFFERENT STYLES OF SINGING IN CRICKETS, KATYDIDS, MOLE CRICKETS, AND GRIGS

by Laure Desutter-Grandcolas

Orthoptera show a high diversity across all facets of bioacoustics, particularly the Ensifera. The evolution of this acoustic diversity, the subject of research for more than a century, focuses mostly on the wing structures used to produce the acoustic signals. It takes into account the crickets (Grylloidea), mole crickets (Gryllotalpoidea, usually not separated from true crickets as far as bioacoustics is concerned), katydids or bush crickets (Tettigonioidea), and the extant species of grigs (Hagloidea, Prophalangopsidae; all referred to here as the Acoustic Four), plus the diverse fossil Ensifera (see chapter 1.2, Ancient worlds).

Two main hypotheses have been proposed to explain the origin and evolution of acoustic communication in the Acoustic Four: (1) acoustic communication evolved only once in Ensifera and is homologous in all acoustically communicating Ensifera (the inheritance hypothesis); and (2) acoustic communication has evolved several times in Ensifera, and the structures involved are not homologous (the convergence hypothesis). However, several problems have biased the analysis of this biological mystery: (i) The structures involved in the production and perception of sounds in Ensifera are actually quite similar in crickets, mole crickets, katydids, and grigs: all these species stridulate with their fore wings raised above their body, rubbing a file (located on one fore wing) on a scraper (located on the other fore wing). In the same way, all these species can hear, thanks to the tympana located on the tibia of their forelegs. Too similar not to be the same? (ii) Most studies have limited their scope to communication with the fore wings, but acoustic communication in Ensifera is much more diverse. The complete story of their evolution needs to take into account all ways of sound production and perception, without choosing a priori one to focus on. (iii) No evolutionary hypothesis can be scientifically tested without a phylogenetic tree and without a reasonable analysis of the homologies of the characters that are compared. Let's discuss these three points and see what they mean in terms of acoustic evolution in Ensifera.

TOO SIMILAR NOT TO BE THE SAME? IN FACT, NOT SO SIMILAR!

The Acoustic Four stridulate with their raised fore wings, but this apparent similarity hides a world of structural and functional differences (fig. 1):

— Forewings are almost symmetrical in crickets, mole crickets, and grigs, but highly asymmetrical in katydids: both wings present a file and resonators in the former, while in katydids, the file is on the left fore wing only and the resonator on the right.

— In singing crickets, the right fore wing is on top of the left (hence it bears the functional file), but this is the reverse in katydids. Both mole crickets and grigs can reverse the fore-wing position during singing.

— Which vein bears the file in Ensifera? Several hypotheses have been proposed over the last century. The recent use of X-ray microtomography allows identifying the veins from their very bases, at which they emerge from the bullae near the thorax, and shows that the stridulatory file is situated on the anterior post-cubital vein (PCuA) in crickets and katydids, and on a branch of the posterior cubital vein (CuP) in mole crickets and grigs (Prophalangopsidae, genus *Cyphoderris*). Hence, the resonant areas of the fore wings are not delimited by the same veins and are not homologous.

— In most crickets, the two fore wings vibrate to produce the call, and the stridulation system is well-tuned, so that functional constraints produce a "musical" sound largely at frequencies audible to human observers (ca. 2 to 8 kHz). By contrast, katydids have asymmetrical fore wings that do not vibrate in phase; the signal is less structured and does not appear as musical as that of crickets, and with a high proportion of ultrasounds (>20 kHz). The grig (*Cyphoderris* spp.) resembles crickets with its symmetrical fore wings, both of which vibrate, and the relatively low frequency of its songs (around 12 to 15 kHz). Mole crickets sing from a burrow that they construct in a specific way to enhance the power of their song.

ENSIFERA, A VERY DIVERSE ACOUSTIC CLADE

All ensiferan groups have members capable of stridulation, but not all use the same mechanisms. The Acoustic Four mostly stridulate, as described above, by rubbing their fore wings against each other (so-called elytro-elytral stridulation). However, there are also many Ensifera, such as Gryllacrididae, Rhaphidophoridae, Schizodactylidae, and Anostostomatidae, that do not stridulate with their wings. Anostostomatidae, Gryllacrididae, and Schizodactylidae stridulate by rubbing their hind femora against spines diversely arranged in rows or fields on the abdomen (so-called femoro-abdominal stridulation). Rhaphidophoridae communicate by vibrations of their abdomen (tremulation).

The behavioral contexts of stridulation may also differ. Intraspecific communication requires a hearing organ, the tympanum, but interspecific communication does not necessarily. Vibrations emitted by stridulation or tremulation can also be perceived by the subgenual organs of the legs. Apart from the Acoustic Four, Anostostomatidae cannot sing with their wings, but they stridulate, and they may have tympana on their fore legs. Also, all Ensifera, whether acoustic or mute, detect vibrations with their legs. As soon as a mode of signal detection has evolved, other organs and behaviors may follow in response, as shown by the ultrasonic communication in Lepidoptera that has been proposed to have evolved as a response to the ultrasonic echolocation of bats.

The groups of the Acoustic Four also differ in the proportion of their member species capable of stridulation. Nearly all katydid species communicate by sounds, whereas many crickets are mute, being apterous, brachypterous, or fully winged but devoid of a stridulum. The evolution of the Oecanthidae crickets (the "fifth family" of the Grylloidea clade) clearly shows how acoustic communication has been lost many times in a single cricket clade, without reversal and with a limited diversification of mute species: crickets may have evolved many times independently toward the world of (relative) silence from a noisy ancestor.

A PHYLOGENY AS A GUIDE TO TEST EVOLUTIONARY HYPOTHESES

Testing an evolutionary hypothesis requires a phylogeny reconstructed from the analysis of a robust data matrix, where all the terminals are documented for all the characters, and the characters in the data matrix or the characters of interest optimized onto the phylogeny must be defined according to reasonable hypotheses of primary homologies. When structures are very similar, but too different to support hypotheses of primary homologies, then a hypothesis of convergence must be proposed. This is the case, for example, with the wings of birds and bats, which are built from different parts of a vertebrate arm, or of a stridulatory file not located on the same vein, as in crickets (PCuA) and mole crickets (CuP). But if one structure can support a hypothesis of primary homology – as with the stridulatory file in crickets and katydids, both located on the PCuA vein – the pattern of the phylogeny will support a hypothesis of inheritance (sister groups) or a hypothesis of parallel evolution (not sister groups).

Let's consider the phylogenetic relationships among Ensifera shown in figure 2. If the file was homologous throughout all Ensifera, which it is not, it would then be more parsimonious to consider a parallel evolution of the file in the Gryllidea, on one hand, and in the Tettigoniidea p.p. (grigs, Tettigonioidea), on the other, as many basal members of Ensifera do not have wings to stridulate.

As the file proved to be non-homologous, the most parsimonious scenario of the evolution of the stridulatory file in Ensifera implies a minimum of four steps – that is, the occurrence of the file on the PCuA in crickets (one step), the occurrence of the file on a branch of the CuP in mole crickets (one step), the occurrence of the file on the PCuA in katydids (one step), and the occurrence of the file on a branch of the CuP (one step) in *Cyphoderris*.

Mapping of the stridulatory apparatuses on the Ensifera phylogeny thus supports both multiple convergent and parallel occurrences of wing stridulation in this clade. The functional constraints resulting from fore-wing shape and the movements necessary to sing with raised tegmina could explain why the different files and plectrums occur close to the base of the fore wings, on a limited number of candidate veins. A similar conclusion may be obtained as femoro-abdominal stridulation becomes better documented in Ensifera.

Figure 1 Wing venation and homologies for the stridulatory file (red arrow) as reconstructed using microtomography in Ensifera, showing the fore wing, the vein reconstruction, and the species for: (A) *Cyphoderris* sp. (Tettigoniidea, Prophalangopsidae; photo: Delise and Matthew Priebe); (B) *Quiva* sp. (Tettigoniidea, Tettigonioidea; photo: Artour Anker); (C) *Brachytrupes membranaceus* (Gryllidea, Gryllidae; photo: Bernard Dupont, Wikimedia Commons); (D) *Gryllotalpa* sp. (Gryllotalpoidea, Gryllotalpidae; photo: Martin Husemann). Vein identity after Schubnel et al. (2019): costal vein, deep blue; posterior subcostal vein, red; radial vein, pink; media vein, lighter blue; fusion of media and anterior cubital veins (apomorphy of Orthoptera), green; cubital vein, yellow (posterior cubital, lighter yellow); postcubital vein, gray; anal veins, black.

Figure 2 Phylogenetic relationships of Ensifera and the evolution of song. The Acoustic Four are highlighted in yellow: Tettigoniidae, Prophalangopsidae, the Gryllidea group, and Gryllotalpidae. The photos show families of Gryllidea; see fig.1 for photos of other families. Blue squares indicate the position of the file: dark blue = on the anterior post-cubital vein of the forewing (PCuA), light blue = on a branch of the cubital vein (CuP). This suggests that all four groups have evolved their singing capacity independently, as groups with similar stridulatory structures are not closely related. Figure by Oliver Hawlitschek. Phylogenetic tree based on Shin et al. (2014) (see chapter 1.3, The phylogenetic history of Orthoptera). Photos: Sylvain Hugel.

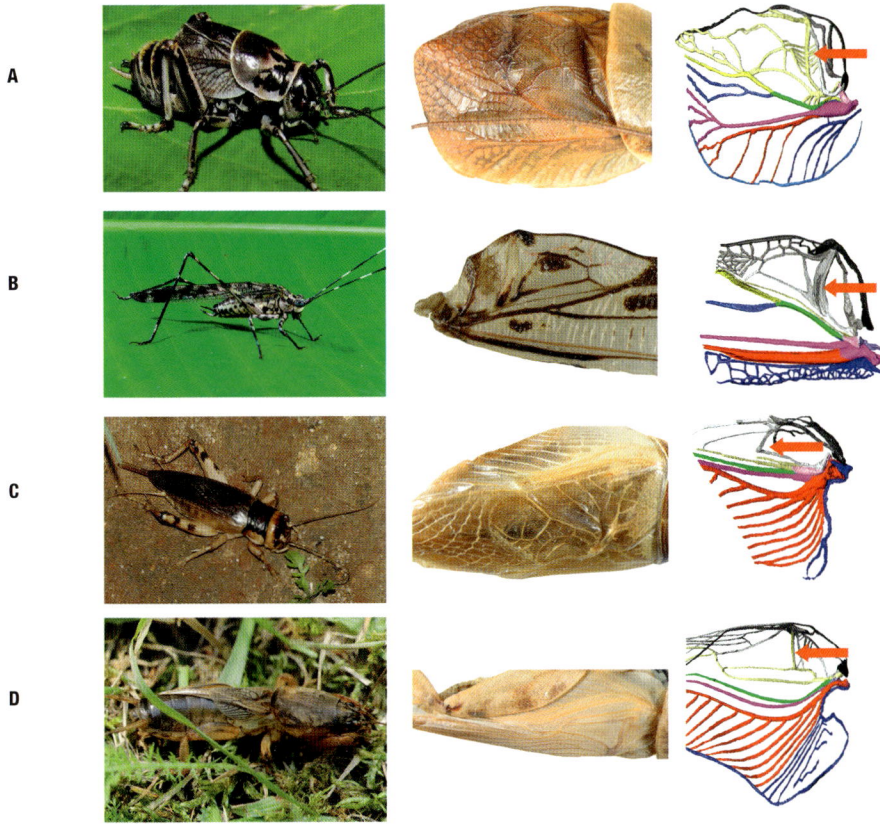

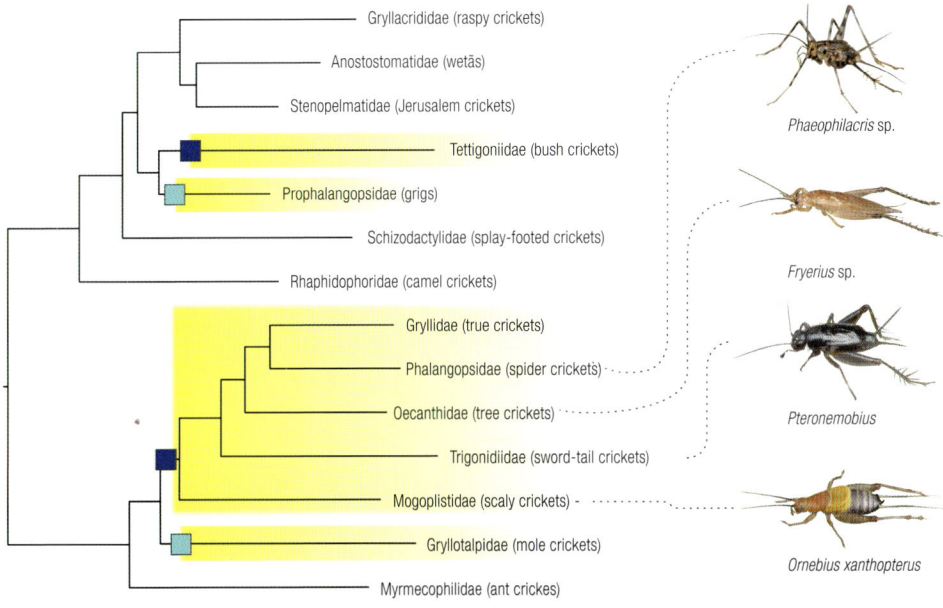

4.4 FIELD GRASSHOPPERS, SOME OF THE MOST VERSATILE SINGERS AMONG ORTHOPTERA

by Varvara Vedenina

Gomphocerinae, or slant-faced grasshoppers, comprise perhaps the most diverse and species-rich subfamily of Acrididae, occurring on all continents except for Australia and Madagascar. Members of Gomphocerinae can be distinguished from other subfamilies of Acrididae by a characteristic structure of the acoustic apparatus – namely, by the presence of stridulatory pegs on the inner surface of the hind femur. Among Acrididae subfamilies, acoustic communication in Gomphocerinae is most developed in terms of complexity of stridulatory leg movements, the number of sound elements, and mating strategies. Notably, acoustic communication was most studied in Palearctic members of this subfamily, and this chapter treats species from only this geographical region.

The song is produced by stroking a stridulatory file of each hind femur across a raised vein on the ipsilateral wing (i.e., the wing on the same body side). In most species, the sound has a broad frequency spectrum, so that the specificity of the calling songs lies not in their frequency band, but almost always in the temporal pattern. However, dominant frequencies of different courtship elements can shift either to lower values (10–15 kHz) or to the ultrasound band (30–40 kHz; see below). Using both hind legs, the grasshoppers have two separate sound-producing devices, in contrast to Ensifera, which produce sound with one stridulatory apparatus. The stridulatory movements of the two legs can differ in amplitude and form, and the legs can exchange roles from time to time.

To analyze the stridulatory leg movements, the German scientists Otto von Helversen and Norbert Elsner developed a unique opto-electronic device in 1977. A piece of reflecting foil was glued to the distal part of each hind-leg femur of a male, and two opto-electronic cameras focused on the illuminated reflecting dots. Each camera was equipped with a position-sensitive photodiode that converted the upward and downward movements of the hind legs into voltage changes. These signals, together with the microphone recordings of the sounds, were A/D-converted with a custom-built PC card. Later, this design was further developed in various laboratories in Germany and other countries.

The opto-electronic device produces a graphical representation of the leg movements that can be analyzed jointly with an oscillogram – that is, the graphical representation of an audio recording of a song. There are three lines on the oscillogram of each song recording: the two upper lines represent movements of the two legs; the lower line represents the temporal sound structure (fig.1). For the song description, we distinguish the elements *pulse, syllable*, and *echeme*. *Pulse* is a structural unit of the lowest level, which is produced by one stroke of a hind leg against a fore wing (fig.1). *Syllable* starts when the legs leave their initial position and ends when the legs return to this position; the syllable can be produced by one complete up-and-down leg movement or by several up-and-down leg movements. *Echeme* is a structural unit of the highest level, representing a series of consistent syllables separated by pauses.

Various Gomphocerinae species demonstrate different degrees of song complexity. One of the simple song patterns may be characterized as a sound produced by straight upward and stepwise downward movements of the hind legs (*Gomphocerus sibiricus*; fig.1A). The stepwise movements produce a series of pulses separated by gaps. The

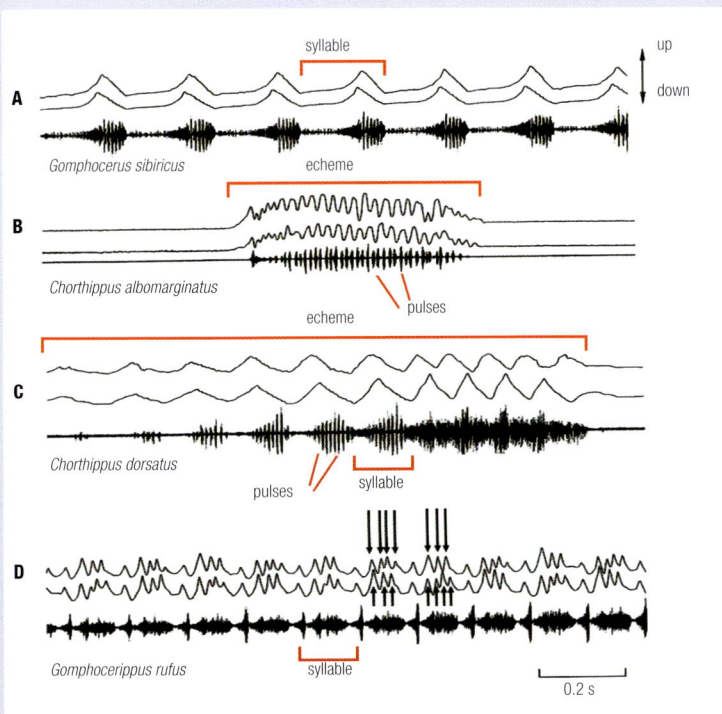

Figure 1 Oscillograms of the Palearctic Gomphocerinae calling song and terminology used in the acoustic analysis. The two upper lines are recordings of hind-leg movements, and the lower line is the sound recording. In *G. rufus* (D), arrows show that the legs exchange roles every syllable so that each leg produces either three or four up- and downstrokes. Image by Varvara Vedenina.

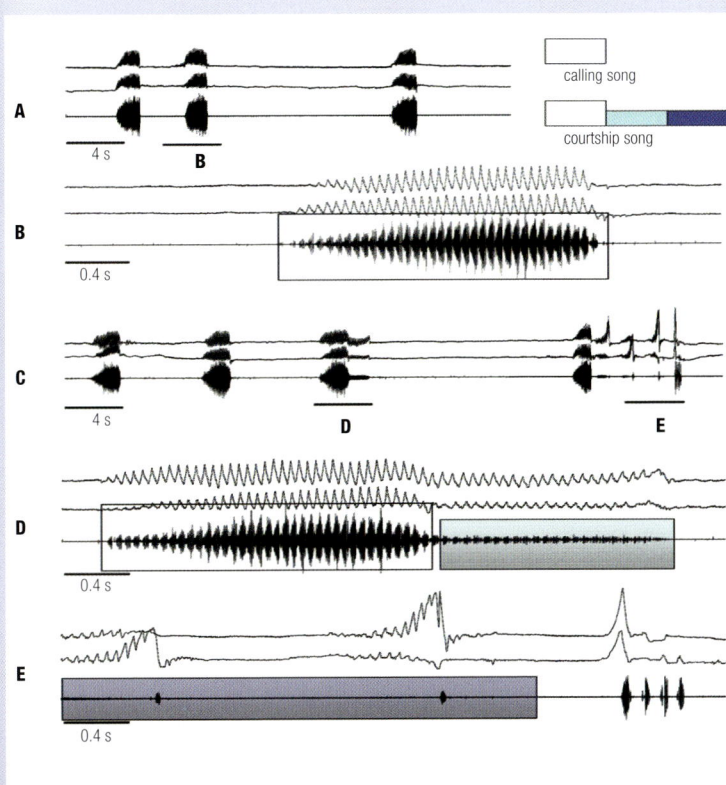

Figure 2 Oscillograms of the calling (A–B) and courtship (C–E) songs of *Omocestus minutus* at two different scales. In all figures, the two upper lines are recordings of hind-leg movements, and the lower line is the sound recording. Rectangles indicate different song elements, showing that the courtship song possesses one calling element and two unique courtship elements. Image by Varvara Vedenina.

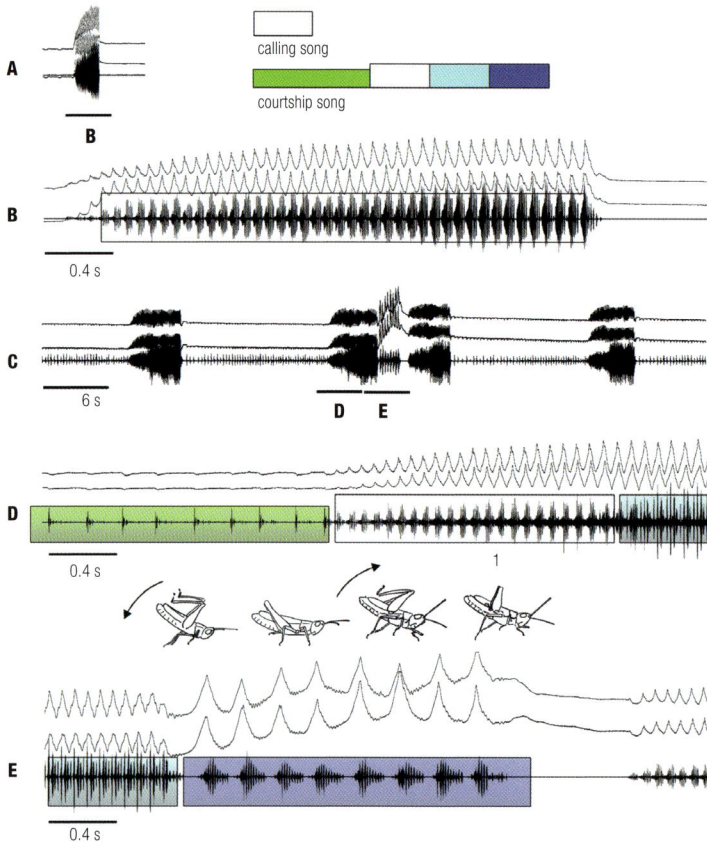

Figure 3 Oscillograms of calling (A–B) and courtship (C–E) songs of *Stenobothrus fischeri* at two different scales. In all figures, the two upper lines are recordings of hind-leg movements, and the lower line is the sound recording. Similar leg movements generate two different sound patterns, shown as 1 and 2 (D). Drawings show the strokes with the hind tibiae and the side-to-side movements of the whole body at the corresponding moments of the song (E). Rectangles indicate different song elements, showing that the courtship song possesses one calling element and three unique courtship elements. Image with drawings by Varvara Vedenina.

Figure 4 (A) This male of *Chorthippus oschei* kicks his legs high in the air, a typical element of the species-specific courtship. (B) This male hybrid between *C. oschei* and *C. albomarginatus* also kicks, but with substantially less fervour. Will the female on the underside of the straw be impressed? Photos: Ján Svetlík.

legs are moved almost synchronously. The song complexity may increase by the changing of the phase shift between the two legs; for example, *Chorthippus albomarginatus* starts singing with synchronous leg movements, but most of the song is produced by movements at antiphase (fig.1B). In *C.dorsatus*, the first four to six elements are produced by synchronous movements with stepwise downstrokes, which generate distinct pulses; a second song part is produced by straight downstrokes of the legs moved at antiphase, which result in a buzzing element without distinct pulses (fig.1C). In *Gomphocerippus rufus*, the legs exchange roles every syllable, so that each leg produces either three or four up- and downstrokes (fig.1D).

The song in Gomphocerinae also varies according to the behavioral situation. A solitary male produces a *calling song* to advertise his position to females over longer distances. A female that is ready to mate approaches him and/or responds with her own song. The response of a female may initiate a duet, during which the female usually sits still while the male approaches her. When a male finds a female, in many species the male begins a distinct courtship song, which just carries over short distances, may reach a high complexity, and may be accompanied by conspicuous movements of different parts of the body, such as the abdomen, head, antennae, or palps. In the conditions of high population density typical for many grasshopper species, a male may occasionally meet a conspecific female by chance. In such a situation, the male may start to court even without producing a calling song first. Courtship songs, therefore, should function both for species identification (otherwise a function of the long-range calling song) and for evaluation of mate quality and sexiness (otherwise a function of the close-range courtship behavior). In most grasshopper species, one courtship element is usually similar to the calling song. For example, in *Omocestus minutus*, the first part of the courtship sequence is similar to the calling song, while the second part contains a much quieter element (fig.2). After that, one leg gradually rises and produces several up- and down-

strokes. The single pulse produced by these strokes is relatively quiet, which indicates that the high-amplitude strokes of one leg serve as a visual signal. In *Stenobothrus fischeri*, courtship starts with small-amplitude movements of the hind legs producing quiet pulses (fig.3). Then the legs generate movements of the calling-song type. However, the sound produced by this movement pattern is similar to the calling song in the first half of the sequence; in the second half of the sequence, a different sound element is produced. We suppose that the legs are pressed to the wings with different forces, which cannot be visualized in the oscillogram (fig. 3D). Then the high-amplitude leg movements follow and produce a fourth sound element. During these high-amplitude strokes of the legs, the tibiae are kept away from the body at an angle of about 20 degrees. This phase of the courtship is accompanied by side-to-side movements of the whole body.

The evolution of male song and female preference in gomphocerine grasshoppers may be faster than in other members of Orthoptera due to their specific mating strategy: during courtship, several competing males often sing around a female, and courtship songs are very long and frequent. When a female refuses the first attempt at copulation, but at the same time does not retreat, a male can continue to sing for two or even more hours. Females can hear and compare different courting males for many hours in the field (fig. 4). This may facilitate female choice and favor competition among males, similar to a lek – that is, an aggregation of males gathered at a site to court a female.

4.5 DOES COMPLEX COURTSHIP PREVENT OR PROMOTE HYBRIDIZATION BETWEEN GRASSHOPPER SPECIES?

by Varvara Vedenina

Hybrid zones – that is, geographic areas where genetically distinct groups, typically species, of organisms interact, leaving at least some offspring of mixed ancestry – are found in all major groups of higher organisms. Scientists regard hybrid zones as natural laboratories for the study of the processes involved in divergence and speciation. Behavioral isolation is a crucial part of speciation, as parental species often evolve and maintain divergent courtship rituals. Behavior may contribute not only to pre-mating isolation, but also to post-zygotic isolation as a result of the reduced mating success of hybrid male offspring. Since the behavior and song of orthopterans are almost completely inherited genetically, hybrid males may suffer reduced fitness because their mating signals are intermediate between the parental signals and therefore unattractive to females of either parental species. Such lack of hybrid fitness is called behavioral hybrid dysfunction (BHD). Among grasshoppers, there are several examples demonstrating BHD. This has been especially well studied in the species-rich Gomphocerinae genus *Chorthippus*, which is widespread in Asia and Europe (see chapter 5.2.1, Grasshoppers of the vast Central Asian steppes).

Females of the closely related grasshopper species *Chorthippus brunneus* and *C. jacobsi*, as well as hybrid females, show reduced responsiveness to F1 (first-generation) hybrid male songs. Strong post-zygotic barriers, such as reduced courtship motivation of F1 hybrid females and behavioral isolation of F1 hybrid males and females, were also found between *C. biguttulus* and *C. mollis*. Hybrid females between *C. biguttulus* and *C. brunneus* preferred songs of only one parental species, *C. biguttulus*, and discriminated against hybrid males. All these species, however, possess a rather simple calling and courtship behavior. Their calling songs are characterized by different degrees of complexity in temporal structure. In *C. brunneus* and *C. jacobsi*, the temporal pattern of the echeme is simple, whereas it is considered more complex and advanced in *C. biguttulus* and *C. mollis*. At the same time, the number of song elements is small in all these species: calling songs consist of one repeating element in three species and of two elements only in *C. mollis*. Courtship behavior can be characterized as simple: courtship songs are very similar to calling songs in all species except *C. biguttulus*, which adds one element during courtship.

In other grasshopper species, courtship songs reach high levels of complexity and are accompanied by conspicuous visual displays. We would expect that hybrids suffer particularly reduced fitness in such systems because complex courtship behavior is highly relevant for mate choice. But is this true? We studied this in *Stenobothrus clavatus* and *S. rubicundus*, which meet in a narrow contact zone on Mount Tomaros in northern Greece (fig. 1A). These species differ in calling and courtship songs and in some obvious morphological characters: *S. clavatus* has conspicuous spatulate and darkened tips of the antennae, whereas the hind wings in *S. rubicundus* are dark and have heavily sclerotized leading edges (fig. 1B). Remarkably, both antennae and wings are used in visual display during courtship behavior. In playback experiments, females from the localities with the *clavatus*-song phenotype demonstrated assortative preferences, choosing songs of *clavatus*-males (fig. 1C). Females from the localities with the *rubicundus*-song phenotype also preferred to respond to conspecific songs; however, the differences between the response frequency to *rubicundus* song, *clavatus* song, and natural hybrid song were not significant. Finally, F1 hybrids bred in the lab did not show significant preferences for any of the four song types (with natural and F1 hybrids treated separately). Thus, hybrid females do not show reduced fitness; they can even have an advantage over parental females because they do not distinguish between the parental species. Hybrid males also do not show any reduced fitness in *rubicundus*-like localities. This is not what we expected. What could be the reason?

We studied a second system to find out more about this question. *Chorthippus albomarginatus* and *C. oschei* hybridize in a contact zone of about 200 km width in Ukraine and Moldova (fig. 2). In contrast to *S. rubicundus* and *S. clavatus* and to the *Chorthippus* species described above, these two species are similar in morphology and in calling songs but quite different in courtship songs, which are accompanied by conspicuous kicks with the hind legs in *C. oschei* (fig. 3). In the courtship song of *C. albomarginatus*, three elements alternate in a characteristic order, whereas five different elements are distinguishable in the more elaborate *C. oschei* song. Natural hybrids, firstly, demonstrated a high degree of inter- and intraindividual song variability compared with the rather narrow range of song variability in the allopatric populations, and, secondly, produced novel elements and novel combinations of elements. It is remarkable that similar song novelty was found in hybrids between *C. albomarginatus* and *C. karelini*, another closely related species, as well as in hybrids between *S. rubicundus* and *S. clavatus*.

In mate choice experiments conducted on *C. albomarginatus* and *C. oschei*, females from allopatric populations demonstrated strong assortative mating (fig. 4). At the same time, females from *albomarginatus*-like localities of the contact zone did not distinguish between conspecific and hybrid males, whereas females from *oschei*-like localities preferred conspecifics. The F1 hybrid females equally often chose males of *C. albomarginatus* and hybrid males and less often males of *C. oschei*, whereas F2 females did not show any selectivity among the three types of males. Thus, similarly to the hybrids between *S. rubicundus* and *S. clavatus*, hybrid females between *C. albomarginatus* and *C. oschei*, due to their little selective mate choice behavior, may even have an advantage over parental females in mixed populations. Hybrid males also do not show reduced fitness in *albomarginatus*-like localities.

A genetic analysis of the courtship-song differences between *C. albomarginatus* and *C. oschei* showed an unusual type of inheritance for most song characters studied involving duplicate pairs of loci and elaborate interactions between non-allelic loci. The homologous elements in the songs of *C. albomarginatus* and *C. oschei* could be controlled by the different copies of the duplicate loci. In the hybrids, both parental copies may be expressed, and this could result in the development

A

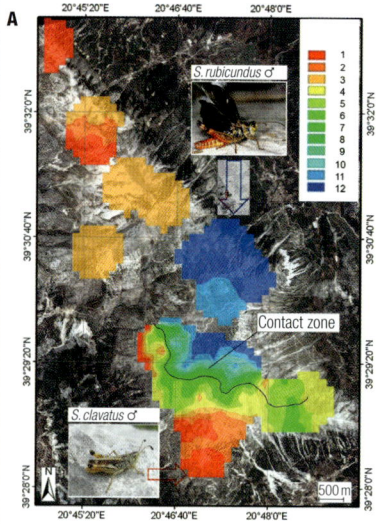

B

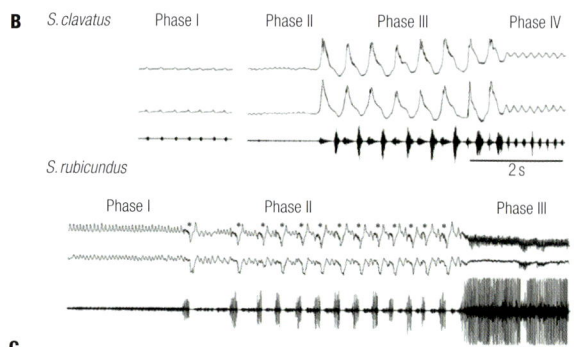

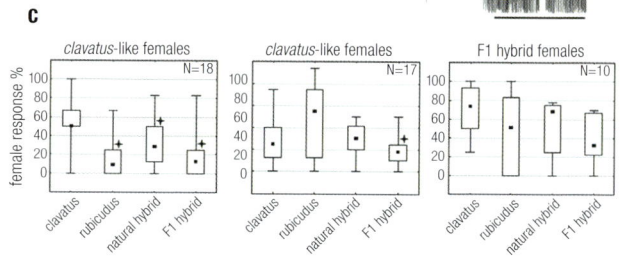

C

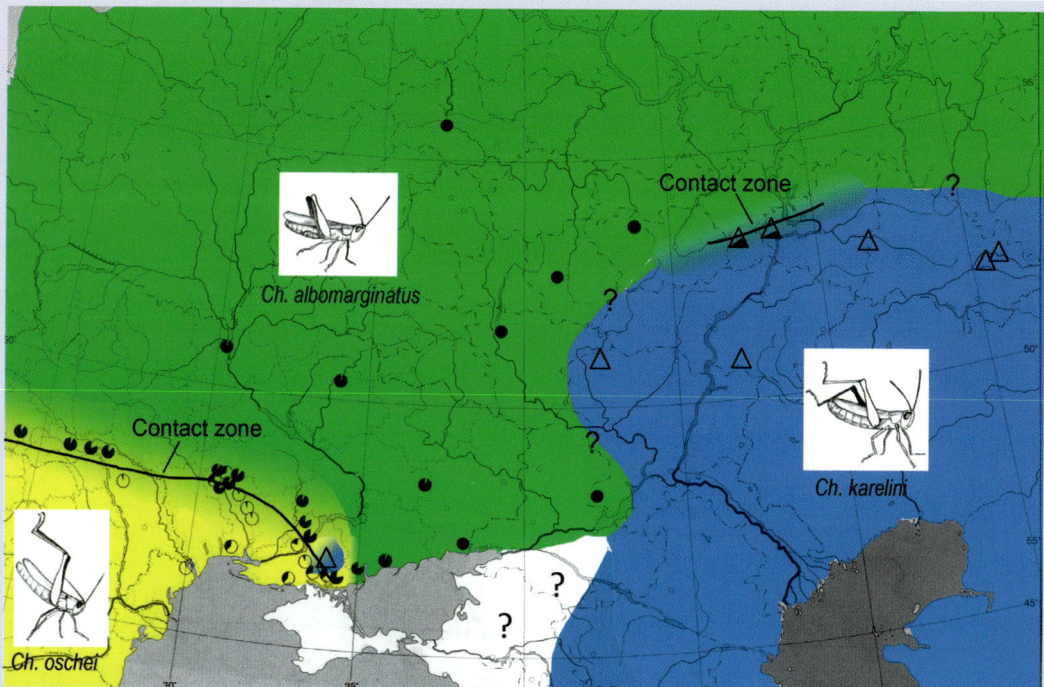

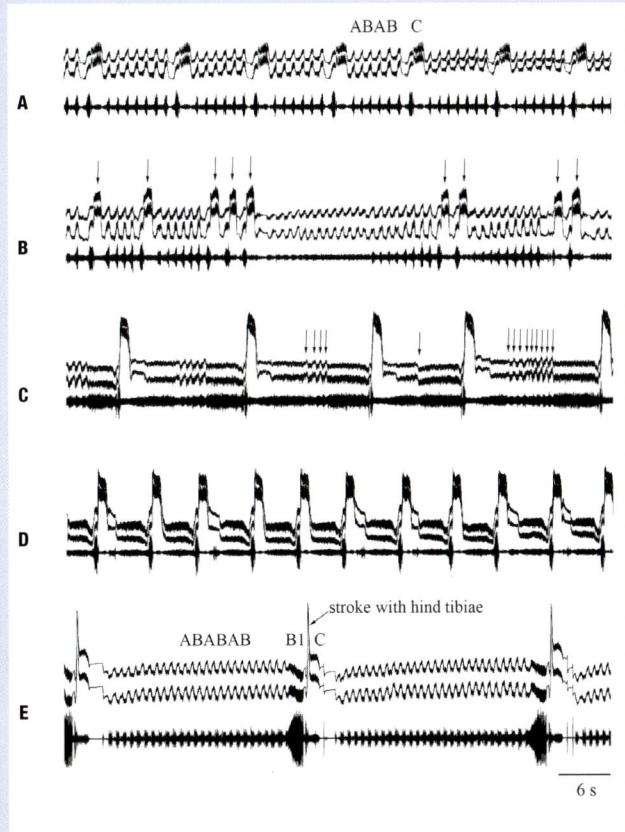

< **Figure 1** (A) Hybrid zone between *Stenobothrus clavatus* and *S. rubicundus* on Mount Tomaros (Greece) based on the courtship-song analysis. Courtship-song phenotypes from 1 (pure *S. clavatus*) to 12 (pure *S. rubicundus*) with intermediates 2 to 11 indicated by different colors, are shown on the right. The center of the contact zone is shown by the black line. (B) Oscillograms of the courtship songs (the two upper lines are recordings of hind-leg movements, and the lower line is the sound recording) in *S. clavatus* and *S. rubicundus*. In *S. rubicundus*, phase II is produced by two different mechanisms, leg stridulation and wingbeats (marked by asterisks); phase III is produced by wingbeats. (C) Responses of females (medians, quartiles, and minimum/maximum) to playback of the courtship songs of *S. clavatus*, *S. rubicundus*, natural hybrids (from Mount Tomaros), and F1 hybrid males. Image by Varvara Vedenina.

< **Figure 2** Map of distribution of three species of the *Chorthippus albomarginatus* group in Russia and adjacent countries. Sampling localities with the courtship-song recordings are indicated by black-filled circles (*albomarginatus*), open circles (*oschei*), and triangles (*karelini*). The centers of contact zones are shown by the black lines. The drawings (made by the author) show the highest positions of the hind legs when producing the C element of the courtship song; additionally, *C. karelini* and *C. oschei* flick the tibiae. The question marks indicate potential contact zones where *C. albomarginatus* and *C. karelini* can hybridize. Image by Varvara Vedenina.

Figure 3 Oscillograms of the courtship songs (the two upper lines are recordings of hind-leg movements and the lower line is the sound recording) in *Chorthippus albomarginatus* (A), *C. oschei* (E), and natural hybrids (B–D). In *C. albo-marginatus*, 3–7 pairs of so-called A/B elements alternate with the C element; in *C. oschei*, a complex of B1-A1-C elements follows after 15–30 pairs of A/B elements; in the beginning of the C element, a male moves the legs into an extra-high position and flicks with the tibiae. In hybrids, the C element is repeated twice or three times (B, indicated by arrows); the number of A and B elements between C elements varies greatly (C, indicated by arrows); and the alternation of several A/B pairs between C elements can be completely absent (D).

105

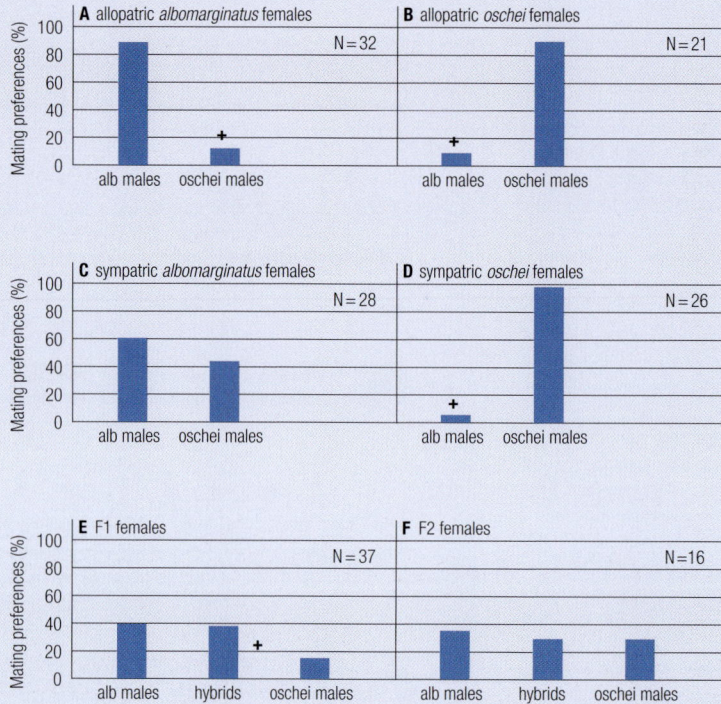

A allopatric *albomarginatus* females — N=32 — alb males, oschei males

B allopatric *oschei* females — N=21 — alb males, oschei males

C sympatric *albomarginatus* females — N=28 — alb males, oschei males

D sympatric *oschei* females — N=26 — alb males, oschei males

E F1 females — N=37 — alb males, hybrids, oschei males

F F2 females — N=16 — alb males, hybrids, oschei males

Mating preferences (%)

of a highly variable pattern. In contrast, most song elements in *S. clavatus* and *S. rubicundus* were found to be non-homologous. Nevertheless, we suggest similar processes: superposition of independent parental song elements in the hybrids could result from formation of two pattern-generating neuronal networks in the hybrid central nervous system. This idea was originally suggested in the study of hybrids between *C. biguttulus* and *C. mollis*, whose songs contain non-homologous elements.

Similarly, an absence of BHD in hybrid females could be explained by an expression of both parental neuronal filters for song recognition. Activation of one of two neuronal filters might be sufficient for a positive response in hybrid females. Studies of birds-of-paradise suggested that increasing the complexity of a suite of male ornaments or displays requires increasingly complex responses of female recognition systems and thus increases the likelihood of mistakes. This theory might also explain the lack of selectivity of hybrid females in both *rubicundus/clavatus* and *albomarginatus/oschei* hybrid zones. We therefore suggest that complex courtship can promote hybridization between closely related grasshopper species.

Figure 4 Mating preferences of female *Chorthippus albomarginatus* and *C. oschei* from allopatric populations (A, B), from the hybrid zone (C, D), and of F1- and F2 hybrid females (E, F). N is the number of females studied, crosses highlight statistically significant differences (binomial test, p<0,05).

Figure > This South American lichen katydid *Markia espinachi* is but one example of the overwhelming diversity of Orthoptera around the World. Photo: Thomas M. Cassar.

5 THE DIVERSITY OF ORTHOPTERA AROUND THE WORLD

5.1 AUSTRALIA AND PACIFIC

5.1.1 *THE ASTOUNDING DIVERSITY OF AUSTRALIAN ORTHOPTERA*

by Matthew Connors

Isolated by both time and space, Australia's Orthoptera are among the most unique and spectacular in the world. Orthoptera inhabit every terrestrial habitat on the continent. A diverse array of weird and wonderful crickets and katydids fill Australia's forests, vast numbers of grasshoppers roam its deserts and woodlands, and a few hardy species even live atop the highest snow-capped peaks. An extensive variety of both niche and habitat preference is reflected in the incredible diversity of their shapes, sizes, and colors, from the aptly named tiny grasshopper *Minyacris nana*, one of the smallest true grasshoppers in the world, to the enormous giant hooded katydid *Siliquofera grandis* (see chapter 8.2, The giant hooded katydid, a pet like no other), one of the largest of all orthopterans.

The vast majority of Australia's Orthoptera are entirely endemic, and many display unique adaptations unparalleled anywhere else. The long-legged sandhopper *Urnisiella rubropunctata*, a desert specialist, uses its long legs to bury itself in sand when danger threatens – one of just a few known examples of grasshoppers engaging in this behavior. The rain-forest-dwelling bulldog raspy cricket *Chauliogryllacris acaropenates* shows an equally unusual adaptation. With a bite force of more than 12 newtons, it has the strongest known bite of any insect; it uses this strength to carve out burrows for itself in wood.

The orthopteran fauna of Australia has been cut off from the rest of the world for so long that several entirely endemic lineages have diversified across the continent. The most ubiquitous and widespread of these are the matchstick grasshoppers of the endemic subfamily Morabinae, part of the family Morabidae. These elongate, wingless grasshoppers are represented by almost 250 species and include the only known obligatorily parthenogenetic grasshoppers, *Warramaba virgo* and *W. ngadju*.

Australia's katydid fauna is perhaps the most distinctive in the world; of the 19 currently accepted tettigoniid subfamilies, five are entirely endemic to Australia. Most spectacular of all are the flower-feeding katydids of the subfamilies Zaprochilinae and Phasmodinae. These katydids are strongly prognathous, and although *Phasmodes* consume entire flowers, the zaprochilines delicately take only nectar and pollen. The subfamilies are equally unique in their morphology; the narrow wings of most zaprochilines are held aloft over the body, and the extraordinarily elongate *Phasmodes* resembles a stick insect far more than a katydid.

Deep in the tropical rain forests of Australia's northeast, a myriad of predators stalk through both the undergrowth and the canopy. It pays to remain unseen, and although for many species this means taking shelter in burrows or retreats, some of Australia's largest orthopterans are able to hide in plain sight. With a dense covering of spines and powerful kicking legs, the spiny rain-forest katydid *Phricta*

Figure 1 A male matchstick grasshopper *Moritala* sp., member of the endemic subfamily Morabinae, is perfectly adapted for blending in with narrow vegetation. Photo: Matthew Connors.

Figure 2 A female crested toothgrinder *Ecphantus quadrilobus* perfectly matches the pale, hairy leaves of her preferred host plant. Photo: Matthew Connors.

Figure 3 Cooloola monsters (genus *Cooloola*) are certainly among the most unusual-looking orthopterans of Australia. Photo: David Rentz.

Figure 4 Male Gondwanan wasp katydids *Veria colorata* call from the tops of trees during the hottest part of the day. Photos: Matthew Connors.

> Figure 5 A male spotted predatory katydid *Chlorobalius leucoviridis* searches for prey at night. Photo: Matthew Connors.

spinosa is a formidable opponent; however, few predators ever get the chance to test its defenses. Although easily located with a flashlight during the night, the katydid's exceptional camouflage allows it to blend in with mossy rocks, logs, and tree trunks during the day, hidden from the eyes of hungry predators.

Venture just a hundred kilometers to the west of these rain forests and the habitat changes dramatically into dry, open savannah dotted with eucalypts. Here too orthopterans rely on crypsis to evade predation. The strange crested tooth-grinder *Ecphantus quadrilobus* belongs to an entire lineage of grasshoppers adapted to feed on the hairy, gray-green shrubs that are prevalent throughout central Australia. Its four-lobed pronotal crest and hair-covered body help it exactly match these plants in both shape and texture. Travel still further inland and the cast of characters changes yet again: gumleaf grasshoppers (*Goniaea* spp.) blend in with dry vegetation, and halgania grasshoppers *Histrioacrida roseipennis* disappear into green-and-white herbs. Most remarkable of all is the living stone *Raniliella testudo*, a flightless, pebble-like grasshopper that lives almost entirely away from vegetation. Combined with its appearance, a

Figure 6 The mountain katydid *Acripeza reticulata* is perfectly camouflaged with its fore wings closed, but can startle a potential predator by displaying its colorful abdomen. Photos: Matthew Connors.

Figure 7 Leichhardt's grasshopper *Petasida ephippigera* (Pyrgomorphidae), here shown on *Pityrodia jamesii*, one of its main food plant, is an emblematic species of Kakadu National Park in Northern Australia. Their bright colors indicate their chemical defense against predators. In the mythology of the indigenous Bininj people living in the region, these insects are called 'alyurr' and believed to be the children of a powerful creation ancestor; when they call out to their father, he answers with storms and lightning. Photo: Thomas Mesaglio.

reluctance to move unless greatly disturbed enables it to all but disappear into the rocky deserts it calls home.

Sometimes, camouflage is not enough to deter predation, and many Australian orthopterans instead rely on bluff to ward off threats. Numerous grasshoppers and katydids show bright patches of color on their legs or wings when threatened, and some nymphs of the katydid genus *Agnapha* are excellent mimics of the aggressive green tree ant *Oecophylla smaragdina*. One of the most accomplished mimics in Australia is the Gondwanan wasp katydid *Veria colorata*, one of the few day-calling katydids on the continent. In the hottest part of the day, males call from the very tops of trees, where they are easily visible to aerial predators. However, their appearance is enough to dissuade most predators; their strongly contrasting black-and-white pattern matches that of a wasp with a painful sting, but it is the katydid's behavior that elevates its mimicry beyond simple color matching. The katydid moves in a jerky, wasplike motion, with its antennae constantly twitching to complete the disguise, fooling any would-be predator.

The spotted predatory katydid *Chlorobalius leucoviridis* is perhaps Australia's most remarkable mimic. By producing clicking noises with their tegmina and twitching their bodies in time, these katydids are able to mimic the courtship responses of female cicadas. Male cicadas, thinking that they are duetting with females, are instead lured into the jaws of the hungry katydid. Most extraordinary of all is that *Chlorobalius* can successfully respond to the calls of species that it has never encountered before, demonstrating an adaptive ability shared by few others.

Some orthopterans have no need for camouflage or mimicry and instead display bright, aposematic colors. None are more spectacular than the iconic Leichhardt's grasshopper *Petasida ephippigera*, a rare species that inhabits sandstone escarpments in northern Australia. These colorful grasshoppers likely use bitter terpene glycosides from their food plants to deter predators. Similar colors are used by the mountain katydid *Acripeza reticulata*, another of Australia's insect emblems. When threatened, the katydid raises its wings to reveal bright red-and-blue stripes on its abdomen. It completes the display by oozing a foul-tasting liquid from between its abdominal segments – more than enough to discourage any potential predator.

The diving grasshopper *Bermiella acuta* of northern Australia is perhaps the most well-adapted of any grasshopper to a semiaquatic life. Both sexes possess dense patches of hair on the abdomen and tegmina to retain air and expanded hind tibiae to aid in swimming. Female *Bermiella* additionally have the base of the tegmina broadly expanded into a dome-like structure that covers the first abdominal spiracle. This structure forms an air chamber that enables the grasshopper to remain underwater for several minutes at a time.

The most unique of all Australia's Orthoptera are surely the Cooloola monsters (*Cooloola* spp.). Once considered a distinct family, these stocky, subterranean predators are now known to be aberrant members of the Anostostomatidae. Although adult males have been found wandering on the surface, females and nymphs spend their entire lives buried in sand and even molt underground. Adept burrowers, they use their elongate palps instead of antennae to feel their way forward and dig using a combination of both legs and highly modified jaws.

There is still much left to discover, not only about *Cooloola* but about all of Australia's orthopterans. The majority of species are undescribed, and many mysteries remain about their biology and ecology. Even some of the most obvious questions remain unanswered – for example, what is the function of the huge, hollow crest of the spectacular crested katydid *Alectoria superba*? These are questions for future generations of orthopterists to tackle.

> **Figure 1** A mating pair of wingless grasshoppers *Phaulacridium vittatum*, with the male (on top) exhibiting the white-striped, wingless morph. Photo: Sonu Yadav.

5.1.2 TWO CONTRASTING ENDEMIC GENERA OF AUSTRALIAN MOUNTAIN GRASSHOPPERS

by Rachael Y. Dudaniec and Sonu Yadav

INTRODUCTION

There are about 3,000 species of Orthoptera in Australia, with at least 500 of them being grasshoppers. Australian grasshoppers are found in almost any terrestrial environment across the continent you can think of, and their relative global isolation has led to high endemicity. Many species remain undescribed or unclassified, and we have varying degrees of knowledge about their ecology and life histories. However, studies that incorporate landscape-wide population sampling, body measurements, and genetic information are providing more and more clues about how grasshoppers are evolving and adapting to diverse Australian environments. With climate change impacts growing across this sunburnt, fire-prone country, grasshoppers play a pivotal role in our understanding of how species will move and adapt under shifting environments. Here we highlight knowledge about the adaptive capacities of two contrasting endemic genera: the habitat generalist *Phaulacridium* (subfamily Catantopinae) and the alpine-restricted group of five species belonging to the genus *Kosciuscola* (subfamily Oxyinae). These genera exhibit vastly different distributions that have major implications for their persistence and adaptive capacity under shifting environments.

THE WIDESPREAD PHAULACRIDIUM VITTATUM

The wingless grasshopper *Phaulacridium vittatum* (fig. 1) is known as an agricultural pest with periodic outbreaks in pastural regions of Australia, where it feeds voraciously on the precious fodder of domesticated livestock. Larvae of this species emerge in the Australian spring, and after laying eggs, adults die in the Australian winter. Wingless grasshoppers have a broad geographical distribution across the southern temperate and subtropical regions of Australia and exhibit morphological variation in body stripe pattern, melanism (i.e., degree of dark coloration), and body size, allowing them to cope with different environments (fig. 2). Experiments and analyses of genetic structure have confirmed that preferred temperature in *P. vittatum* is flexible and influenced by the effect of melanism and stripe pattern on body temperature. At least in the Australian alpine region of New South Wales, stripe pattern and body-size variation are associated with solar radiation: darker *P. vittatum* individuals warm up faster when basking.

Using environmental and genetic data, we found that dispersal of individuals and, therefore, genetic connectivity is high among populations of *P. vittatum* and increases with temperature, while other landscape features such as forest and urban cover factor little in their distribution. Genes that confer potential survival benefits (i.e., are locally adapted) at higher temperatures were also found. These genes have functions relating to body size and stripe morphology, but also to olfaction, chemical signaling, UV shielding, and pigmentation. Overall, these findings suggest that climate strongly influences the adaptive capacity of *P. vittatum*, as well as their abundance and frequency of morphological traits. Therefore, the combined effects of climate warming and land clearing throughout this species' distribution are likely to facilitate the spread of *P. vittatum*, potentially leading to more frequent and severe outbreaks of this agricultural pest. This contrasts with many Orthoptera species that may not fare so well in Australia under global warming, including representatives of cold-adapted species such as the alpine genus *Kosciuscola*.

THE ALPINE SPECIALIST KOSCIUSCOLA

While appreciating the uniquely contrasted warm, gray and brown tones of the Australian Alps in summer, a flicker of turquoise or bright green may catch a hiker's eye as they navigate the steep, rocky terrain. More than likely, this is a grasshopper of the genus *Kosciuscola*, which are restricted to elevations of 700–2,200 m in the Australian Alps of New South Wales and Victoria and can be found at lower elevations only in Tasmania (fig. 2). Currently, scientists recognize five species and one additional subspecies, but a recent phylogenetic analysis suggests up to 14 evolutionary lineages that show geographic and morphological divergence. These lineages may therefore constitute a much larger species group within the genus. Interestingly, members of *Kosciuscola* show elevational partitioning in their distributions, and this has been linked with corresponding, species-specific thermal tolerances.

Figure 2 A *Kosciuscola tristis* female (above) and male (below), known as the "chameleon", color-changing grasshoppers. Photos: Sonu Yadav.

With increasing temperatures and decreasing suitable habitat, like many other alpine taxa, species of *Kosciuscola* are susceptible to population decline and loss of genetic variation that may be vital for adapting to warming alpine regions. To add further fuel, these grasshoppers seem to be very bad at maintaining gene flow even across short distances: populations of *K. usitatus* and *K. tristis* co-occurring on mountains only 20 to 30 km apart were found to exist in virtual genetic isolation from each other. When examined for adaptive genetic variation, *K. tristis*, which has a narrower elevational niche breadth of ~ 1,600–2,000 m, showed stronger genetic structure and more evidence of local adaptation than *K. usitatus*, which occupies a broader elevational range of ~ 1,400–2,200 m. Patterns of selection on genes also varied spatially in both species toward higher elevations, but was more pronounced for *K. tristis*. Genes under potential selection from climatic variables were involved in lipid metabolism, which is associated with cold acclimatization, and other genes found to be related to development have consequences for fecundity. The two species varied markedly in their evolutionary sensitivity to elevation and temperature, with *K. tristis* likely to be more sensitive to warming temperatures given its narrower niche breadth and adaptive flexibility.

The microgeographic evolutionary responses of *Kosciuscola* to subtle changes in elevation and temperature and the group's high evolutionary diversity place this genus as a potential sentinel for understanding climate change impacts – a somber role that is rarely attributed to an insect and, moreover, a humble grasshopper.

CONCLUSIONS

Despite this brief introduction to two contrasting Orthopteran genera, the vast diversity of grasshoppers and katydids in Australia hold great potential for assessing species' adaptive processes, particularly in relation to climate and land-use change. From endemic, widespread pests of agriculture, like *Phaulacridium*, to habitat-restricted and thermally sensitive fauna like *Kosciuscola*, orthopterans have many lessons to offer about how human impacts are disrupting biodiversity at landscape scales.

5.1.3 SANDGROPERS: A UNIQUE GROUP OF UNDERGROUND ORTHOPTERA

by Terry Francis Houston

Cylindrachetids (or sandgropers, as they are called in Australia) are very atypical orthopterans. Their strange appearance is associated with their wholly subterranean existence: they are adapted for tunneling in sandy soils. True to their name, they have a cylindrical body form, the head and prothorax being strongly sclerotized like a beetle, while the abdomen is relatively soft and wormlike (fig. 1). Fully grown, they range in length from 35 mm to 88 mm, depending on the species. Wings are never developed, and all appendages are reduced, streamlining the insects so they can slip through the soil. The forelegs are extremely short, broad, and modified for digging in such an extreme way that they take no part in walking. Powered by muscles in the enlarged prothorax, they serve to part the soil ahead of the insect, creating a space. It is the puny mid and hind legs that enable the insects to shuffle along in their tunnels. Compound eyes are replaced with simple eyes, and the antennae are relatively short and simple.

Some similar adaptations for a subterranean life are also found in mole crickets (Gryllotalpidae) and Cooloola monsters (Cooloolidae), but they are not as extreme, and these insects are only distantly related to cylindrachetids, being members of the "long-horn" suborder Ensifera. Cylindrachetids are classified within the "short-horn" suborder Caelifera. Their nearest living relatives are the tiny (and unfortunately named) pygmy mole crickets and mud crickets (families Tridactylidae and Ripipterygidae, superfamily Tridactyloidea), which have nothing to do with real crickets or true mole crickets.

The world distribution of cylindrachetids is quite odd, the insects being recorded only from Australia, South America, and (putatively) New Guinea. Given that cylindrachetids are flightless, they must have evolved in Gondwanaland prior to its breakup in the Eocene.

The most recent revision of the family Cylindrachetidae is that of Günther (1992). This work revealed that the family is most diverse in Australia. The author created the genus *Cylindraustralia* for 13 species inhabiting the Australian mainland (excluding the southeastern portion) and one species in Papua New Guinea, while restricting the genus *Cylindracheta* to one species inhabiting the "Top End" of Australia's Northern Territory.

Cylindrachetids inhabit diverse environments. The Argentinian *Cylindroryctes spegazzini* is recorded from high-rainfall areas in sandy shores of freshwater lakes and streams in mountainous areas and in nearby rain-forest soils. In Australia, sandgropers are found in sandy terrain ranging from well-watered coastal dunes and sand plains to dune systems of the arid interior. Even in desert environments, water is important to the survival of the insects. They are found near the surface of the ground only after it has been moistened by a substantial rainfall. They have been excavated from depths down to 1.9 m in deep sands, confirming that they are able to retreat to cool, moist soil during dry periods.

The habits and life history of only one cylindrachetid species, *Cylindraustralia kochii*, have been studied intensively, although comparative notes were also recorded for the related *C. tindalei*. Both species are inhabitants of southwestern Australia and occur commonly in both natural and altered habitats, including farmlands. The presence of the insects is revealed after rain in the cooler months of winter and spring, when they create surface trails by tunneling just beneath the surface of the ground, pushing up long, raised ridges (fig. 2). Beneath each ridge is an open tunnel into which the maker may reverse at the

Figure 1 Adult female of the sand-groper *Cylindraustralia kochii*. Photo: Terry F. Houston.

Figure 2 Trails of sandgropers (across middle of view). Photo: Terry F. Houston.

Figure 3 Eggs of the sandgroper *C. kochii* showing attachment pads and pedicels. Photo: Terry F. Houston.

first sign of danger. Additionally, it may backtrack if it strikes an impenetrable barrier, then tunnel off to one side or go deeper.

Examination of gut contents of a large series of specimens revealed that both *C. kochii* and *C. tindalei* are omnivorous: the insects consume a wide array of plant, fungal, and animal materials. Plant materials comprised root, leaf, floral, and seed tissues. While most of their food is consumed within the soil, the insects evidently access some food items lying on or touching the ground surface. Animal tissues consumed included mostly those of soil-dwelling insects (including cutworms, well-known agricultural pests) and arachnids. In very dense sandgroper populations, cannibalism was evident.

During the wet winter months, females of *C. kochii* lay relatively large eggs (ca. 7.5 mm long), each placed singly in a separate chamber, suspended from the ceiling by an adhesive pad and thin, flexible pedicel (fig. 3). The eggs hatch en masse in summer, and the first instars take at least 12 months to reach the molting stage. Adults and all nymphal stages are found year-round, so it is evident that these insects are long-lived.

>**Figure 1** New Zealand alpine grasshoppers vary in size and ecology. (A) Adult male *Sigaus minutus* on top of an adult female *S. villosus*, illustrating size differences between the smallest and largest species within this radiation. (B) The largest, *S. villosus*, lives in rocky outcrops at the highest elevations on the Southern Alps, whereas (C) the smallest, *S. minutus*, is found only in low-elevation, semi-arid habitat in central South Island. Scale bars = 10 mm. Photos: Fabio Leonardo Meza-Joya.

5.1.4 SPUR-THROATED GRASSHOPPERS OF NEW ZEALAND'S MOUNTAINS

by Fabio Leonardo Meza-Joya, Mary Morgan-Richards, and Steven A. Trewick

The mountains of Aotearoa/New Zealand are home to an endemic radiation of short-horned grasshoppers of the genus *Sigaus*. Fossil calibrated phylogenetic analysis indicates that this monophyletic group radiated about 12 MYA, before the orogenic activity that gave rise to the dominant mountain terrain in New Zealand (~ 5 MYA). This suggests that *Sigaus*, which now consists of 13 species, diversified in the absence of montane or alpine environments. *Sigaus* is assigned to the tribe Catantopini and the subtribe Russalpiina, which is limited to temperate New Zealand and the Australian island of Tasmania.

These diurnal Acrididae live in open habitats, where they rely on solar basking for thermoregulation. Prior to the arrival of humans about 800 years ago, the vast majority of the New Zealand landscape was covered in wet forest, so open habitat was mostly above the elevational tree line on mountains of Ka Tiritiri-o-te-Moana (Southern Alps), and this is where most of these species persist today, with many areas and habitats occupied by multiple sympatric species. Ten species are alpine specialists, but some have expanded their ranges to lower elevations, and three species live only in lowland habitats.

All of these grasshopper species are brachypterous, with only reduced tegmina visible in adults. As a result, they are flightless and silent. The species and sexes differ considerably in size; the

smallest individuals are male *S. minutus* (~10 mm), and the largest are female *S. villosus* (~ 50 mm, fig. 1). *S. villosus* lives in rocky outcrops at the highest elevations of the Southern Alps, while *S. minutus* is restricted to low-elevation, semi-arid habitat in central South Island. Most widespread are *S. australis*, *S. nitidus*, and *S. nivalis*, while most restricted are the endangered and localized *S. childi* and *S. robustus*. The large, rugose species *S. robustus* lives on low-elevation braided riverbeds, where it suffers disturbance and predation by introduced birds and mammals.

These grasshoppers are generalist herbivores, and their occurrence is not driven by the presence of particular plant species in their environments. Across their overlapping ranges, the most widespread species (*S. australis*, *S. nitidus*, and *S. nivalis*) occupy distinct microhabitats within scree-shrub-herbfield mosaics. Habitat isolation and limited auditory

Figure 2 Mating and laying eggs occurs mostly during the New Zealand summer (December to March), when they are most active. Mating pairs of *S. nitidus* (a; photo: Christopher Stephens) and *S. nivalis* (b; photo: Fabio Leonardo Meza-Joya). Ovipositing female *S. australis* (c; photo: Cheryl Dawson) and *S. piliferus* (d; photo: Leon Perrie)

> **Figure 3** Predicted shifts in suitable native area and sampled intraspecific diversity for three New Zealand alpine grasshoppers under the warmest future climate scenario (RCP8.5) published by the Intergovernmental Panel on Climate Change (IPCC). Faded lineages in the phylogenies correspond to populations surveyed that are predicted to be lost (×) under global warming (image from Meza-Joya et al. 2023). Many other populations not included in the analysis are likely to have the same fate.

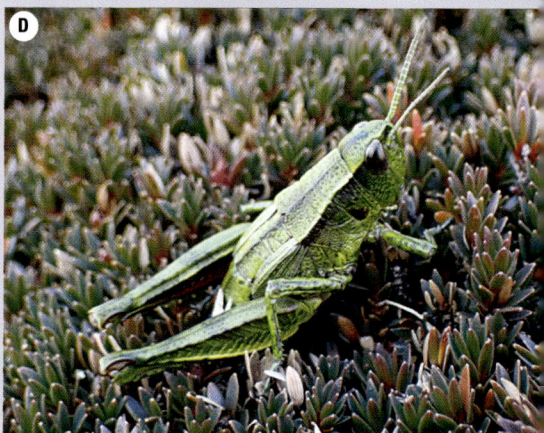

(a) *Brachaspis nivalis* (b) *Paprides nitidus* (c) *Sigaus australis*

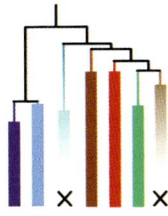

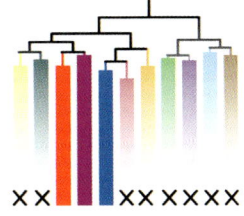

Suitable habitat
- Current
- Future

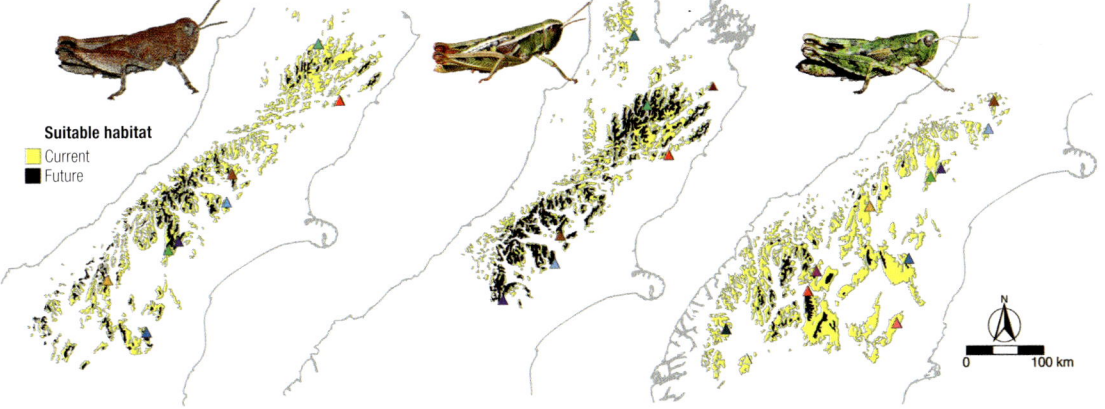

0   100 km

signaling suggest that mechanical or chemical cues are key for food and mate selection, yet visual signals may also play a role. In addition to consuming a wide range of native shrubs and herbs, these grasshoppers have recently been shown to eat a number of invasive European weeds, but they make little use of the native tussock-forming grasses (*Chionochloa* spp.) that are prominent in alpine habitat.

In some species, such as *S. australis*, *S. campestris*, and *S. nitidus*, color patterns are highly variable, even at the same site; this is most apparent in well-vegetated areas. Others, including *S. nivalis* and *S. villosus*, that live primarily on rocky habitat are less patterned and typically appear drab gray or brown with a rugose surface. The phenotype of some species also varies greatly; for instance, two distinct ecotypes are evident in *S. piliferus*, depending on habitat type: alpine and shrubland.

New Zealand's temperate climate is particularly erratic and changeable on the mountains, where alpine snowfall and freezing conditions can arise at any time of the year, even during summer, and the longevity of winter snowpack is variable. *Sigaus*

grasshoppers have evolved to tolerate this by being able to freeze at any life stage and reanimate as their environment warms. Breeding takes place mostly in summer, with multiple egg clutches per season if the spring–summer–autumn transition is particularly long (fig. 2). The unusual ability of these grasshoppers to overwinter at any life stage allows for multiple overlapping generations to be present at a given time and for relatively long lifespans (three to four years).

The future for alpine biology is bleak (see chapter 7.3, Alpine grasshoppers of the Mediterranean: Isolated in refuges shrinking from global warming). Population genetic diversity and climate niche modeling indicate range contraction following the last Pleistocene glacial cycle, with a legacy of spatially separate intraspecific lineages. *Sigaus* species each have distinct environmental characteristics, but ecological niche models predict that all will suffer habitat reduction and fragmentation as global heating from anthropogenic activity reduces their habitat to higher elevations (fig. 3).

123

5.1.5 WETA AOTEAROA: DIVERSITY OF NEW ZEALAND'S ENDEMIC ANOSTOSTOMATIDAE

by Steven A. Trewick and Mary Morgan-Richards

Anostostomatidae are a family of Ensifera represented around the Southern Hemisphere and Asia. Their nocturnal habits and concealment in holes during the day make the group rather cryptic and unknown. In New Zealand, they are known as wētā, a word from the language of the indigenous Maori people of Aotearoa, but elsewhere they are referred to as king crickets. The term wētā has also sometimes been applied to another family of Ensifera, the Rhaphidophoridae, usually in the form "cave wētā." More recently, a second term, Tokoriro, has become better recognized as applying to species of New Zealand Rhaphidophoridae that comprise the camel crickets in other parts of the world (see chapter 1.6., The evolutionary radiation of camel crickets).

The anostostomatid wētā in New Zealand include at least 43 endemic species, making up a quarter of the recognized Orthoptera in the country. They are generally subdivided into a number of informal groups based on key ecological or morphological features. There are 7 species of tree wētā, 11 giant wētā, more than 22 species of ground wētā, and 3 tusked wētā. These different groups are distinguished, respectively, by their different features: they roost in tree holes, are relatively large and robust, live in holes in the ground, and their males develop prominent mandibular tusks when adult. All are wingless and nocturnal.

Together, these endemic wētā are an ecologically diverse group of insects, and many regions and habitats are occupied by multiple sympatric species. This appears to be achieved through a wide range of size, diet, and reproductive behavior. Most species live in forest habitat, but several species occupy other

Figure 1 Adult tree wētā *Hemideina crassidens* form "harems" with many females and one adult male in tree cavities, such as this artificial roost. Photo: Steve Trewick.

habitats, including higher mountains – for example, the alpine ground wētā *Hemiandrus focalis*, alpine scree wētā *Deinacrida connectens*, and stone wētā *Hemideina maori*, which have independently evolved freeze tolerance.

TREE WĒTĀ

Most members of the genus *Hemideina* hide in tree cavities during the day, but one species in this radiation (*H. maori*) lives in the alpine zone above the tree line in the Southern Alps and hides under and between rocks. All species emerge at night and feed primarily on leaves, supplemented by fruit. They also scavenge animal remains. By hiding in tree holes, they escape daytime predators, but adults also have

Figure 2 The biggest of the giant wētā is *Deinacrida heteracantha*, which survived on a single offshore island until translocation to other islands. Photo: K. Bolstad.

potential mates nearby. In this setting, large-headed males with elongated mandibles compete for and attempt to monopolize access to females in harems (polyandrous mating system); however, females often mate with more than one male (polygynandry).

The species commonly known as the Auckland tree wētā *H. thoracica* has a range that extends from the warmer northern tip of North Island southward, where it meets the more cold-tolerant Wellington tree wētā *H. crassidens* (fig. 1). *H. crassidens* lives in cooler southern and higher-altitude habitats. Analysis of population genetic diversity and modeling of the climate niches of these two species help explain their latitudinal and elevational distribution. Pleistocene climate cycling caused the conditions to which each species was adapted to shift repeatedly in latitude and elevation. Along with the connection of the main islands of New Zealand due

to lower sea level during cold phases, this led to the current distribution.

GIANT WĒTĀ

Members of this group are characterized by their large size but do not display sexual head dimorphism. Female *Deinacrida heteracantha* have bodies up to ~70 mm long. Adult males are smaller than females, but have relatively long legs that are useful as the males engage in scramble competition for mating (fig. 2). Whereas *Hemideina* tree wētā have proved quite tolerant of habitat modification and even thrive in some urban situations, species of *Deinacrida* have suffered extreme population and range reduction since the arrival of humans and the predatory mammals they brought with them. Today, only species living in the Southern Alps persist on mainland New Zealand, while relict populations of others survive only in small refugia, primarily on offshore islands. The most widespread species, *D. connectens*, lives among scree above the tree line on several mountains on South Island (fig. 3). Here, along with a number of other endemic insects, including *H. maori*, it survives the winter by freezing and "reanimating" in the spring.

The three largest *Deinacrida* species are a closely related group in northern New Zealand; today they are restricted to offshore islands or, in one case, a patch of remnant habitat where it appears that prickly gorse *Ulex europaeus* plants afford them some protection from mammal predators (fig. 3). These forest-living *Deinacrida* do not usually roost in cavities during the day but shelter under leaves; this habit made them vulnerable to introduced predators such as possums *Trichosurus vulpecula* and rats *Rattus* spp.

TUSKED WĒTĀ

Three phylogenetically related species are most readily distinguished from other wētā by the prominent pair of "tusks" that extend from the mandibles of adult males. The largest species is also the rarest: just a single population has ever been found on the small arid island of Atiu. Atiu and other islands in the Mercury group would have stood in a coastal plain during the lowered sea level of Pleistocene glacial phases; thus *Motuweta isolata* very probably once occupied a wider range throughout the region (fig. 4).

Only a few individuals were ever found, but an audacious captive-breeding program resulted in several hundred individuals being translocated to other nearby islands, and reproduction in the wild has been documented. This species, like most large wētā species, clearly did not fare well in the presence of mammal predators brought to mainland New Zealand. However, two other tusked wētā do survive on North Island. One, *M. riparia*, is found in remote northeastern forests, where it lives close to small streams where individuals hide during the day in cavities beneath stones, whereas the other (*Anisoura nicobarica*)

Figure 3 Alpine scree wētā *Deinacrida connectens* are cold-adapted and live among rocks on mountains. (a) Adult females are much bigger than males. (b) Males lack the enlarged head and mandibles of male tree wētā but have relatively long legs. Photos: Euan Brook.

Figure 3 Alpine scree wētā *Deinacrida connectens* are cold-adapted and live among rocks on mountains. (a) Adult females are much bigger than males. (b) Males lack the enlarged head and mandibles of male tree wētā but have relatively long legs. Photos: Euan Brook.

Figure 4 The smallest tusked wētā species, *Anisoura nicobarica*, has the proportionally longest tusks. This species hides in holes of tree and shrub branches during the day and caps the opening with a mixture of wood dust. Photo: Tom Musson.

hides in tree holes of northern forests. Males of this small species have proportionally the largest mandibular tusks. The curious species name referencing the Nicobar Islands resulted from historic mislabeling of the type specimen when taken to Europe.

GROUND WĒTĀ

This is the most species-rich group of New Zealand Anostostomatidae wētā comprising two distinct genera. In addition to the 22 described species of *Hemiandrus* and *Anderus*, at least seven further species await description. New species continue to be discovered, including the large and brightly colored *H. jacinda*, named in 2021 for the then prime minister Jacinda Ardern (fig. 5). Most parts of the country are home to at least two species of ground wētā, usually displaying contrasting appearance and habits.

Adult females of many ground wētā, including *H. jacinda*, *A. brucei*, and *A. maculifrons*, appear to use a lay-and-leave approach to their eggs, which is typical of Anostostomatidae and many other Orthoptera (fig. 5). In contrast, species of the group including *H. pallitarsus* have almost no ovipositor. This species also has a peculiar sternite extension on the underside of the abdomen; it is used during copulation for the transfer of a nuptial gift from the male, and this provisioning of nuptial gifts in some *Hemiandrus* is associated with short ovipositors and maternal care of the young (fig. 6).

Figure 5 A male of the large and boldly colored Jacinda's wētā *Hemiandrus jacinda*. Photo: Steve Trewick.

Figure 6 (A) Short-ovipositor ground wētā. (B) An adult pair of *H. bilobatus* remain attached by genitalia as the female starts to eat the nutritious spermatophylax (nuptial gift) from the male. Photo: Steve Trewick.

5.2 ASIA

5.2.1 GRASSHOPPERS OF THE VAST CENTRAL ASIAN STEPPES

by Nikita Sevastianov

Central Asia, despite its topographical complexity, is dominated by vast stretches of relatively uniform habitats that are the home of many species with extremely extensive global distribution ranges. One of the most extensive biomes characterizing Central Asia are steppes. Not only are the steppes of the region the largest and the oldest of Eurasia; Central Asia is also suggested to be the place of origin for many characteristic steppe plant species, such as needle grasses (*Stipa*) and fescues (*Festuca*). These genera originated and spread during the Oligocene Epoch (around 33 to 20 MYA). Phylogenetic studies on the Gomphocerinae, a subfamily of Acrididae and the most common group of Eurasian grassland grasshoppers, indicated an origin and diversification in the same period, possibly also in Central Asia.

As described for Europe, Central Asian biomes and habitats were subject to profound changes in the course of Pleistocene glaciations (the Ice Age). During the glacial and interglacial periods, the boundaries of natural zones of altitudinal zonality shifted up and down, while, probably, a variety of biotopes remained habitable for Orthoptera throughout the entire process. Thanks to paleoclimate reconstructions, we can compare the modern climate with the climate of the Pleistocene, specifically during the last glacial maximum (LGM), 21,000 years ago, to infer the approximate biotope distribution of that time. According to these models, the ecological conditions of the LGM were potentially suitable for a wide variety of grasshoppers, including species adapted either to a humid continental climate or to Mediterranean and semi-arid climates. Modern areas with these climates have high orthopteran species-richness, and the same could be expected dur-

ing the Ice Age. However, vast expanses of Siberia and Europe were covered by cold tundra-steppes, which were also inhabited by grasshoppers, but were far less species-rich than warmer habitats.

After the end of the LGM, the tundra-steppes of the lowlands were largely replaced by forests. While Orthoptera adapted to warmer climates could expand their ranges, those of the tundra-steppe, typically grasshoppers of the family Acrididae, declined substantially. Today, isolated patches of this ancient steppe fauna can be found in various mountain systems, the Alps being the best-known example, but also the Tien Shan and many other Asian high mountain ranges. Some typical relic species of the tundra-steppes among Gomphocerinae, such as *Aeropedellus variegatus* and *Gomphocerus sibiricus*, inhabit huge but extremely patchy ranges across all these widely separated mountain ranges.

Areas like the Ukok Plateau in the north-eastern part of the Tien Shan system (the Sayan-Altai) are very similar to the Pleistocene tundra-steppes and could be considered a real relic of the Ice Age. Some data indicate the role of Central Asia as an important refugium for amphibians, mammals, and mollusks. The study of the role of this region in the formation of the Eurasian Orthoptera fauna is a very promising and intriguing task. The combination of several features of the Central Asian acridid fauna makes this region very interesting for future investigation.

— On closer inspection of the species-rich Acrididae groups (Gomphocerinae, Oedipodinae), Central Asia appears to be an important center of biodiversity. The best example is *Chorthippus*: 46 of 221 valid species were described from Central Asia. Most of them are brachypterous

forms with high endemism: almost every mountain range has its own unique species (fig. 1). However, many species have never been recorded after their original description.

— The Gomphocerini tribe unites the most widespread Gomphocerinae grasshopper species. According to recent reconstructions, some Gomphocerini genera (*Aeropedellus, Mesasippus*) represent basal lineages of this enormously species-rich tribe. Together with the common Eurasian *A. variegatus*, all species of these genera live in Kazakhstan and adjacent areas except the American *A. clavicornis*. If we combine that with dozens of endemic *Chorthippus* species, it could indicate the origin of this tribe.

— Not only Acrididae comprise unique components of the Central Asian fauna. There are also many endemic short-winged genera within Gomphomastacidae, Dericorythidae, and Pamphagidae. They are often species-rich but poorly studied. The case of the genus *Conophyma* (Dericorythidae) is particularly complicated (fig. 2). Morphological conservatism makes using morphological identification keys very difficult,

while the absence of complex behavior precludes acoustic identification. Brachypterous phenotypes and high-mountain habitats correspond with high endemism: more than a hundred species were described, and their identification is an almost impossible challenge (see also chapter 5.5.1, Sky islands: Hot spots of endemic grasshopper diversity in the Americas). Several closely related genera (125 species in total) combined in the subfamily Conophyminae are endemic to Central Asia. In this regard, Conophyminae can be compared with the West Asian shieldback locusts Nocarodeini:

Figure 1 The first photo of *Chorthippus oreophilus*, endemic of the Dzungarian Alatau, (Kazakhstan, Tekeli). Photo: Nikita Sevastianov.

Figure 2 An unidentified species of the diverse and complex genus *Conophyma* (Kazakhstan). Photo: Oliver Hawlitschek.

Figure 3 *Mesasippus kozhevnikovi iliensis* (Kazakhstan, Ili River Valley). Photo: Nikita Sevastianov.

same phenotype, same ecology, and the same identification issues.

— Other genera are less species-rich, but vary considerably in their ecology. *Mesasippus* (Acrididae) inhabit low-elevation biotopes spanning a range from river floodplains (*M. kozhevnikovi iliensis*, fig. 3) to deserts (*M. ammophilus*).

— Central Asia is considered a part of the Palearctic biogeographical realm, but due to its vast extent, it has close contact with the Paleotropics and includes the northern outposts of several subtropical or tropical groups. One of them is Eumastacoidea, known as monkey grasshoppers (see chapter 5.5.6, The Neotropical monkey grasshoppers). These groups are common and comprise many colorful and conspicuous species in tropical and subtropical areas of Asia, Africa, Australia, and America. Surprisingly, they can also be found in South Kazakhstan (Trans-Ili Alatau) and Kyrgyzstan, where they inhabit not only humid meadows (*Gomphomastax*) but also arid semi-deserts (*Phytomastax*, fig. 4). Another example are very specific gaudy grasshoppers – the Chrotogonini. They are common in tropical savannahs of Africa and India. Outside the tropics, they can be found only in Central Asia.

Finally, many characteristic elements of the pan-Eurasian fauna are also common in Central Asia. The widespread Italian locust *Calliptamus italicus*, Moroccan locust *Dociostaurus maroccanus*, and migratory locust *Locusta migratoria*, as well as many species of *Chorthippus*, *Bryodemini*, and *Sphingonotini*, can be found in Central Europe, the Mediterranean, and the Volga steppes. To be comprehensive, any study on the distribution of these species will only be complete with a visit to Central Asia!

Figure 4 *Phytomastax* cf. *artemisiana,* one of the northernmost monkey grasshoppers (Kazakhstan, Bartogay Lake). Photo: Nikita Sevastianov.

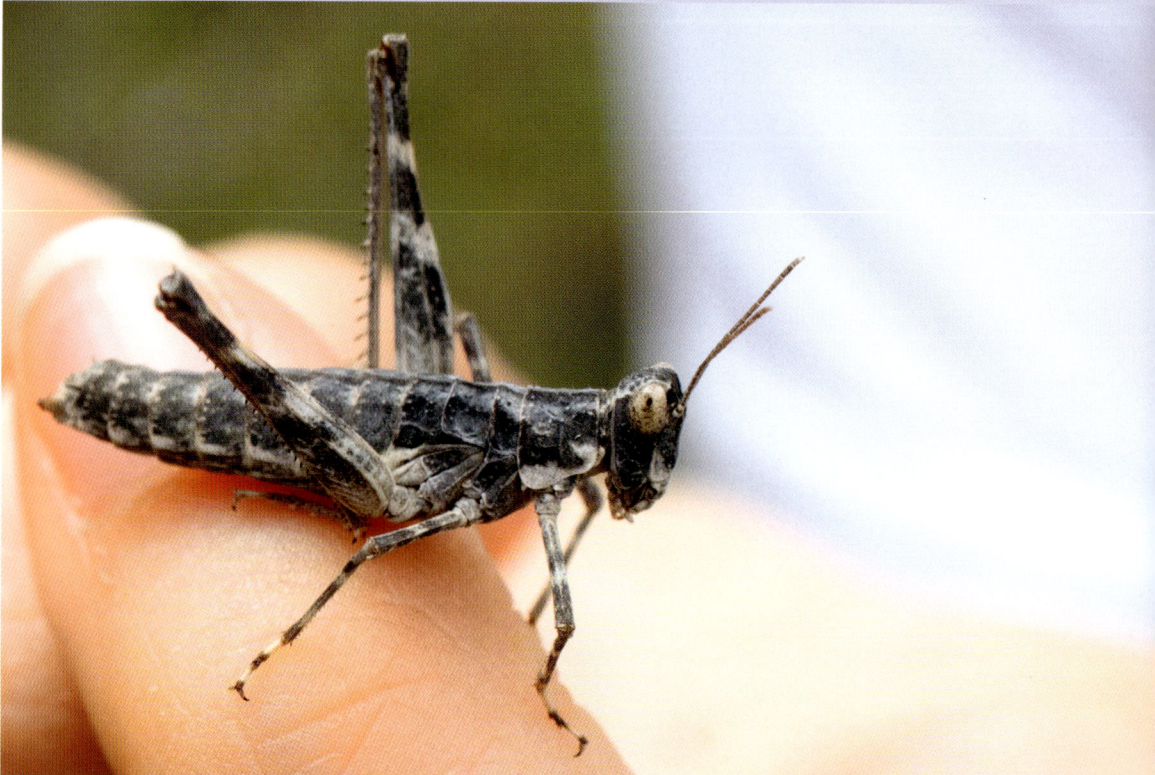

5.2.2 SONG AND SIGNALING BEHAVIOR IN AN INDIAN WĒTĀ

by Swati Diwakar

First described as early as 1888, the genus *Gryllacropsis* was erected to include a single Indian species, *G. magniceps*, the Indian wētā. It is yet another member of the interesting family Anostostomatidae (see also chapter 5.1.5, Wētā Aotearoa). Indian wētās have large bodies and wings and also long, threadlike antennae. Females are generally larger than males, but as females lack an external ovipositor, distinguishing the sexes in the field can be difficult. Indian wētās are nocturnal and are mostly seen calling high up on trunks of evergreen trees. They are dark brown and well-camouflaged in their natural habitat.

While in many Orthoptera, only males produce sounds, females of the Indian wētā were also reported to stridulate. Unlike many ensiferans (see also chapter 4.3, Many different styles of singing in crickets, katydids, mole crickets, and grigs), these species produce low-pitch, broadband calls by rubbing pegs on the underside of the femur with the row of pegs on the lateral side of the abdomen. The stridulatory structures of the Indian wētā are arranged in a more systematic and organized manner, as opposed to the femoro-abdominal stridulatory structures of wētā species from New Zealand. The inner surface of the hind femur has between 5 and 7 transverse ridges and 2 parallel rows of pegs above the transverse rows. The second abdominal tergite on each side has between 20 and 28 ridges. Each ridge is made of a row of rounded, bead-like pegs.

Males produce four-syllable chirping calls at regular intervals, whereas female calls were typically two-syllable chirps produced either by single individuals or in groups. Males call with regular chirp periods and durations, whereas chirp rates in the female calls were more variable, owing to groups of chirps produced randomly. Dominant frequencies for the male and female calls were 1.71 + 0.05 kHz and 1.68 + 0.1 kHz, respectively. The average bandwidth of the frequency spectrum was reported to be 2.0 + 0.3 kHz for males and 2.19 + 0.15 kHz for female calls. Acoustic duetting between the sexes, otherwise predominant in the Tettigoniidae, has previously been reported in some wētā species, but has so far not been detected in Indian wētā. Field observations showed that males initiated calling after sunset, followed by females. While both sexes were then calling simultaneously, fewer females than males would be active throughout the peak time of calling activity.

The calling activity of *G. magniceps* is largely limited to the post-monsoon season between October and December. The animals were observed calling from tree trunks, most commonly from *Terminalia* species, the most abundant tree genus in the area, but otherwise showed no preference for any particular plant species.

Courtship involves antennation by both sexes. During this phase, other conspecific individuals are often spotted nearby. Subsequently, the male positions behind the stationary female and continuously vibrates his abdomen (courtship drumming). This is followed by the female mounting the male, as is typical for many ensiferan species. The male raises his wings during mating and transfers the spermatophore to the female, which immediately starts eating this nuptial gift. Post-copulation, the male stays with the female and grooms her (fig. 1). The specific behavioral mechanisms involved in the responses of females to courtship drumming by males and the aggregation of multiple individuals at the pre-mating stage are still subject to research. They present a case for investigating the evolutionary costs-benefits of bi-directional and bi-modal acoustic communication strategies.

Our understanding of the Indian wētā is limited, as this species is arboreal, well camouflaged, extremely sensitive to artificial light, and restricted

Figure 1 Female wētā with spermatophore attached. The male stays with the female after transferring the spermatophore. Photo: Manisha Tomar.

Figure 2 Unidentified species belonging to genus *Diaphanogryllacris*. Photo: Manisha Tomar.

to tropical evergreen forests. In addition to *G. magniceps*, a variety of non-singing wētā species have been reported from the Western Ghats and northeast Himalayan biodiversity hot-spot regions in India. Species belonging to the genera *Eugryllacris*, *Diaphanogryllacris* (fig. 2), and *Gryllacris* have been documented, but species-level identification is yet to be carried out. It is very likely that new species will still be found.

5.2.3   LITTLE MONSTERS: SPLAY-FOOTED CRICKETS OF PAKISTAN

by Riffat Sultana and Santosh Kumar

Pakistan, a country in southern Asia, features a diverse array of ecological regions across its large latitudinal and altitudinal extent. Spanning over 882,000 km^2, the country is roughly the size of Chile or of all Central European countries combined. The local Orthoptera fauna is very diverse; currently, 208 taxa are recognized in Pakistan. The superfamily Schizodactyloidea, the splay-footed crickets, stands out as one of the most ancient and distinctive taxonomic groups; it diverged from crickets over 200 MYA and comprises only 21 extant species. Within this superfamily, there are only two known genera: *Comicus*, found in southwest Africa, and *Schizodactylus*, found in West and South Asia.

Four *Schizodactylus* species are found in Pakistan: *S. monstrosus*, *S. minor*, *S. hesperus*, and *S. sindhenesis*. Schizodactylidae possess several peculiar features that distinguish them from other orthopterans. Particularly, their very long hind wings, which, at rest, are coiled into a tight spiral, make them stick out. The tarsal segments are strongly developed and broadly expanded into lateral lobe-like or finger-like processes. The latter represents the most characteristic feature of flightless forms.

S. monstrosus, the maize cricket, is a large species, commonly known by many local names, such as *bherwa* in Bihar (India) and *pani tharo* or *muccher* (Pakistan). Their large mandibles make them a scary sight (fig. 1). Indeed, species of *Schizodactylus* have an intimidating habitus and clear predatory adaptations, including raptorial prothoracic legs and the aforementioned powerful, enlarged mouthparts. Sub-social behavior and cannibalism have been observed in populations of *S. monstrosus* and *S. inexpectatus* (Werner 1901).

Schizodactylids are nocturnally active on the surface of sandy and arid environments. Their peculiar tarsi permit them to run on sand with ease. During the day, they live inside burrows, preferably built in moist places on slopes or level ground in the deep layers of fine sand (fig. 2). Both adults and nymphs construct individual burrows and are equipped with several morphological adaptations for digging in sandy environments. They dig with their mandibles and anterior legs to scrape and loosen sand, then push the sand behind them out of burrows, using the abdomen and hind tibiae. The width of burrows varies from nymphs to adults, measuring between 35 mm and 55 mm; the depth ranges from 9 cm to 30 cm.

They never keep their burrow open, and once they leave it, they never return to the same burrow. When they enter their burrow, they throw back the sand to close the opening. In summer, the entrance remains closed, while in other seasons, it remains mostly open during daytime. When threatened inside, they dig a new exit. Hence, the burrowing behavior is also considered a defense strategy.

Splay-footed crickets can fly and swim. They never use their wings during swimming, but stretch their fore and hind legs. For faster travel, they open their rolled wings and take a short flight of 10 to 20 seconds at 3 m to 5 m height. Their limited flight ability leaves them vulnerable to many natural enemies. Being carnivorous themselves and feeding on some well-known agricultural pests, they play a key role in the food chain and can contribute to natural pest management.

Despite their importance in pest control and for ecosystem functioning, schizodactylids are under threat. Their habitats have been severely impacted by various factors, including road construction, overgrazing, rising temperatures, drought, and the devastating flood in 2022 in Pakistan. Hence, conservation measures should be undertaken to conserve these beautiful little monsters.

Figure 1 A specimen of *Schizo-dactylus monstrosus*. Photo: Riffat Sultana.

Figure 2 A splay-footed cricket at the entrance of its burrow. Photo: Maleeha Jamil.

5.2.4 CAVES WITHOUT CAVE CRICKETS IN BHUTAN

by Cheten Dorji and Steven A. Trewick

Bhutan is a small Himalayan country of about 38,394 km^2 nestled between two giants: India in the south and China in the north. Perched on the south face of the eastern Himalayas at the contact point of the Indian and Asian continental tectonic plates, Bhutan exhibits elements of oriental and Palearctic ecoregions dominated by forest, steeply grading from snow-capped mountain peaks in the north to southern plains. The elevation ranges between 150 m and 7,500 m, with a latitudinal distance of just 150 km; many rivers and tributaries criscross the slope. This altitudinal range and associated climatic variation hosts a wide range of poorly studied species within a small biodiversity hot spot.

A very typical feature of the mountainous landscape of Bhutan are natural caves with extensive dark zones (fig. 1). Despite a scarcity of limestone, many smaller caverns and rock shelters have been formed either by weathering or by fracturing from active tectonic movement. Human inhabitants have used these caves for domestic and religious purposes for millennia, and many of them still constitute key elements in the sacred geography of Buddhism today. However, only two such cave systems have been thoroughly explored and mapped while investigating for the mining of limestones for cement production: the Dung Phug "Ghost cave" in Bumthang and the Khaling cave in Trashigang. As with many habitats of Bhutan, the biodiversity of its cave systems remains poorly known. While more than 40 species of bats have been documented there, very limited data on invertebrates are available.

Camel crickets (Rhaphidophoridae) are common inhabitants of cave systems, but the camel cricket fauna of Bhutan remains virtually unknown, with just two species documented: *Rhaphidophora angulata* and *Diestramima tsongkhapa*. Therefore, we conducted surveys of some better-known cave systems for rhaphidophorids and found them to be unexpectedly scarce. This surprised us, as caves of neighboring countries are inhabited by camel crickets, such as species of *Tachycines*, and there is no obvious biological reason why they should be absent from Bhutan.

However, there was plenty to be discovered around the cave entrances, where almost no biological exploration had been conducted since the 1972 expedition by the Swiss Natural History Museum of Basel. In addition to the two known species of camel crickets already mentioned, we encountered many undocumented Rhaphidophoridae in wet and moist forest habitats, under rocks, cracks, and crevices at the base of cliffs, and among leaf litter and fallen branches. These discoveries clearly indicate a substantial hidden diversity of this family in Bhutan. Preliminary morphological and molecular analyses suggest more than 15 species new to science belonging to four genera.

The two subfamilies found in Bhutan appear to be at their western range limits. Aemodogryllinae, such as *Tachycines*, appear to be more closely associated with cave systems in East and South Asia; *Diestramima*, with currently 41 species, is recognized throughout south and southeast Asia (fig. 2). Within Rhaphidophorinae, *Rhaphidophora* comprises over 109 species with a distribution extending south to New Guinea and the Solomon Islands. Our recent sampling in Bhutan suggests a position of the country at the western nexus of these two subfamilies that here occupy the compact habitat gradient ranging from moist tropical forest to cool broadleaf forest between 500 m and 2,500 m above sea level.

Bhutan has so far escaped intense landscape change and retains about 70% of its land under native forest cover. About 52% of the country is

137

designated as protected areas, spanning the habitat from high alpine to lowland rain forest. Here, many more species may await discovery, especially in areas that are hard to reach, and further systematic exploration of caves throughout the country may still yield true cave crickets and help our understanding of the overlapping biology and traditional culture.

Figure 1 (above) Unexplored Rangtse cave system with bats, Chego village in Gakiling Geog, Haa (900 m above sea level). (below) Typical habitat type used by camel crickets: the rocky entrance of Minjiwoong cave. Photos: Cheten Dorji.

Figure 2 Examples of the genera belonging to two subfamilies found in Bhutan. (A) Aemodogryllinae: *Diestramima* sp. newly discovered from Bhutan. (B) Genus *Tachycines*, common in caves of neighboring regions. (C) *Tachycines* sp. from China. (D) *T. asynamorus* from South Korea. (E) Rhaphidophorinae: *Rhaphidophora taiwana* from Ishigaki Japan. (F) *Rhaphidophora* sp. from China are likely relatives of Bhutan camel crickets.: Photos: (a, b) Cheten Dorji; (c) Zhou Yu; (d) Hyun-tae Kim; (e) Do-yong Kim; (f) Fan Gao.

5.2.5 SINGAPORE: A MICROCOSM OF HIGH ORTHOPTERA DIVERSITY IN A DENSE URBAN ENVIRONMENT

by Ming Kai Tan

The Republic of Singapore, located at the southern tip of the Malay Peninsula, is a tropical island city-state. In bird's-eye view, the landscape presents itself as a network of tree- and shrub-lined roads intersecting a conglomerate of urban high-rise buildings and skyscrapers with scattered pockets of green spaces – both natural and managed. Rapid urbanization over the last 80 years developed the Republic from a rural swampland to a commercial and economic hub and has led to massive deforestation and species extinction. Yet, situated in the heart of a biodiversity hot spot in southeast Asia, Singapore still harbors a rich diversity of flora and fauna, including many orthopterans.

URBAN DIVERSITY

At least 120 species of crickets, katydids, and relatives and 30 species of grasshoppers have been documented in the largest nature reserves in Singapore (namely the Central Catchment and Bukit Timah Nature Reserves). Most impressively, 23 species and 1 genus new to science were discovered between 2010 and 2016 (fig.1). This new species discovery rate accounted for around 15% of all species documented. Additionally, the remaining mangrove forests along parts of the island's coastline also harbored numerous new species: *Ornebius tampines*, *Svistella chekjawa* (fig.2), *Amusurgus caerulus*, and *Ornebius lupus*. These species are so far restricted to mangrove habitats and have not yet been reported elsewhere outside Singapore – hence they are endemics.

Owing to the biodiversity surveys and taxonomic research on the orthopterans in Singapore, we now have a relatively good understanding of their true diversity. This is unlike the situation in most parts of southeast Asia, where vast areas of forests have been surveyed only fragmentarily or not at all. The foundation laid by this taxonomic work serves as a crucial basis for further research on ecology and behavior.

Orthoptera are one of the insect orders capable of producing sound for communication. However, the calling songs of most species in southeast Asia are unknown, a far cry from the bioacoustics data available from other sound-producing animals such as birds, bats, and even frogs. In a first of its kind for southeast Asian taxa, recent studies used ultrasonic-sensitive devices to record and describe

Figure 1 *Singapuriola separata*, a new genus and species of crickets discovered in recent surveys in Singapore. Photo: Ming Kai Tan.

Figure 2 *Svistella chekjawa*, a mangrove-restricted sword-tailed cricket described and so far found only in Singapore. Photo: Ming Kai Tan.

Figure 3 *Asiophlugis temasek*, a crystal predatory katydid described from Singapore, can produce a calling song with a frequency of nearly 60 kHz. Photo: Ming Kai Tan.

Figure 4 A male (left) and female (right) *Lebinthus luae* exhibiting pre-mating rituals. Photo: Ming Kai Tan.

Figure 5 *Phaneroptera brevis* feeding on the flower of *Neptunia plena*. Pollen grains can be found on the antennae, legs, and face. Photo: Ming Kai Tan.

Figure 6 Despite its small size and the immense pressure from urban development, Singapore hosts a remarkable variety of habitats of Orthoptera. Photo: Ming Kai Tan.

the previously unknown calling songs of 19 katydid species from Singapore. Many of these katydids have peak frequencies in the ultrasonic range and songs that are inaudible to humans, such as *Asiophlugis temasek* described from Singapore (and named after the old Javanese name of Singapore; fig. 3). This member of the Phlugidini tribe of Meconematinae, commonly known as crystal predatory katydids, produces a song with energy peaking at 58 kHz, far above the human audible limit of 20 kHz. Without sound taxonomy (no pun intended), it would not have been possible to identify these katydids accurately and precisely to species level.

HOW THE PANDEMIC AFFECTED ORTHOPTERAN COMMUNICATION

The COVID-19 pandemic was found to also affect crickets in Singapore. *Lebinthus luae*, an eneopterine cricket commonly observed in the understory of the coastal forest parks of Singapore (fig. 4), was found to change its calling song parameters in response to the change in human activity during the lockdown. Specifically, crickets in parks, where fewer people visited during the lockdown, started producing longer calls, whereas crickets in parks more accessible to people produced shorter calls. It represented one of the first studies that examined how the global pandemic affects animal activity. Moreover, from recordings under laboratory conditions and in the field, this eneopterine consistently follows a complex circadian cycle in its calling activity, peaking it thrice daily – at dawn, in the afternoon, and at night. These findings highlight that it would be too simplistic to consider an orthopteran as strictly either diurnal or nocturnal in its activity. Once again, owing to our knowledge of the local species diversity and calling songs of orthopterans, it is far easier to distinguish the calling songs of *L. luae* from the other species.

ORTHOPTERA AS POLLINATORS

Orthopterans are generally not considered to be flower visitors. Nevertheless, a study based mainly on Singaporean species demonstrated that orthopterans are, in fact, more frequent flower visitors than

previously thought. Contrary to the popular generalization that orthopterans are gregarious pests, the katydid *Phaneroptera brevis* in Singapore feeds gently on the nectar and petals of flowers – hence, rarely damaging the reproductive parts (fig. 5). In the process, pollen is attached to the legs and antennae and subsequently transferred to other flowers (see chapter 3.5, Raspy crickets as pollinators of orchids). Consequently, *P. brevis* can actually help with pollination and hence can be considered an ecosystem service provider. Further behaviorial studies on this species also revealed that katydids (and probably many orthopterans) exhibit animal personality and are capable of efficient foraging, but they may suffer from poor decision making when presented with an overwhelming number of choices. These behaviors are not too different from us, humans!

CONCLUSIONS

Even on a small urbanized island like Singapore, there are still many recent new species discoveries and even more new knowledge generated on the ecology, behavior, and natural history of orthopterans. The same is likely to be said for the rest of southeast Asia if and when more effort to study the orthopterans is made. That the studies of the orthopteran taxonomy in Singapore have facilitated research on orthopteran ecology and behavior emphasizes the pressing need to appreciate taxonomy, train more taxonomists, and offer more funding for taxonomic research. However, with the worsening of the anthropogenic-induced climate change and biodiversity loss, exacerbated by the taxonomic crisis in the region, the future does not necessarily appear bright for the orthopterans and the rest of the biodiversity in the region.

5.2.6 WHERE CAN'T YOU FIND THEM? ECOLOGY OF PYGMY GRASSHOPPERS IN SOUTHEAST ASIA

by Niko Kasalo and Josip Skejo

Alfred Russel Wallace spent a fair amount of time collecting all kinds of animals in southeast Asia. Having seen an astonishing number of specimens on countless islands, Wallace could not help but notice the patterns of species distributions and the varying degrees of similarity between them. Eventually, those observations coalesced into one of the most important theories in science, the theory of evolution. More than 150 years later, the birthplace of modern evolutionary thought remains an inexhaustible spring of information on biological interactions. With hundreds of endemic species of pygmy grasshoppers, Southeast Asia provides a perfect microcosm to examine the wealth of their ecological diversity (fig.1).

Pygmy grasshoppers (family Tetrigidae) are a large group of tiny grasshoppers; scarcely any are more than a couple of centimeters long. This historically neglected group is currently going through a renaissance, and only recently have we found out how unique these barely perceptible critters are. Pygmies feed on decaying organic material, mosses, lichens, algae, and fungi, which associates them more with decomposers, such as earthworms, than with their sister group, the largely herbivorous grasshoppers. They occupy strange and diverse microhabitats, and the ways they interact with the environment have been glimpsed only barely.

Perhaps the most awe-inspiring group is Scelimenini, which includes species that dive underwater. At first glance, it is apparent that they are no strangers to water: their bodies are smooth, with a few sharp protrusions, and their hind legs are shaped like oars. When they jump into the water, the oars go into action, and the insect is quickly out of sight. Some like to swim on the surface, while others can submerge and stay underwater for more than half an hour, and some may even feed during that time.

Some researchers have noted that these grasshoppers can use air bubbles trapped under their pronotum – the shield covering their body – as a kind of "scuba tank". Interestingly, the behavior of jumping into the water has also been observed in another group, the Tetriginae. When in danger, some individuals jump into a nearby stream or puddle, but they are not good swimmers, and it is not uncommon for them to drown.

On the other extreme lie the cryptic species, which belong to many different groups. Discotettigini, close relatives of the swimming Scelimenini, are usually covered with spikes and spend their time among tangled tree roots. A cursory look will surely mistake them for tree bark (a so-called corticolus lifestyle). Cladonotini hide using the same principle but in different forms; some have a single vertical protrusion on their pronotum, making them resemble tiny twigs, while others carry thin crests on their backs and look exactly like dead leaves. They also seem quite sluggish, with some reports claiming that the same individual was spotted in the same place days apart.

As a bonus, the cryptic species are often covered with a mat of algae. This unique interaction has never been examined, leaving future researchers with an interesting question: how far does this symbiotic relationship go?

Yet another type of cryptic morphology can be found in Batrachideinae, an ancient group of pygmy grasshoppers that can often be recognized by the spiky "helmet" that hangs over their head. They seem to spend most of their time among vibrantly colored plant parts and are often spotted on leaves. Appropriately, they are often bright and intricately patterned. This, in combination with their comparatively large size, makes them irresistible to photographers.

a) phytophilous

b) corticolous

c) leaf-litter dwelling

d) amphibious

e) ground-dwelling

Figure 1 The ecological diversity of pygmy grasshoppers: (a) *Camelotettix curvinotus* (Metrodorinae), photo: Philipp Hoenle, iNaturalist, CC0 licence. (b) *Discotettix kirscheyi* (Scelimeninae, Discotettigini), photo: Jonghyun Park (iNaturalist clurarit), cropped, CC-BY licence. (c) *Holoarcus* sp. (Cladonotinae), photo: Philipp Hoenle, iNaturalist, CC0 licence. (d) *Paramphibotettix sanguinolentus* (Scelimeninae, Scelimenini), photo: Wildan R. Ardani (iNaturalist), CC-BY licence. (e) *Paratettix* sp. (Tetriginae), photo: Philipp Hoenle, iNaturalist, CC0 licence. Jungle photo in the middle: Gido CC BY Available at https://www.flickr.com/photos/103499652@N04/11888308085/.

Southeast Asia is well known for its overwhelming number of islands. Island biogeography is both a blessing and a curse for the same reason – incredibly high endemism. Nearly every island harbors species unique to it. Sometimes it is easy to place those endemics into the wider biogeographic context, but some strange ones keep boggling our minds. Tetrigidae in their basic shape are relatively competent fliers, which provides an opportunity to colonize new habitats. However, when they manage to secure themselves in a new and isolated environment, their progeny shows a tendency for the loss of wings. With wings out of the way, their pronotum can change drastically, obscuring the hints about their descent. How pygmy grasshoppers dispersed and populated the southeast Asian islands is a question that may continue to bear partial answers in perpetuity.

One possible and fascinating answer to the dispersal patterns of pygmies could be birds. Japanese cranes were found to feed on *Tetrix japonica*, which has interesting implications. For example, the eggs of stick insects can survive the bird's digestive system and hatch after being expelled, and even some adult beetles, it has been suggested, may be able to survive this kind of transport. If this is true for Tetrigidae as well, bird migrations might explain the incredible number of separate colonization events on some islands. It could perhaps be the key to explaining how some wingless species have close relatives hundreds or thousands of kilometers away.

Unfortunately, the life cycles of the vast majority of species are completely unknown, with a few European species providing most of the data. The sensory structures are similarly unexplored. Recently, there has been some research on the antennal sensilla of a few species, which revealed a great diversity in the shape and distribution of these structures. Tetrigidae are known to communicate through vibrations produced by beating on the substrate with their legs, but we do not even know where their auditory organs are placed, if they possess them at all.

How pygmy grasshoppers perceive the world and how it perceives them is a mystery that we have only begun to untangle. The research on these fascinating animals is slowly ramping up, but we have a long way to go. Just as Wallace had his revelations, so will we have our own.

5.3 EUROPE

THE MEDITERRANEAN AND THE BALKANS: HOW GEOLOGY SHAPED BIOGEOGRAPHY

by Oliver Hawlitschek, Paolo Fontana, Bruno Massa, and Dragan Chobanov

The biogeography of the Mediterranean region, defined as the area of lands around the Mediterranean Sea that are characterized by a Mediterranean climate, with hot and dry summers and cool and rainy winters, is the product of a complex and active geological history. Many islands are scattered among the deep basins of the Mediterranean Sea, and its shores are flanked by high mountain ranges. This geological variety at many scales has given rise to an enormous biodiversity with high rates of regional and local endemism, making the Mediterranean a global hot spot of biodiversity. While traditionally among the areas best studied by biologists, new species – as demonstrated in this book – are still being discovered regularly.

The geological history of the Mediterranean Sea is closely connected with the Alpine orogeny, the process that began in the late Mesozoic Era, more than 65 MYA, and gave rise to many of today's mountain ranges of Europe, northern Africa, and western Asia, all the way to the Himalayas. In this process, the African and Eurasian tectonic plates converged and finally collided, closing the Tethys Sea, which separated the ancient landmasses of Gondwana and Laurasia over most of the Mesozoic. The Alpine orogeny formed the mountain ranges of the Pyrenees, the Atlas, the Alps, and the Apennines of the western Mediterranean, and the Dinaric High-

lands, the Hellenides, the Rhodopes, and the Anatolian Highlands of the eastern Mediterranean. The orogeny is still ongoing: the tectonic activity causes volcanism, and the mountain ranges keep growing. The resulting dramatic topography of summits, plateaus, and valleys, each with their own unique set of environmental conditions, is matched by numerous larger and countless smaller islands scattered across the sea.

Global climate changes have also been of pivotal biogeographical importance. About 5.8 to 5.6 MYA, during what is called the Messinian crisis, the Mediterranean almost dried out. Islands and landmasses that had previously been isolated by the sea were then connected by dry land. The basin was flooded again after the crisis, and many islands, such as the Balearics and Crete, remained isolated until the present day. Others were newly connected through the recurring sea-level drops beginning at the onset of the Pleistocene glaciations, the Ice Ages, around 2.6 MYA. In addition to causing the sea level to fall and rise again, the glacial cycles induced severe shifts in vegetation, both latitudinal and elevational. Mountaintops became inhabitable, valleys became isolated, and forests disappeared, to regrow cycle after cycle. Species that were adapted to temperate climates retreated to refugia in the Mediterranean. Glaciers carved out valleys, created moraines, and formed the

147

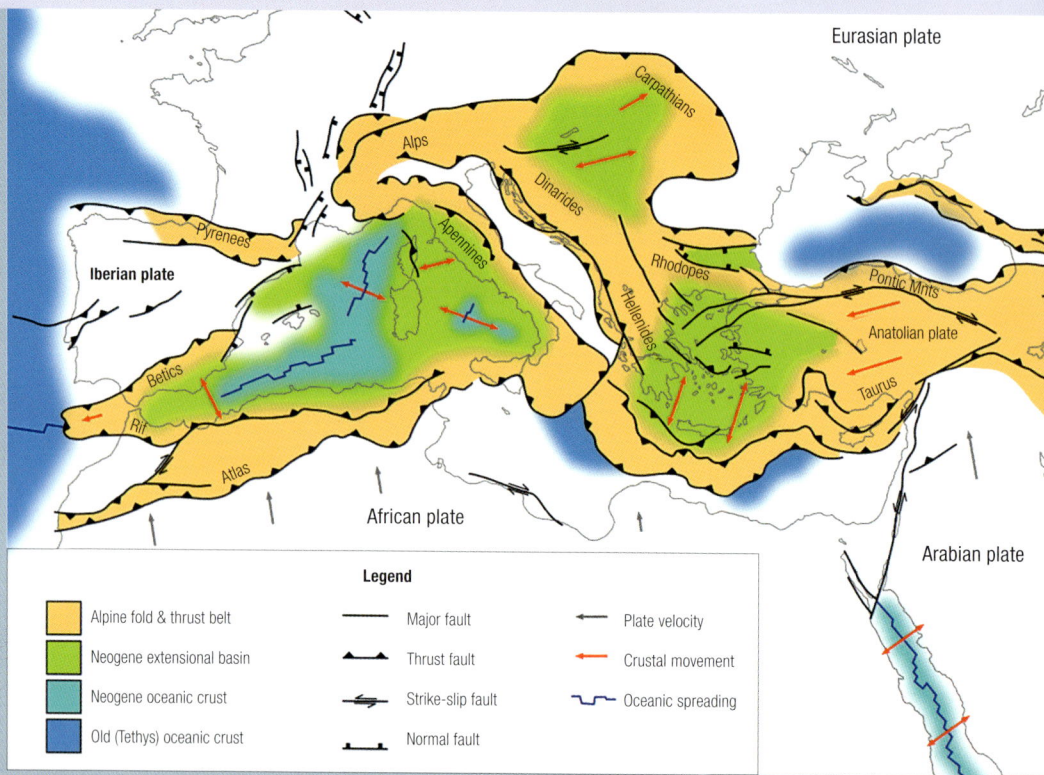

Figure 1 Geological map of the Mediterranean region. Arrows show the ongoing movement of tectonic plates through geological history, which has led to the uplift of mountain ranges. Source: Creative Commons License (modified).

topography of the existing alpine landscapes. The weight of the ice sheets caused the Earth's crust to sink and then to rebound with the melting of the ice. Changes in elevation and drainage patterns and the formation of new habitats shaped the distribution and evolution of many alpine species. Finally, karst formation – that is, the chemical dissolution of limestone rocks – created caves and access points to the subterranean ecosystems, promoting the evolution of highly specialized cave species.

Most recently in geological time scales, but over many millennia of civilization history, human activities have deeply altered the Mediterranean ecosystems. Especially in the Mediterranean lowlands, complex rural civilizations have deeply impacted and shaped the natural landscapes and species communities since prehistoric ages. In the last century and, most intensely, over the last decades, deforestation, intensification of agriculture, urbanization, tourism, and anthropogenic climate change have placed additional pressure on the fragile biodiversity of the region.

5.3.1 THE MEDITERRANEAN HOT SPOT I: THE BALKANO-ANATOLIAN REGION OF THE EAST

by Dragan Chobanov, Slobodan Ivković, and Luc Willemse

In southeastern Europe, cradled between the Danube River and its tributaries, and the Black, Mediterranean, and Adriatic Seas, lies the Balkan Peninsula. The peninsula derives its name from a Turkish word for "forested mountain." In the east, just few hundred meters separate the Balkans from Anatolia, also known as Asia Minor, via the Bosporus Strait. The two regions have been connected for long stretches of geological time, most recently since around 7,000 years ago, and form a geological and biogeographic unit.

During the formation of the Mediterranean basin via the separation of the ancient landmasses of Gondwana and Laurasia ca. 150 MYA (Late Jurassic), the area that later became the Balkans was represented by scattered archipelagos within the Tethys Sea. Only 120 million years later, in the Oligocene Epoch, these islands began merging with Anatolia into a subcontinental landmass called the Aegean. The Aegean landmass broke up for a short period 23–18 MYA during the Miocene into at least three fragments constituting the Hellenides in the west, the Rhodopes and Dinarian Highlands in the northwest, and the Anatolian Land in the southeast, and then reunited. With the incursion of Mediterranean waters into the Aegean and the fast development of a strait connecting the Mediterranean Sea with the Paratethys, the Balkan Peninsula was disconnected from Anatolia around 8 MYA, the result of a formation referred to as the Aegean Barrier. Crete remained isolated since the time of the Messinian crisis, but the Cyclades Islands retained connections with the Balkan landmass until the Early Pliocene. The position of the Balkan Peninsula as an important refugium during the glacial periods as well as the multiple connections and disconnections with other landmasses during geological history, as described above, have likely contributed to a diverse fauna of various origins, including possible multiple exchanges with Asia (via the so-called Lut Block) and pre-consolidated European archipelagoes.

The Balkan Peninsula has the highest diversity of Orthoptera in Europe, with 45% of its species diversity (555 out of 1,229 species). Anatolia is even richer in species, with nearly 700 species recorded in Turkey (Türkiye). Both regions are characterized also with exceptional endemism rates. The high diversity may be explained by the complex paleogeographic and paleoclimatic history of the region, which represented repeatedly a "bridge" between Europe and Asia or was isolated for long periods. Recent scientific studies on the origins and phylogeographic patterns of certain Orthoptera groups of eastern Mediterranean and Aegean origin have shown that the orthopteran faunas of the Balkans and Anatolia have many things in common, suggesting a joint "Balkano-Anatolian" hot spot. Phylogenetic studies allow us to trace the evolutionary origin of faunal elements of this hot spot through geological history. After their establishment in the area, such groups exhibited explosive diversification into genera and species, showing examples of phylogenetically "old" (paleo-) and "young" (neo-) endemics. Though placing the timeline between these groups is subjective, there are some clear examples from both phylogenetics and ecology for paleo- and neoendemic taxa.

Phylogenetic signs for pre-Eocene common ancestry with Laurentian fauna can be traced in the origin of tribe Barbitistini (bush crickets), suggesting ancestral migrations via North America–Europe land bridges. Examples of more recent exchanges between Eurasian and Gondwanan faunas due to the connection of the African and Arabian landmasses around the Paleogene-Neogene border

< Figure 1 **Figure 1** The Balkan predatory bush cricket *Saga ephippigera* is one of Europe's largest orthopterans and prefers prey of the same insect order. Photo: Nikolay Simov.

< Figure 2 **Figure 2** Like other members of its family, the Bulgarian stone grasshopper *Nocaracris bulgaricus* is flightless and sluggish, relying instead on its perfect camouflage as a pebble. Photo: Dragan Chobanov.

Figure 3 *Anadolua schwarzi* (left) inhabits semi-desert-like habitats with sparse vegetation (right). Photos: Dragan Chobanov.

Figure 4 The southern barbed-wire bush cricket *Onconotus servillei* is a specialist of steppe habitats, of which only small fragments remain in the Balkans. Photo: Dragan Chobanov.

(23 MYA) are the Pamphagidae, which have a center of diversity in Anatolia and another one in northwestern Africa, and the Saginae, with two centers of diversity in the Balkano-Anatolian region and in southern Africa. The common ancestor of the tribe Pholidopterini of Tettigoniidae bush crickets originated in this area 17.9 MYA in the Early Miocene. This matches quite well with the origin of some endemic and western Palearctic genera (e.g., *Anterastes*, *Eupholidoptera*, *Pholidoptera*, *Psorodonotus*, *Isophya*, *Poecilimon*) during the Late Miocene between ca. 12 and 5 MYA. The latter speciation events correspond to the Middle Miocene Climate Transition (MMCT; 14 MYA) and subsequent climate cooling and aridification that led to the replacement of the earlier megathermic flora with mesothermic floral elements and steppe vegetation, followed by the appearance of new habitats in the process of the Alpine orogeny.

In addition to phylogenetic affinity, it was adaptation to the high diversity of habitats that led to the evolution of numerous distinct genera and species (see chapter 8.3, On the hunt for the best photo). Today, we often see specific communities of orthopteran species characterizing certain habitat types. Open and semi-open Mediterranean xerophytic habitats, predominated by shrubs and diverse herbaceous vegetation that have established after the MMCT, are home to a highly diverse orthopteran community, including groups that diversified between the Late Miocene and Early Pleistocene with terminal lineages (species) of the *Poecilimon jonicus* group and the *Isophya straubei* group as old as 6 to 4 MYA. Of a similar age may be the main lineages of the carnivorous genus *Saga* (except for the siblings, like *Saga pedo/campbelli*), including the largest insects in Europe (*Saga natoliae*) and in the Holarctic (*Saga ephippigera*; fig. 1). Such groups should be considered Balkano-Anatolian taxa, as lineages inhabiting the Balkans and Anatolia are often closely related.

The Alpine orogeny formed a vast mosaic of mountains and valleys, including typical Mediterranean mountain habitats. These "sky islands" (see chapter 5.5.1, Sky islands) gave rise to some diverse

montane genera endemic to Balkano-Anatolia up to the Caucasus, like *Anterastes*, *Psorodonotus*, and *Nocaracris* (fig. 2). The first two genera have had their own common ancestors dated between 8 and 5 MYA, an age similar to the age of species groups inhabiting the older Mediterranean xerophytic habitats. Other genera are confined to one side of the Aegean Sea, either the Balkans (*Parnassiana*, *Oropodisma*, *Peripodisma*) or Anatolia (*Anadolua*, *Rammepodisma*; fig 3). The evolution of these genera, similar to the younger mountainous groups within the hyperdiverse genera *Poecilimon* and *Isophya*, happened later, so that they could not "escape" their centers of origin. Within those taxa, speciation seems to have been reinforced by the Mid-Pleistocene transition to longer glacial cycles (1.25 to 0.7 MYA) and the mid-Brunhes transition to larger-amplitude cycles (0.43 MYA), as in the *Poecilimon ampliatus* and *Isophya rectipennis* groups.

The alpine "tundra" is a habitat typical of the highest northern mountains of Balkano-Anatolia that dominate the landscape over 2,200–2,500 m. Presently, these habitats are considered interglacial refugia inhabited by recent Pleistocene migrants or the so-called glacial (neo-)relics. These can either have wider distribution divided between the southern high mountains and the Eurasian far north, such as *Aeropedellus variegatus* and *Melanoplus frigidus* (see chapter 5.2.1, Grasshoppers of the vast Central Asian steppes), or they can be restricted to isolated fragments, like *Stenobothrus cotticus*.

Another recently established habitat specific to the Balkan-Anatolian hot spot and less common or absent from the central and western Mediterranean is the steppe. Zonal steppes stretch from

Figure 5 *Phonochorion uvarovi* is an endemic of the eastern Black Sea region of Anatolia. Photo: Dragan Chobanov.

Figure 6 The Serbian stick grasshopper *Pyrgomorphula serbica* is known to inhabit only a very small area in Serbia and is Critically Endangered. Photo: Laslo Horvath.

northeastern China to the northwestern corner of the Balkan Peninsula, whereas so-called extrazonal steppe patches persist along the Danube Plain, in Central Anatolia, and also in other parts of Europe. The origin of the Asian steppe is linked to the tectonic uplift of the Tibetan Plateau and the Pamir and Tian Shan ranges around the Paleogene-Neogene border. The Balkano-Anatolia steppes appeared in the Late Miocene. During the Pleistocene climate cycles, they periodically expanded or contracted, thus connecting or disconnecting from the central Asian steppes. After the LGM, the eastern Balkans retained dry conditions until the Mid-Holocene, which could have favored the incursion of eastern fauna from the Eurasian steppes. The low genetic diversity of species with wide distribution in steppe-like habitats, like *Celes variabilis* and *Arcyptera microptera*, suggest a very recent, possibly post-LGM, expansion over the dry-grass habitats of the Balkan Peninsula. On the other hand, species with very specific habitat requirements, like *Onconotus servillei* (fig. 4) and *Montana medvedevi*, retained significant genetic differentiation between and even within isolated-occurrence sites. Climate aridification possibly contributed to speciation and out-of-the Balkans dispersal of certain lineages like *Poecilimon intermedius* and *Saga pedo*.

Some types of forest habitats house unique orthopteran communities. The Euxine-Colchic broadleaf forests stretching along the southern Black Sea coast are characterized by high precipitation and constantly high air humidity. At their western edge, the topography is mostly low and hilly, with many neo-endemics of the *Poecilimon bosphoricus* group, as well as the characteristic paleo-endemic *Pholidoptera brevipes* and *Paranocarodes straubei*. At its eastern edge, this ecoregion stretches high in the mountains and is characterized by some endemic species of the genera *Isophya*, *Poecilimon*, and *Psorodonotus*, and the endemic genus *Phonochorion* (fig. 5).

Other interesting examples of specific habitat types that are inhabited by local endemics are the flooded riparian forests along the lower course of the Danube and Sava (*Zeuneriana amplipennis*);

the small remnant of ca. 10 km² of a seasonally flooded forest in southeastern Bulgaria (*Isophya gulae*); mid-latitude fens (marshes; *Conocephalus ebneri*); and sparse black pine forests on serpentine rock in western Serbia and eastern Bosnia and Herzegovina (*Pyrgomorphula serbica*). While the first three examples seem to represent neo-endemics due to the existence of closely related taxa inhabiting different habitats, the case of *P. serbica* (fig. 6), representing a monotypic genus, requires special phylogeographic attention.

With ca. 1,000 taxa of Orthoptera, the region of the Balkans and Anatolia, covering about 1 million km², is comparable in species-richness with some tropical areas. Considering the overall organismal diversity and habitat deterioration (though of a lesser scale than in the rest of Europe), it clearly matches the criteria for a biodiversity hot spot. Recent estimates show that the Balkans include the highest concentration of orthopteran diversity and number of endemic species in Europe and large areas with high numbers of threatened taxa. All of this underlines the importance of this region for the conservation and general knowledge of Orthoptera. The phylogeographic data known to us show the integral history of the paleo-(sub-)endemic lineages of the Balkans and Anatolia, which is partially preserved in the neo-endemic lineages occurring along the Black Sea coast, while mountainous neo-endemics tend to accumulate on either side of the Aegean Sea. Even today, new faunal elements enter via the dry grassland habitats from the north, enhancing the diversity and conservation importance of "Balkano-Anatolia" as a unique and remarkable cradle of orthopteran diversity.

> **Figure 1** A map of the western Mediterranean region, showing the main mountain ranges. Map by Roberto Battiston, based on Nature Earth 2 basemap, public domain.

THE MEDITERRANEAN HOT SPOT II: MOUNTAIN RANGES OF THE WEST

by Paolo Fontana, Bruno Massa, and Roberto Battiston

The mountains of the western Mediterranean – namely, the Alps, the Apennines, and the Pyrenees – have played a pivotal role in shaping the regional biodiversity (fig. 1). These mountains are characterized by high-altitude landscapes and harsh climates, and their unique biodiversity is deeply intertwined with the process of orogenesis, the formation of mountain ranges.

The resulting climatic patterns have influenced the development of characteristic vegetation types, such as maquis and garrigue, and shaped the life strategies of numerous plant and animal species. Elevation gradients of temperature, precipitation, and atmospheric pressure from the Mediterranean to the Alpine zones helped create a variety of microclimates.

Alps
Central Massif
Pyrenees
Apennines
Cantabrian Range
Sierra Nevada
Atlas Mountains
Anti Atlas

0 ___ 200 km

Figure 2 Unusual-looking and bulky for a bush cricket, these members of the genus *Eugaster* can sometimes be observed in high numbers in the Atlas Mountains. Photo: Oliver Hawlitschek.

Figure 3 Alpine bush crickets are all inhabitants of alpine grasslands and shrublands of higher elevations of the Alps. The Mercantour Alpine bush cricket *Anonconotus mercantouri* shown here is endemic to a small area at the very western end of the Alpine Arc. Photo: Paolo Fontana.

Figure 4 An endemic radiation of grasshoppers of the genus *Italopodisma* inhabits mountains of the Apennines in the Italian peninsula. Photo: Paolo Fontana.

Figure 5 *Melanoplus frigidus* has a wide distribution across Eurasia, but is largely limited to high mountains. Photo: Dragan Chobanov.

Concerning the biogeography of Orthoptera, we recognize three zones in the western Mediterranean: (1) the Iberian Peninsula, including the Pyrenees (Spain, Andorra, and France); (2) the Italian Peninsula (including the Alps and the Apennines), a bridge between Europe and northern Africa; and (3) the Maghreb region, including Morocco, Algeria, and Tunisia, separated from the Sahara by the Atlas Mountains. All these regions are highly diverse and rich in endemic species and genera, especially the high-elevation habitats.

The tettigoniid subfamily Bradyporinae is represented by many endemics in various mountain ranges. *Baetica ustulata* is endemic to the Sierra Nevada massif (Granada and Almeria), where it lives between 2,200 and 3,400 m above sea level. The genus *Ephippiger* is particularly species-rich in Italy, both in the Alps and on the peninsula, with even an endemic species in Sicily, *E. camillae*. To name a few more, *E. ruffoi* is widespread in the highest altitudes of the Central and Southern Apennines, while *E. carlottae* is typical of the northern parts of the same range. The closely related genus *Dinarippiger*, recently described, lives from coastal areas to mountain and alpine environments between northeastern Italy and the Balkan Peninsula. Morocco, especially its mountainous areas, is characterized by numerous endemic species of the genus *Uromenus*, a genus that is also well represented in western Europe and that reaches as far as the Balkans. Somehow similar and related to the Bradyporinae are the Hetrodinae, which spread across Africa and the Middle East. Among these, the species of the genus *Eugaster* often also live in mountainous areas of Morocco (fig. 2).

The alpine mountain habitats of southern Europe host many further endemic tettigoniids belonging, among others, to the genera *Poecilimon* (more widely distributed in the eastern regions), *Leptophyes*, *Decticus*, *Metrioptera*, *Platycleis*, and *Tettigonia*. Several species of the genus *Tettigonia* are typically mountainous, such as the widely distributed *T. cantans*, *T. hispanica* from the mountains of the Iberian Peninsula, *T. silana* from southern Italy, and *T. longispina* of the highest peaks in Sardinia. The genus *Anonconotus* has a very limited distribution, including the central-western Alps and the northern Apennines, and includes 10 species living exclusively in high-altitude habitats (fig. 3). The genus *Amedegnatiana* is endemic to southern France, while *Sardoplatycleis* is endemic to Sardinia, with only one species, *S. galvagnii*, known from the highest peaks of Sardinia.

Among acridids, the perhaps most characteristic group of western Mediterranean mountain ranges is the tribe Podismini of the subfamily Melanoplinae. The highest altitudes of the Pyrenees (Spain, France, and Andorra) are inhabited by *Cophopodisma pyrenaea*. The Alps and the Apennines are home to numerous genera of Podismini, some of which are endemic. There are species of modest altitudes, such as *Odontopodisma schmidtii* and *Micropodisma salamandra*, species widely spread between northeastern Italy and the Balkan Peninsula. Species that are typical of higher elevations, but do not reach the alpine areas include *Pseudopodisma fieberi*. Endemic to the eastern Alps are two genera with only one species each, namely *Chorthopodisma cobellii* and *Pseudoprumna baldensis*, while *Epipodisma pedemontana* is endemic to the western Alps between Italy, France, and Switzerland. The genus *Podisma* has a complex distribution: central-northern Europe is populated by a single species with a very wide distribution, *Podisma pedestris*, while in southern Europe there are many endemic species. *P. cantabricae* and *P. carpetana* are endemic to the Iberian Peninsula, *P. amedegnatoae* to southern France (Central Massif), and *P. dechambrei* to the southwestern Alps (France and Italy). The Italian Alps are also inhabited by some populations currently considered subspecies of *P. pedestris* (whose identity is under investigation), whereas the Apennines are the home of several endemic species of the genus: *P. emiliae*, *P. goidanichi*, *P. magdalenae*, and *P. ruffoi*. The Apennines are also characterized by the genus *Italopodisma*, with nine species and some subspecies, similar to the Pyrenean genus *Cophopodisma* and to the Balkan genus *Peripodisma* (fig. 4). Other Podismini from the mountainous areas of

southern Europe are the species of the genera *Kisella*, *Miramella* and *Nadigella* (all considered subgenera of *Miramella* by some authors). A particularly interesting species due to its Holarctic distribution, although fragmented, is *Melanoplus frigidus* (fig. 5), present from Alaska to Kamchatka, through Central Asia, the Scandinavian Peninsula, and the Alps. Also endemic to the central Apennines (Italy) is the genus *Italohippus*, with three species.

Concerning wing length, many taxa among montane Orthoptera are micropterous or brachypterous, and this may have increased the degree of geographic isolation during their evolution. In total, 378 and 380 taxa are known from Greece and Italy, respectively; a high percentage of them (ca. 35%) are endemic. Similar diversity is found in Morocco, which represents the westernmost part of northern Africa and at the same time reaches high altitudes (High Atlas) that have allowed the speciation of many taxa. In particular, the Moroccan Atlas, as well as other Maghrebian mountains, host a large number of endemic species and even some genera exclusive to those mountains; many of them belong to micropterous Pamphagidae, but there are also the southernmost representatives of the genera *Stenobothrus* and *Chorthippus*. Overall, the mountains of northern

Africa represent the meeting point of high-altitude Palearctic species and those specialized to the dry, arid climates of the southernmost latitudes. At least 95 species of Pamphagidae belonging to 17 genera are known in northern Africa. The squamipterous species of the subfamily Pamphaginae are subject to isolation and local differentiation, due to low or absence of gene flow, genetic drift, and founder effects. In many cases, mainly in mountain taxa, it is possible to find cryptic species living side by side. Many species are present within Morocco, Algeria, and Tunisia, belonging to the genera *Pseudoamigus*, *Pamphagus* (fig. 6), *Paracinipe*, *Acinipe*, *Eurypa-ryphes*, *Paraeumigus*, and *Eunapiodes*. The Pamphagidae are also present on the European continent, especially with a high contingent of species in the Iberian Peninsula, Italy and the Balkans and the Balkan Peninsula.

This chapter describes just a few examples of the high species-richness of western Mediterranean Orthoptera. Many other groups show similar patterns of distribution and diversity in the area.

Figure 6 Stone grasshoppers of the genus *Pamphagus*, such as this *P. sardeus* of Sardinia, occur in southern Europe and northern Africa. Photo: Paolo Fontana.

5.3.3 THE MEDITERRANEAN HOT SPOT III: A PLETHORA OF ISLANDS AND ISLETS

by Bruno Massa and Paolo Fontana

During the cold periods of the Pleistocene glacial cycles, many species otherwise adapted to temperate climates retreated to the Mediterranean peninsulas. These refugia became hot spots of species-richness, leading to the evolution of many rare and isolated taxa. Apart from the mountain ranges, the islands were important areas conserving Orthoptera diversity and contributing to the proportion of endemic taxa in the Mediterranean. There, the most relevant process was the change in sea levels: many islands were connected by land bridges to other islands or were connected to the continent at lower sea levels but isolated at higher sea levels. Sicily and the Maltese Islands formed a joint mass of dry land, the Hyblean Plateau, which was connected to the Italian peninsula during the Last Glacial Maximum around 18,000 years ago. Other islands – for example, the Balearics, Crete, and Cyprus – have been isolated for much longer.

We here look at the Orthoptera species present on the major Mediterranean islands (Sicily, Sardinia, Corsica, Crete, the Balearic Islands, Cyprus, and the Maltese Islands) and give the number of known species for each island based on the historic and recent literature and open databases. Overall, we found 254 species, with the highest species-richness on Sicily (islands of the Sicilian Channel excluded) with 117 species, followed by Sardinia with 115, Corsica and Cyprus with each 77, Crete with 73, the Balearics with 52, and the Maltese Islands with 46 species (fig.1). The inferred numbers can be interpreted with the background of the theory of island biogeography and the related species–area relationship, which implies that species-richness increases as a power function of the surface area. This analysis reveals that species diversity is well correlated with island size: a 13-fold increase of the island area is necessary to double the number of species. It further indicates that Cyprus is

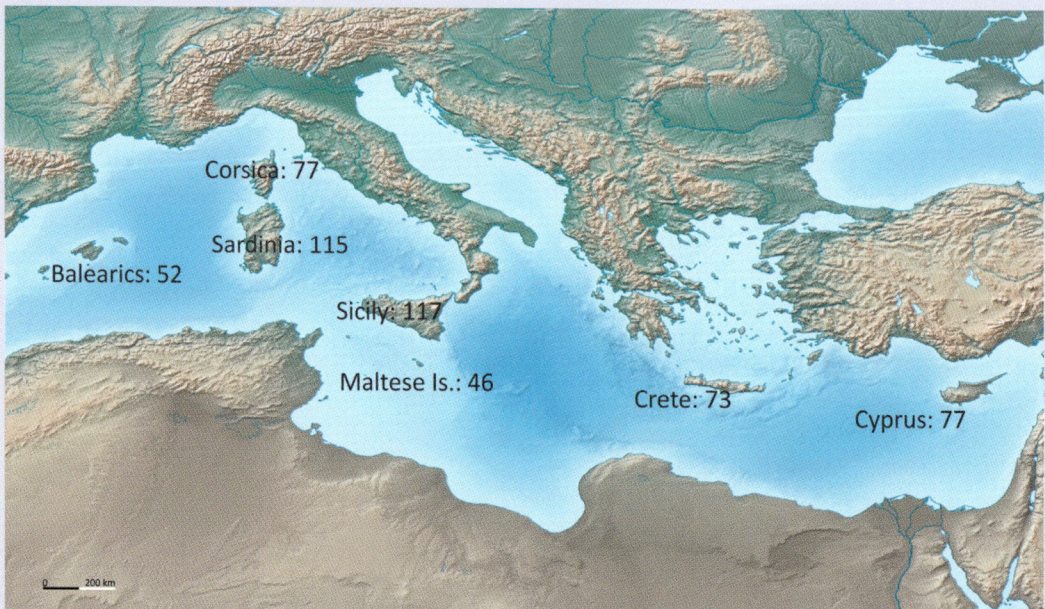

Corsica: 77

Sardinia: 115

Balearics: 52

Sicily: 117

Maltese Is.: 46

Crete: 73

Cyprus: 77

0 ____ 200 km

poorer in species compared with the other islands, probably as a result of its eastern Mediterranean position, much influenced by the Asiatic steppe climate.

Interestingly, many taxa are endemic to a single island: Sicily (islands of the Sicilian Channel excluded) has 22 endemics, translating to 18.8% endemism of its fauna; Sardinia: 15 (13.0%; fig. 2, 3); Corsica: 11 (14.3%); Crete: 26 (35.6%; fig. 4); Cyprus: 19 (36.5%); and the Balearics: 2 (3.8%). The Maltese Islands have no endemics. The highest percentage of endemic taxa is found on Cyprus and Crete – for the latter, thanks mainly to the incredible number of *Eupholidoptera* species, which have undergone a local radiation. We even find endemic genera on some islands.

Many endemic species of Mediterranean islands are characterized by brachypterism – a short-winged and flightless state – compared with other species of the same genus (e.g., *Platycleis concii, P. ra-*

gusai, P. monticola, P. kibris, Sardoplatycleis galvagnii, Tessellana lagrecai, Dociostaurus minutus, Chorthippus corsicus, C. pascuorum, C. biroi, and *Euchorthippus sardous*). Brachypterism is associated with limited dispersal abilities and is often found on mountaintops or islands; very likely it is the result of isolation not only territorially, but also within specific habitats.

Some habitats, like sand dunes, are highly threatened and will disappear if they are not seriously protected from human-induced degradation. Species linked to dune formations – such as *Brachytrupes megacephalus* on Sicily, Malta, and Sardinia, *Ochrilidia sicula* on Sicily, *Ochrilidia nuragica* on Sardinia (fig. 5), and *Dociostaurus minutus* on Sicily – may disappear due to environmental degradation and the severe reduction of populations. These dunes are probably among the most degraded coastal ecological systems

< **Figure 1** A map of the largest islands of the Mediterranean with the number of Orthoptera species. Map by Bruno Massa and Roberto Battiston.

Figure 2 The speckled Sardinian bush cricket *Rhacocleis maculipedes* is endemic to the island of Sardinia, where it mainly inhabits maquis shrubland. Photo: Paolo Fontana.

Figure 3 The short-backed saddle bush cricket *Uromenus brevicollis* inhabits shrublands of several islands of the Mediterranean and the North African coast. Photo: Paolo Fontana.

for various reasons, such as greenhouse cultivation, tourist exploitation, and even, in recent times, recreational activities (e.g., concerts). At least three types of dunes in Italy are considered priority habitats by the Habitats Directive 92/43; as they are in danger of disappearing, the European Union has a special responsibility. Many of these are included in the Sites of Community Interest (SCI) now definitively declared Special Areas of Conservation (SAC). Putting acronyms on environments would serve no purpose if behind them there was no commitment of member states to protect those habitats; yet many dunes are placidly leveled by bulldozers every year, commissioned by municipalities that have an interest in growing local tourism. As in many regions worldwide, conflicts of interest delay the implementation of

conservation laws. For some species, these delays may make the difference between survival and extinction.

< **Figure 4** *Eupholidoptera francesia* (above: male; below: female) and *Leptophyes axeli* (left: male, right: female), described as recently as 2022, are both endemic to Crete and, in the case of *E. francesiae*, the nearby islet of Antikythira. Photos: Paolo Fontana.

Figure 5 *Ochrilidia nuragica* is an endemic of a small area of coastal dunes in Sardinia. Photo: Roberto Scherini.

NATURALISTIC SERENDIPITY: THE DISCOVERY OF A NEW CRICKET VIA ENVIRONMENTAL BIOACOUSTICS WHILE STUDYING PELAGIC BIRDS

by Paolo Fontana and Bruno Massa

In late April 2022, while listening to audio files from an unsupervised bioacoustic assessment of shearwater populations (Aves, Procellariiformes) conducted by Camillo Cusimano, Tommaso La Mantia, and Bruno Massa on the coast of Pantelleria Island (Sicilian Channel, Italy), researchers heard sounds similar to a cricket song. To understand whether the recorded sounds could really be attributed to some type of cricket and, above all, to understand which species could live in those extreme environments of lava cliffs from which they must necessarily have been recorded, two scholars of Orthoptera bioacoustics, Cesare Brizio and Paolo Fontana, joined the research on these mysterious sounds.

The original recordings were obtained by environmental recorders, which are controlled by an application on a smartphone that allows recording only at night or only during the day or at defined times (e.g., half an hour every hour from sunset to sunrise). The sound quality is not particularly suitable for in-depth analysis, but, in this case, it was sufficient to confirm that these songs were of some kind of cricket. Applying more refined bioacoustic techniques and analyzing various parameters of the recorded songs, and after in-depth research among bibliographic sources and sound archives, it was possible to establish quite soon that these songs could not be attributed to any Italian or Mediterranean species of cricket whose song was known.

In the ensuing weeks, physically challenging field research down the cliffs of the southern coast of Pantelleria Island, at some of the original sound recording stations and at further localities, provided photographs, living specimens, and further audio records. As soon as the photos were shared among all the researchers involved, it became clear that the species belonged to the genus *Acheta*. Further bioacoustic analyses and morphological comparison with type specimens of Mediterranean and North African congenerics in insect collections and the scientific literature were carried out: they confirmed that the findings could only be attributed to a still-undescribed species that had previously escaped detection, even on a small island that had already been the target of several entomological surveys.

The peculiar song of the new species has never been heard in the inland part of the island, despite several research projects carried out by the authors in 2021, 2022, and 2023. The newly described *Acheta pantescus* lives mainly among the fissures of the lava rocks a few meters above sea level and in the steep sandy slopes, from which, well hidden, the males emit their song at night. These habitats, nearly impervious to human explorers, are apparently restricted to the effusive coastal cliffs of the island of Pantelleria. This very limited extent of occurrence and the vulnerable habitat suggest that *A. pan-*

> **Figure 1** *Acheta pantescus* male from Pantelleria, a new species discovered serendipitously during nocturnal bioacoustics surveys searching for seabirds. Photo: Paolo Fontana.

> **Figure 2** The peculiar habitat of *Acheta pantescus* on the southern coast of Pantelleria island. Photo: Paolo Fontana.

tescus requires protection actions, such as inclusion in a special Red List by the Italian Committee of the International Union for Conservation of Nature.

A further curious aspect relating to the description of this new species of cricket lies in the fact that the name of the species was chosen through a survey launched on the official website of the Pantelleria Island National Park, as well as on the Forum entomologiitaliani.net, offering the public the opportunity to choose the name of the newly discovered species from among four alternatives: *pantescus* (meaning "from Pantelleria"; 396 votes, 56.6%), *phantasma* ("ghost"; 146 votes, 20.8%), *petrosus* ("of stones"; 136 votes, 19.4%), and *marinus* ("marine"; 22 votes, 3.1%). The discovery of *A. pantescus* is a perfect case of naturalistic serendipity, in which research in one field produces unexpected results in another.

5.3.4 UNRAVELING THE LIFE CYCLE OF THE ATLANTIC BEACH CRICKET

by Karim Vahed

Members of the genus *Pseudomogoplistes* occupy a highly unusual habitat for a cricket. They occur on beaches, among shingle and under cobbles, close to the upper strand line. They are small (body length around 10 mm), gray, and wingless (fig. 1) and, like other members of the family Mogoplistidae, are covered in minute scales. The Atlantic beach cricket *P. vicentae* is known to occur in scattered populations along the East Atlantic coast, from Morocco to Wales, and is classified as Vulnerable on the IUCN Red List. I was intrigued when I read that knowledge of its ecology, including its life cycle, was still sketchy, and I wondered how a cricket manages to survive in such a harsh environment. Hence, in 2011, I began my studies of this species.

Where the females laid their eggs was unknown (beyond a suggestion that it was probably in sand), so I collected adult females and kept them in a terrarium with a choice of likely oviposition sites, including a layer of fine, damp sand and some soft driftwood, and observed them at the same time each night. To my surprise, I only observed females laying eggs in the driftwood, which became packed with eggs over time. Fortunately, females would also lay their eggs in damp cotton wool, so I was able to collect over 200 eggs to examine their development. If the eggs are damp, the pale egg shell becomes transparent enough to be able to observe the embryo under a dissecting microscope. By doing this at intervals, I found that eggs laid in July and August showed no signs of embryonic development until the next May, following a period of cooling, and took nearly a year to hatch (in July).

The number of nymphal instars and the time nymphs take to become adult was also unknown. To solve this riddle, I used baited pitfall traps to sample populations in England and Wales repeatedly across all seasons and also reared individuals from egg to adulthood in the laboratory. This revealed that the majority of eggs hatch from June to August, and the resulting nymphs tend to reach the seventh instar by autumn. They resume development in spring (April) and reach adulthood by July to August, after completing 11 instars in males and 12 in females. Like many species, the adult females seem to live longer than the males. In my captive populations, most of the males died off by the end of November, but a third of the adult females survived until March of the next year. On a visit to Guernsey one April, however, members of La Société Guernesiaise and I managed to catch one adult male that had survived the winter, along with 13 adult females. In its lifetime, an Atlantic beach cricket may live through up to three winters: the first as an egg, the second as a mid-stage nymph, and sometimes a third as an adult.

But how do they survive inundation by storm waves? Part of the answer seems to lie in their behavior. In 2016, I was setting pitfall traps on Marloes Sands beach in Wales during an incoming tide when stormy conditions set in. As the wind began to build and crashing waves started to pound the beach, I noticed swarms of hundreds of sand hoppers (Crustacea, Talitridae) hopping up the beach away from the sea. Then I saw around 100 Atlantic beach-cricket nymphs doing the same thing before scaling the rocky cliffs. One nymph climbed to about 6 m above the beach on a partly vegetated portion of cliff before retreating into a crevice. Similar mass migrations of Atlantic beach crickets away from storm waves have been observed on the Island of Sark.

That females will lay eggs in driftwood may provide a further answer to how the species can survive winter storms. Such eggs will be more resistant to being washed out of the substrate than eggs laid in sand. The driftwood itself may be in danger of being washed away, but if it floats, the eggs inside might survive. That the eggs show no embryonic development until after winter backs up this idea: undeveloped eggs are more resistant to the desiccating effects of saltwater. The long time spent in the egg stage raises a further possibility: eggs in driftwood may raft to new locations. This idea still has to be tested, but it may explain how members of this genus can be found on the Canary Islands, Madeira, the Channel Islands, and numerous small and isolated Mediterranean islands. It could even be that, with such a dispersal mechanism, the Atlantic beach cricket constantly expands its range. After all, the two populations I studied were only discovered in the late 1990s. It is certainly worth looking for the species in suitable habitat in areas where it has not previously been recorded.

Figure 1 An Atlantic beach-cricket nymph in its natural habitat (Chesil Beach, Dorset, UK). Photo: Alex Hyde.

5.3.5 THE PALMENHAUS CRICKET, A MYSTERY LOST TO SCIENCE?

by Oliver Hawlitschek

Wolfdietrich Eichler must have fallen in love.

It all started in the 1930s in the Palmenhaus, the large and famous greenhouse of the botanical garden of Dahlem, a part of the German capital Berlin. At the eve of the Second World War, Eichler was a young researcher specialized in parasites that attacked humans and livestock. His love, however, was a small and inconspicuous insect without any relevance to human health or agriculture. It was a "pretty and most fragile-looking cricket-like creature" that was known from no place except the Palmenhaus.

All kinds of organisms have been intentionally transported, or have hitchhiked, with human travelers throughout history. The higher frequencies of land, sea, and air traffic in the first half of the 20th century consequently led to an increase in the number of individuals and species transported around the globe. Zoological and botanical gardens had a particular demand for importing living organisms from the remotest parts of the planet. In many cases, a multitude of insects, other arthropods, worms, fungi, lichens, and microorganisms inadvertently traveled from their tropical origin to a European or North American greenhouse with plants that were actively imported.

For young biologists, like Eichler, who dreamt of nothing more than joining an expedition to the Malayan rainforest, the Mexican deserts, or the African savannahs, the greenhouses must have been a giant playground where the students could investigate the fauna and flora of these exotic regions without having to undertake an arduous and costly journey. Wolfdietrich Eichler certainly capitalized on this opportunity. He meticulously studied every detail of the cricket's biology, formally named and described it as *Phlugiola dahlemica*, and even made it the topic of his doctoral dissertation. Over just a few years, and in the middle of World War II during which some of them had to serve, Eichler and his colleagues authored 14 further publications on this species. Few other species of Orthoptera have been studied with similar intensity.

We may speculate that Eichler's choice of this topic led to some conflict with his superiors, who, after all, employed him to combat the vermin attacking crucial food supplies and livestock. Perhaps the "strong character" and "assertiveness" that his peers attributed to Eichler ensured that no harm was done to his career. He became a full pro-

1cm

fessor of parasitology and continued his research actively through his retirement until his passing at the age of 81 years.

The Palmenhaus cricket *P. dahlemica* was subject to a less fortunate fate. A bomb hit the Palmenhaus during an air raid in the later stages of the war. While some plants could be transferred to other buildings, the remaining inhabitants were exposed to the cold Berlin winters without any protection. Wolfdietrich Eichler's love, which he had studied with so much dedication, was seen no more.

To date, *P. dahlemica* has never been found again – neither in Berlin nor in any other place around the world. Obviously, it must have colonized the Palmenhaus from somewhere, but its origin remains in the dark. The placement of the species by Eichler in the genus *Phlugiola* – at that time represented uniquely by the species *P. redtenbacheri*, known from a single specimen collected in Suriname and described in 1907 – was later questioned by other authors, who tentatively assigned it to the African genus *Phlugidia*. However, between 2002 and 2018, six further species of *Phlugiola* were discovered and described, resulting in the current consensus that sees the Palmenhaus cricket as part of this genus, confirming the South American origin that Wolfdietrich Eichler had proposed.

The Palmenhaus cricket, for now disappeared from our sight, may surprise us again. In the worst case, it has long been extinct from its natural range without ever being discovered there. Since we do not know the environment it lived in, its habitats may already have fallen prey to deforestation and land conversion, or they might yet do so before their fauna can be conclusively studied. On the other hand, the cricket could still reappear in some greenhouse around the world. Less desirably, it might also show up as an introduced or invasive species in a tropical area of Africa or Asia, where the lack of natural predators and (Eichler would be amazed) of parasites allow its rapid expansion. In the best case, an expedition of field biologists or naturalists will stumble upon *P. dahlemica* in the South American rain forest that is its true home and evolutionary cradle and thus close the circle that was opened by research done generations of scholars ago in a place half a globe away.

Figure 1 (left side) Female holotype of *Phlugiola dahlemica*, one of the few extant specimens of this species. Photo: Birgit Jaenicke.

Figure 2 The historical Dahlem Palmenhaus. Photo: Wissenschaftshistorische Sammlung, Botanischer Garten Berlin, Freie Universität Berlin.

5.3.6   BEI-BIENKO'S PLUMP BUSH CRICKET, ONE OF EUROPE'S RAREST INSECTS

by Soňa Svetlíková

Bei-Bienko's plump bush cricket is an endemic insect known only from one place in the world – the Slovak Karst National Park, located in the very heart of Europe, in southern Slovakia close to the border with Hungary.

The story of *Isophya beybienkoi* (fig. 1) began as recently as 1951, when it was first discovered by the Czech entomologist Josef Mařan from the National Museum in Prague. Mařan recognized that he had found a new species and conducted further surveys in the following years on the Zádielska and the Plešivská Plateaus. He realized that this was the first Slovak endemic bush-cricket species and formally described it in 1958. Mařan named the new species *beybienkoi* in honor of the very famous Russian entomologist Professor Grigory Yakovlevich Bey-Bienko, who devoted his whole life to the studies of the taxonomy, biology, and ecology of grasshoppers and crickets, especially of the European part of Russia.

Over the decades following its discovery, scientists learned more about the biology of the new species. True to its English name, Bei-Bienko's plump bush cricket moves very slowly and, when disturbed, prefers to fall down or hide in the lower vegetation rather than jump away. It is most active at dusk and night, when its very inconspicuous but complex mating song can be heard. The song resembles a falling ping-pong ball and is sometimes answered by beautiful, short, pulsating responses from females. But it is not easily observed: the species is linked to a very specific environment that is exclusively found on huge karst plateaus that are usually inaccessible to humans (fig. 2). There, it inhabits patches of fringe vegetation, rich in a number of flowering plants that occur in or near karst grasslands.

But these habitats are threatened. For many centuries, European wildlife had adapted to a landscape formed by farming and pasture, where herds of livestock gradually took on important ecological roles and, after the extinction of the original megafauna, were crucial for maintaining a mosaic of various types of woodland and grassland vegetation. The advent of industrialized agriculture changed this system and led to the abandonment of traditional grazing practices in the span of just a few decades, causing very fast changes in land cover. Much like many parts of Europe, the Slovak Karst National Park is experiencing these dramatic changes right now. The abandonment of pastures and meadows, once used for grazing animals, has led to the nearly complete overgrowth of the karstic grasslands by shrubs. The habitats of *I. beybienkoi*, which formerly covered wide areas of the karst, are now reduced to isolated patches. Unfortunately, Bei-Bienko's plump bush cricket is not only a slow walker, but also completely unable to fly. As a result, the migration of individuals between habitat fragments is almost impossible, habitats are not re-colonized after extinction, and the overall population declines. The last few isolated subpopulations are disappearing right in front of our eyes.

In 2021, an extensive and detailed research and conservation program was launched to increase the hope of saving this globally threatened endemic from extinction. *I. beybienkoi* was recognized as part of the Slovak national heritage, whose protection and management are of global interest. Very intensive surveys in 2021 and 2022 detected the presence of the species on some remaining habitat patches of most karstic plateaus of the Slovak Karst National Park, but overall, in very low numbers: no more than 200 individuals could be found. As all patches are very small, one of the greatest conservation challenges will be the improvement of the quality, extent,

Figure 1 Female of *Isophya beybienkoi*. Photo: Ján Svetlík.

and connectivity of habitats for this species. If this is successful, we believe that Bei-Bienko's plump bush cricket may become a flagship for the protection of the threatened biodiversity dependent on traditional land use.

Figure 2 Karst landscape rich in fringes as the main habitat of this species. Photo: Ján Svetlík.

5.4 AFRICA

5.4.1 ORTHOPTERA OF THE SAHARA AND THEIR ADAPTATION TO DESERT LIFE

by Michel Lecoq

In the Sahara, the world's largest hot desert, Orthoptera are confronted with extreme ecological conditions almost everywhere. Daytime temperatures sometimes exceed 50 °C, and the temperature range between day and night is often in excess of 35° or 40°C. During the hottest hours of the day, the soil can reach critical temperatures of 60° to 70 °C. Annual rainfall is scarce and irregular and is the lowest on the planet. Most regions receive an average of less than 130 mm of precipitation per year, and some go several years without any rain at all. Nevertheless, quite a few species of Orthoptera live in this hostile environment. The most emblematic is certainly the desert locust *Schistocerca gregaria*, (fig. 1) whose range, in its so-called solitary phase, coincides fairly closely with the extent of the desert zones stretching from

Figure 1 Solitarious adult of *Schistocerca gregaria* (Acrididae) in desertic areas of Mauritania. This locust can be considered a model of the adaptation of Orthoptera to the Saharan desert climate, withstanding both the extreme temperatures and the low amount and random nature of rainfall. Photo: Antoine Foucart, CIRAD, France.

Mauritania to western India. The remarkable adaptation of these Orthoptera is the result of a series of morphological, behavioral, physiological, and reproductive adjustments.

Saharan Orthoptera have various morphological adaptations that help them survive and even thrive in the arid conditions of the desert. Their bodies are generally enveloped in a robust, thick cuticle covered in waxes and lipids that reduce water loss through evaporation. A fairly thick covering of long hairs sometimes helps further to increase this resistance to desiccation. Several species have long legs that keep their bodies at some distance from the hot substrate (fig. 2). In addition, Saharan grasshoppers often display cryptic coloration, helping them to blend into their desert environment and avoid predation (fig. 3).

Saharan grasshoppers also display specific behaviors to help them survive in the desert. They are able to modify their activity according to ambient conditions, avoiding periods of intense heat. The desert locust, for example, can regulate its temperature by a series of daily vertical movements between the vegetation and the ground, and by adopting various thermoregulatory postures. The orientation of the body in relation to the sun's rays varies according to the temperature. To warm up on a relatively cool morning, locusts orientate themselves perpendicular to the sun's rays, maximizing the surface area of their bodies exposed. When the temperature rises, the insects take refuge in the shade or orient themselves parallel to the sun's rays, standing high on their erect legs to keep away from the overheated ground,

thus minimizing heat gain and reducing overheating, which could be lethal.

To further escape the high temperatures and dehydration, some desert Orthoptera bury themselves. This is, for example, the case of *Eremogryllus hammadae* (Eremogryllinae), associated with sandy and fine gravel soils. This small species flies well, making short zig-zag flights, and buries itself very quickly in the sand as soon as it lands, only the antennae, the top of the head, and the eyes remaining visible (fig. 4). Another species found in desert and semi-desert areas, *Poekilocerus bufonius hieroglyphicus*, presents yet another effective adaptation against heat. This grasshopper is almost exclusively associated with the plants of the milkweed family (Asclepiadaceae) and is common on *Calotropis procera*, all parts of which are toxic, particularly the latex. However, this grasshopper is able to withstand the toxicity of its food plants and therefore has access to an abundant water supply. In very hot weather, it is able to lower its internal temperature by evaporative cooling.

The adaptability of the annual biological cycle also plays an important role. Some species, such as *Dericorys albidula*, survive the dry season in the form of diapausing eggs. In *P. bufonius hieroglyphicus*, the eggs develop if there is sufficient humidity, but can withstand a certain amount of desiccation. They can wait in quiescence for the arrival of rain for a year or more. Other species survive in quiescence in the imaginal state, reproduction being delayed until favorable conditions appear, as in the case of the desert locust. In addition, the polyphagy of certain species – that is, their ability to feed on a wide variety of plants – enables them to adapt to seasonal variation in desert vegetation and to the various plant species they are likely to encounter during their migrations.

Indeed, many grasshoppers are also known for their ability to migrate over long distances. This ability is particularly important in desert environments. For the desert locust in general (fig. 1), more than 20 mm of precipitation is required for the maturation of females, egg laying, and the development of eggs and larvae. These conditions rarely persist long enough in any single area of the Sahara to allow

Figure 2 *Truxalis* sp. (Acrididae). Note the length of the legs, which enable the individual to perch away from the heat of the ground, and the pale color, which allows it to hide on sandy soils. Photo: Antoine Foucart, CIRAD, France.

Figure 3 An example of camouflage in this nymph of *Tuarega insignis* (Pamphagidae). This deserticolous grasshopper – the Sahara's largest one – lives on extremely rocky ground and is present throughout the Sahara. Photo: Antoine Foucart, CIRAD, France.

for the development of another generation. This is where – driven by the prevailing winds – the ability to migrate in search of food resources and suitable habitats for reproduction plays a fundamental role. Locusts migrate across the Sahara following the seasonal rainfall, solitary individuals at night and gregarious swarms during the day. Under unfavorable conditions, the imagos remain in quiescence without reproducing, often for several months, migrating several tens or hundreds of kilometers a day. This allows them to search for favorable conditions across large ranges and long times. The seasonal nature of rainfall means that breeding takes place in the northern part of the Sahara in winter and spring, thanks to the rains that hit the Mediterranean basin at that time of year, and in the southern part of the Sahara in summer and early autumn, where the monsoon rains fall. In this way, the desert locust adapts perfectly to local variation in the availability of resources, making the best of the few areas where conditions are most favorable.

These examples show that Saharan Orthoptera have developed an impressive range of adaptations that enable them to thrive in one of the most hostile environments on the planet. Their ability to survive and adapt to the extreme conditions of the Sahara Desert illustrates the remarkable plasticity of these organisms in the face of environmental pressures.

Figure 4 *Eremogryllus hammadae* Krauss, 1902 (Acrididae) buried in sand. The grasshopper first perches on its two pairs of front legs, while the hind legs are alternately shaken backward, throwing sand from under the abdomen and digging a deep hole where the grasshopper eventually burrows. In the end, only the vertex, eyes, and antennae remain visible above the sand. The whole "digging" process takes less than a minute. Photo: Antoine Foucart, CIRAD, France.

> Figure 1 Map of West Africa with a schematic representation of the seasonal migrations of the Senegalese grasshopper relative to the position of the intertropical convergence zone (ITC) in January (red dotted line) and August (blue dotted line). G1, G2, G3: main locations of the first, second, and third generations; G1D: diapausing eggs during the dry season. Blue and red arrows: main population movements (blue: during the rainy season; red, at the start of the dry season) resulting from seasonal movements of favorable habitats (around 25–50 mm of rain/month). Main vegetation zones: 1, desert; 2, semi-desert; 3, steppe; 4, savannah; 5, wooded savannah; 6, tropical forest. Map by Michel Lecoq, with data from Think Africa (2020), Exploring Africa (2025).

by Michel Lecoq

The savannahs of West Africa are rich in a diverse and abundant grasshopper fauna. All these species live in a seasonally highly variable environment marked by the alternation of a short, intense, rainy season and a long dry season – longer the farther north you go toward the limit of the Saharan zone (less than 3 months in the south, 5 to 8 months in the north). In the south, where the climate is Sudanian, rainfall varies from 750 to 1,500 mm/year, and from 250 to 750 mm farther north, in the Sahelian zone. In these regions, the intertropical convergence zone (ITC) plays a key role marking a discontinuity between the warm, dry continental tropical air of the harmattan and the cool, wet maritime air of the monsoon. It moves seasonally from south to north and back again, determining the alternation and duration of the dry and rainy seasons (fig. 1).

Adaptations of life cycles. Grasshoppers in these areas have developed a number of strategies to cope with such seasonal conditions. Some species reproduce continuously throughout the year, taking advantage of the rare points of humidity that remain in the dry season, adapting to a variety of ecological conditions; individuals of all stages can be observed at all times, even in the midst of the unfavorable season. This is the case, for example, for *Acrida bicolor, Pyrgomorpha vignaudii,* and *Gastrimargus*

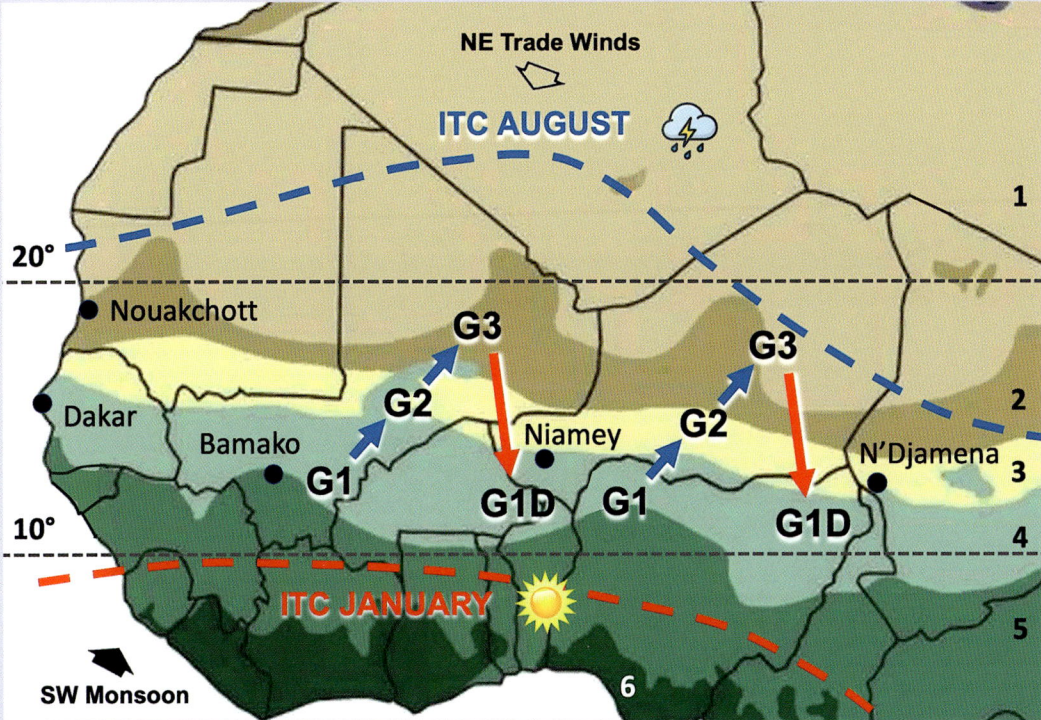

africanus. Other species survive the dry months as immature imagos; in such a case, the first rains trigger sexual maturation and egg-laying, as for *Acanthacris ruficornis* and *Ornithacris cavroisi* (fig. 1). Finally, some species spend the dry season as eggs in diapause, such as *Kraussaria angulifera* and *Oedaleus nigeriensis*, and the first rains trigger the start of embryonic development followed by the hatching of the young nymphal instars.

Following suitable conditions. In addition to these various forms of life-cycle adaptations to survival in the dry season, many grasshopper species in these regions have developed a seasonal migration strategy that allows them to follow the movement of the most favorable conditions as the rainy season progresses. These migrations can be illustrated by the case of the Senegalese grasshopper *Oedaleus senegalensis*, a very regular pest of cereal crops, millet and sorghum in particular (fig. 3). This grasshopper has a life cycle characterized by an embryonic diapause during the long dry season (roughly October through June) and three generations during the rainy season. Eggs normally develop within 12–15 days; there are five nymphal instars, which normally complete their development in three weeks; then

adult females begin to produce eggs about 8–15 days after molting. Thus, under the most suitable conditions and in the absence of diapause, the life cycle of *O. senegalensis* lasts less than two months. Throughout the rainy season, adult populations undertake extensive seasonal migrations (fig. 1). They fly long distances at night, up to 350 km in a single night, to follow the ITC and the associated rains and habitats suitable for breeding. As a result of these migrations, successive generations of Senegalese grasshoppers (G1 to G3) tend to occur in different areas. The northward shift of the ITC at the onset of the rainy season leads to a gradual increase in biotope humidity. Senegalese grasshopper populations are thus obliged to move as the momentarily suitable environment gradually becomes too humid. Southwesterly air currents carry these insects (G1 and G2, first and second generations) to more northern regions where there is less rainfall and ecological conditions are more suitable. This movement gradually accompanies the slow northward shift of the ITC. This phenomenon is reversed at the end of the rainy season. *O. senegalensis* populations (generally G3, third generation) are gradually forced out from the northern areas of their range as the biotopes dry out

Figure 2 Orthoptera have adapted to the climate of the Sahel with very divergent life cycles. (left) *Acanthacris ruficornis citrina*, a species associated with shrubby savannahs, reproduces a single time per year, passing through the dry season in the form of sexually immature nymphs. (right) *Acrida bicolor* (green morph shown), a species of grassy formations, breeds continuously and may produce three generations per year. Photos: Michel Lecoq.

and are drained by northerly winds, and they accompany the ITC in its rapid southward retreat. Some adults stop in the middle zone and lay, while others (depending on the speed of the returning ITC) continue migrating south to the initial distribution zone at the beginning of the season; this is where the majority of egg pods laid by G3 adults are deposited; these eggs spend the dry season in diapause before giving rise to G1 nymphs and imagos in the following rainy season. From one year to the next, the spatial and temporal distribution of rainfall determines the extent and timing of migration. The degree of synchronization between the distribution of favorable habitats, in space and time, and the phenology of Senegalese grasshopper populations will determine the reproductive success of each generation and, consequently, the probability of outbreaks threatening crops.

These long-range migrations are an adaptation to the fragmented and temporary aspect of suitable habitats in semi-arid areas where the spatial and temporal rainfall distribution is shifting and erratic, especially at the beginning and end of the rainy season. This general pattern of migration is characteristic of many grasshoppers in the savannahs of West Africa, although this pattern may vary in detail, depending for each species on their life cycle and ecological requirements. Thus, although the main migrations take place during the rainy season, for species with imaginal diapause, such as *Acanthacris ruficornis*, *Acorypha clara*, and *Ornithacris cavroisi*, imagos continue to be mobile during the whole dry season and to move according to the daily movements of the ITC. Overall, these migratory movements in the savannah regions are so important that the density of a grasshopper population depends on both local reproduction and long-distance migrations, which result in the arrival and departure of imagos at an often very rapid rate.

Figure 3 Brown and green morph of *Oedaleus senegalensis* (Krauss, 1877), one of the most important crop pests for millet and sorghum throughout the Sahelian zone. Associated with grassy formations on sandy or sandy-clay soils, it spends the dry season in diapausing eggs and produces three generations during the rainy season. Photo: Michel Lecoq.

5.4.3 AFRICAN JUNGLES: A WHOLE WORLD OF UNDISCOVERED GRASSHOPPER DIVERSITY

by Jeanne Agrippine Yetchom Fondjo, Martin Husemann, and Oliver Hawlitschek

HISTORY OF RESEARCH ON TROPICAL WEST AFRICAN ORTHOPTERA

Most global insect diversity is found in the tropics. However, compared with the Americas and Asia, the orthopteran fauna of the African rain forests appears very poor in species. But is this really true? Or is it rather an artifact of the different intensity with which we have studied tropical regions? Did we simply not look at the African rain forest closely enough? A dive into the history of the biological research of the region may provide some insight.

We will use the fauna of Cameroon and Togo, especially the Caelifera, as an example. The bulk of existing literature, in which a large number of species were scientifically described, was published by the Swiss Karl Friedrich Brunner-von Wattenwyl and the German Ferdinand Karsch, mostly in the 1890s, and later by Willy Ramme, Lucien Chopard, and Ignacio Bolívar in the early 20th century. European researchers had been active in South America and Asia for a long time already, but only recently in Africa, especially south of the Sahara. For many centuries, Europeans had restricted their involvement with the region to coastal trading stations. The late 19th century then marked the beginning of what historians call the "scramble for Africa," during which western European powers annexed and colonized large parts of the continent. Colonization, in turn, paved the way for European collectors and researchers.

Cameroon was under colonial control by the German Empire from 1884 to 1919. Most taxonomists of that epoch conducted their work in European natural history museums and never ventured into the field. The people collecting samples in the field, however, were in many cases either directly involved in the infrastructure that allowed the European empires to exert their military control and economic exploitation, or they were its beneficiaries. The same persons who were condemned by parts of the European public for their ruthlessness toward the African populace were rewarded by taxonomists who named new species in their honor.

One of the most prominent orthopterists involved in these activities was Ferdinand Karsch, a German entomologist working at the Natural History Museum of Berlin and later at the Agricultural University of Berlin from 1881 to 1915. He lived his whole life in Germany, never traveling to Africa, but described 54 genera and 152 species of African Orthoptera. He named a genus of phaneropterine bush crickets *Preussia* after Paul Rudolph Preuss, a botanist who claimed to be on good terms with the African residents, but also participated in military actions of the colonial forces. Karsch also honored the general of the infantry, Curt von Morgen, with the genus name *Morgenia*, which the general had found time to collect during his military expedition to the Douala and Adamawa regions.

Figure 1 The first author of this chapter, J. Fondjo (left), with an assistant in search of unknown species of Orthoptera in Cameroon's forests. Photo: Jeanne Agrippine Yetchom Fondjo.

Figure 2 The spine-kneed grasshopper *Oxya hyla* typically occurs in open habitats. Photo: Jeanne Agrippine Yetchom Fondjo.

Figure 3 *Apoboleus degener* is typically found in dense forests with a closed canopy. Photo: Jeanne Agrippine Yetchom Fondjo.

The specimens collected during the colonial period came mostly from the southwestern and northwestern regions of Cameroon. After the end of the colonial period, successive missions of various French and Belgian biologists, such as M. Descamps, A. Descarpentries, A. Villiers, M. Donskoff, and J. Le Breton, resulted in around 1,200 collected specimens. These samples, most of which are stored in the entomological collection of the Muséum National d'Histoire Naturelle de Paris in France, led to the publication of Donskoff's work in 1981. More work on this material followed in the 21st century: Mestre and Chiffaud reported the presence of 447 Caelifera from 20 West African countries in their 2009 catalog of Acridoidea.

More recently, Orthoptera research in Cameroon gained new momentum (fig. 3) when native Cameroonian researchers finally took the lead with the contributions of Professor Richard Seino, Professor Sevilor Kekeunou, Dr. Charly Oumarou Ngoute, Dr. Alain Christel Wandji, Dr. Alain Simeu Noutchom, and Dr. Jeanne Agrippine Yetchom Fondjo. Numerous species were subsequently recorded and new species described from this country, and to date the Cameroonian Caelifera fauna encompasses 240 known species. Many more are currently described and are still awaiting their discovery. After almost 100 years of colonial and post-colonial control, Cameroonian researchers have finally become equal partners in the study of the Orthoptera fauna of their own country.

Figure 4 *Hintzia squamiptera* is a short-winged grasshopper so far only known from Cameroonian forests. Photo: Jeanne Agrippine Yetchom Fondjo.

Figure 5 The black-and-green color pattern of *Afromastax zebra zebra* serves as excellent camouflage in the light and shadow of the rain forest. Photo: Premaphotos.

Figure 6 *Stenocrobylus festivus* prefers shrubs and trees. Photo: Jeanne Agrippine Yetchom Fondjo.

Despite this recent fast progress, the West African fauna of Orthoptera remains poorly studied compared with other tropical areas. A great deal of inventory, faunistic, and taxonomic work remains to be done. The occurrence of many species has yet to be confirmed, especially given the ongoing forest degradation in this part of the world. Even more than other groups, the highly diverse rain-forest species are understudied and highly threatened by deforestation.

THE DIVERSITY AND DISTRIBUTION OF AFRICAN RAINFOREST CAELIFERA

Within the tropics, rain forests are the most diverse habitats. This is also true for the rain forests of Africa, yet these have been so far less well investigated compared with other regions of the world. With the little existing data, the Caelifera composition of tropical Africa seems, at the first look, to be relatively similar to that of the savannahs of East Africa. The suborder Caelifera is represented in both regions by five superfamilies: Acridoidea, Eumastacoidea, Pyrgomorphoidea, Tetrigoidea, and Tridactyloidea, and some species indeed occur in both savannahs and rain forests – for example, the Acrididae *Coryphosima stenoptera*, *Morphacris fasciata*, and *Trilophidia conturbata* and the Pyrgomorphidae *Taphronota calliparea*.

Nevertheless, despite the limited data, it is very clear that the Western African rain-forest Caelifera represent a unique community, including many endemic species. Several genera are specific to the African rain forests – for example, the Acrididae belonging to the genera *Serpusia*, *Pteropera*, *Pterotiltus*, *Cyphocerastis*, *Hintzia*, *Digentia*, and *Holopercna*, along with the Chorotypidae *Hemierianthus* and the Pyrgomorphidae *Parapetasia*.

The genus *Hemierianthus* comprises about a dozen species, with a distribution extending from the Ivory Coast to the Congo. Species of this genus are restricted to forest habitats and are very scarce. The very strong sexual dimorphism in this genus makes it difficult to match sexes, even if males or females of a species have already been described and can be recognized.

The genus *Parapetasia* contains two described species to date, *P. rammei*, and *P. femorata*. Both are restricted to the wet forests of West and Central Africa, but they occupy different habitats. *P. rammei* is limited to the highlands areas of Cameroon, whereas *P. femorata* inhabits forested lowlands in Western and Central African countries, such as Cameroon, Equatorial Guinea, Gabon, and Nigeria. Although the species is widespread in Cameroon, its distribution in neighboring countries may be underestimated due to a lack of sampling. Both species are present in their natural habitats throughout the year.

The genus *Pterotiltus* is more diverse and contains 14 described species, 13 of which are from wet forests of West and Central Africa; one (*P. hollisi*) is restricted to Uganda and Western Kenya. Species of this genus are forest-dwelling and mostly found in the herb layer of forests, forest edges, and paths. The wings are completely reduced, which may explain its species-richness and the high degree of endemism.

A high diversity of other genera and species can be found throughout sub-Saharan Africa, ranging from low to high altitudes, from tropical zones to deserts, cultivated areas, bare ground, and woodland. These species may serve as indicators for different habitat and forest types. For example, *Acrida bicolor*, *Acridoderes strenuus*, *Atractomorpha acutipennis*, *Catantops stramineus*, *Chirista compta*, *Coryphosima stenoptera*, *Eyprepocnemis plorans*, *Oxya hyla* (fig.2), *Spathosternum pygmaeum*, and *Zonocerus variegatus* can only be found in open habitats. These heliophilous species need the light, space, diverse food sources, and favorable climatic conditions provided by their habitats. In turn, other species prefer a closed canopy – for example, *Apoboleus degener* (fig.3), *Bunkeya congoensis*, *Euschmidtia congana*, *Hintzia squamiptera* (fig. 4), *Holopercna gerstaeckeri*, *Serpusia opacula*, and *Afromastax zebra zebra* (fig.5). There is also segregation at the vertical level: while some species – for example, *Chrotogonus senegalensis*, *Gastrimargus africanus*, *Heteropternis thoracica*, and *Morphacris fasciata* – generally can be considered geophilous and spend most time on the ground, others, like *Oxycatantops congoensis*, *Stenocrobylus festivus* (fig. 6), and *Taphronota calliparea*, live on shrubs and small trees.

All these taxa display specific adaptations to the environments they live in. These species can only survive if their habitat requirements are met and hence may quickly disappear in the face of ongoing forest destruction.

5.4.4 AFRICAN GAUDY GRASSHOPPERS: PRETTY POISONOUS PESTS

by Ricardo Mariño-Pérez and Jeanne Agrippine Yetchom Fondjo

The family Pyrgomorphidae (Orthoptera: Caelifera) contains some of the most colorful grasshoppers in the world (hence, they are often called gaudy grasshoppers). Currently, there are 488 valid species in 148 genera in this family. Pyrgomorphs are easily diagnosable by the presence of a groove in the fastigium of the vertex and the very distinctive characteristics of the male genitalia. The family is specifically diverse in Africa.

African Pyrgomorphidae (excluding Madagascar, but including Socotra) include 56 genera, of which 45 are endemic to the region. The 11 non-endemic genera are also distributed in Israel, Madagascar, the Arabian Peninsula, the continent of Asia, southern Europe, and Australia.

The family Pyrgomorphidae includes species with different levels of wing development, from apterous (wingless), passing through micropterous (presence of tegmina that are nonfunctional and unable to open), and brachypterous (tegmina reduced but still able to open) to fully winged. Some species even have different levels of wing development in a single population, such as Maura rubroornata (fig. 1), Zonocerus elegans, and Z. variegatus.

CHEMICAL DEFENSES

Many Afrotropical members of the family Pyrgomorphidae are stunning examples of aposematic warning coloration. Some feed on poisonous plants and retain the toxic compounds from the plants in their own bodies as a defense mechanism against predators. For this reason, they are commonly called the toxic milkweed grasshoppers. Members of many genera of this family, such as Phymateus, Phyteumas, Poekilocerus, Zonocerus, and Colemania, possess a mid-dorsal abdominal gland that discharges highly repugnant substances, such as cardenolides and pyrrolizidine alkaloids. Both nymphs and adults can use this mechanism when disturbed, often involving an arched stance to increase hemostatic pressure and hissing noises.

Another chemical defense mechanism is the secretion of foam. Taphronota species emit these secretions through openings on the pronotum and abdomen, whereas Dictyophorus (fig. 2), Loveridgacris, and Parapetasia (fig. 3) produce the toxic foams by combining haemolymph with air through the spiracles. Sometimes the foam can be very colorful, and the color depends on the pyrgomorphid diet.

Their chemical defenses do not protect these species from widespread consumption by humans. The seven species most consumed are the abundant Chrotogonus senegalensis, Occidentosphena uvarovi, Phymateus viridipes, Pyrgomorpha cognata, P. vignaudii, Zonocerus elegans, and Z. variegatus. The grasshoppers are usually collected manually using nets or baskets and then boiled or fried to (if done correctly) remove any toxins. In many cases, the nymphs are softer and hence pricier.

PLAGUE SPECIES AND THREATENED SPECIES

Nymphs of certain genera, such as Phymateus, Zonocerus, Poekilocerus, and Taphronota, tend to form aggregations, possibly to enhance protection against predators. It has been debated whether these aggregations can be considered gregarization, which implies changes in behavior or color. In some cases, nymphs march in bands, and adults sometimes fly in groups. In the case of Zonocerus elegans (fig. 4), they can also be found in large numbers, but adults aggregate less, probably due to their low mobility. Even the macropterous forms are poor flyers. Others, such as Phymateus viridipes, have shown much more sustained flight. For Taphronota calliparea (fig. 5), it has been reported that there is a difference in color when

Figure 1 *Maura rubroornata*, long-wing morph, South Africa, long-winged adult (left) and nymph (right). Photo: Ricardo Mariño-Pérez.

Figure 2 *Dictyophorus spumans*, South Africa. Photo: Ricardo Mariño-Pérez.

Figure 3 *Parapetasia femorata*, Equatorial Guinea. Photo: Ricardo Mariño-Pérez.

Figure 4 A nymph of *Zonocerus elegans*, South Africa. Photo: Ricardo Mariño-Pérez.

Figure 5 *Taphronota calliparea*, Mozambique. Photo: Ricardo Mariño-Pérez.

nymphs are solitary (green) compared to when they are aggregated (black with yellow markings changing to green when molting to last instar). Adults disperse soon after. Some of these species damage crops and are considered plagues. This has been studied mostly for *Z. variegatus*, but 27 further species belonging to 11 genera are also recognized as pests of cereals, coffee, beans, corn, tobacco, and cotton, among other farm crops.

Yet other species are of conservation concern. Five species found in Somalia are potentially endangered due to their narrow distributions. *Paraphymateus roffeyi* is known from only four or five specimens from three localities in central Somalia. *Megalopyrga monochroma* is known from one unique female holotype. *Parorthacris somalica* is known from male material only from the type locality. Similarly, *Vittisphena somalica* is known only from its type locality. *Xiphipyrgus tunstalli* has been collected only at two localities. These few records could be due to the lack of recent collecting trips or due to political reasons. Nevertheless, in Kenya, which has been better explored, there is a single male specimen known for *Marsabitacris citronota* from Mount Marsabit. Two species endemic to Socotra Island (part of Yemen, but African in nature) are endangered due to habitat fragmentation (*Physemophorus sokotranus* and *Xenephias socotranus*). We still have time to reverse this situation, as *P. sokotranus* was recorded in iNaturalist many times in 2024.

THE AFRICAN PAINTED GRASSHOPPER, A BEAUTIFUL PEST AND DANGER

by Ricardo Mariño-Pérez and Jeanne Agrippine Yetchom Fondjo

One iconic species of pyrgomorphids widespread across Africa is the painted grasshopper *Zonocerus variegatus*. It is one of the most beautiful, but also one of the most feared grasshopper species on the continent. It passes through the larval stage in a gregarious form, causing substantial damage to many cultivated crops.

In addition to its pest status, the variegated grasshopper has a very strong chemical defense. A repellent gland that opens in the dorsal midline between the first and second abdominal segments produces a liquid that has an unpleasant smell. Some of the common names for *Z. variegatus* are *criquet puant* (French: "stinking cricket") and *Stinkschrecke* (German: "stink bug"). The liquid can be ejected up to 20 cm. *When ingested, the toxins in the liquids are harmful to predators, including humans. Symptoms – including* cramps, stomach pain, paralysis, allergy, diarrhea, nerve pain, abdominal pain, and vomiting – have been reported by several ethnic groups in southern Cameroon. Curiously, these well-known cases of intoxication do not discourage people from consuming it; they continue to eat this insect for its delicious taste (after correct preparation) and for its use in traditional medicine to treat diseases such as splenomegaly and umbilical hernia, and to prevent rheumatism and benign prostatic hypertrophy. *A* typical recipe would be to place the grasshoppers first in boiling water to kill them and then to remove the head (with the gut), legs, and wings. After that, salt is added, and later they are sun-dried. Finally, they are roasted, fried (often with a mixture of spices), or boiled again.

Figure 1 A mating couple of *Zonocerus variegatus*, contributing to ensure the next generation of this versatile species. Photo: Igor Siwanowicz.

5.4.5 THE BALLOON BUSH CRICKETS: MYSTERIOUS DENIZENS OF EAST AFRICAN FORESTS

by Claudia Hemp

At night, East African submontane and montane forests are pervaded by a chorus of animal sounds. Many of these can be attributed to a variety of forest birds, most remarkably the evening spectacle of the silvery-cheeked hornbills *Bycanistes brevis*. However, there are also certain loud and curious cricket- or bird-like chirps that have remained a mystery for a long time, as any search for the singer ended in vain: they stopped as soon as the inquisitive scientist approached.

Mount Kilimanjaro has attracted the interest of researchers for a long time, and the invertebrate fauna was thoroughly sampled in an expedition led by the Swedish explorer and scientist Yngwe Sjöstedt at the beginning of the last century (fig. 1). But it was only in 1996 that Andreas Hemp, while cutting a liana on the southern slopes of this enormous volcano, accidentally discovered a vivid green, medium-sized katydid with curious, balloon-like inflated wings. Grabbing this delicate-seeming insect with his hand, Andreas quickly found out that it was not only armed with long raptorial spines on its fore and mid legs but also – while emitting angry chirps – knew how to use its mandibles to best effect. We brought our amazing discovery home to our scientific station in Old Moshi. Right after nightfall, and even in its cage, it readily commenced a noisy and incessant chirping that would later be declared the loudest song of any bush cricket, solving the enigma of these striking nocturnal sounds. We coined the name balloon bush cricket for our new discovery because of its ostentatious, balloon-like inflated wings that we observed in this and all other males. Once we knew how to locate the continuously chirping insects, which were well camouflaged under broad leaves in the bush and canopy layer, we found many more of them, as well as the wingless and silent females (fig. 2). Our field surveys showed that the balloon

bush cricket was a sensitive species with narrow habitat demands that were apparently only fulfilled in the canopy of certain forest communities on Mount Kilimanjaro, where we considered it endemic.

A search of various entomological collections in Nairobi and around Europe confirmed that nothing like this bush cricket had been collected before. We therefore formally erected a new genus *Aerotegmina*, meaning "inflated wings" with the – at that time – single species *Aerotegmina kilimandjarica*. However, surveys of other montane forests of East Africa in the following years discovered balloon bush crickets on the North Pare Mountains, Mount Meru, and the Manyara Escarpment in northern Tanzania, and on Mount Sabuk and Mount Kenya in the Kenyan highlands (figs. 3, 4).

As many grasshopper and bush-cricket taxa of East Africa are represented by endemic species in different mountains and ranges, we expected to find the same pattern in *Aerotegmina* and were therefore surprised not to see any differences among the populations. A flightless canopy species – restricted to closed montane forest and unable to migrate under current climatic conditions, but occurring as far as Mount Kenya in the north and the Kilimanjaro area in the south – raised thrilling questions. Did this bizarre insect indicate that, not very long ago, closed forests resembling those that are today only found on mountains covered a vast expanse of East Africa, even the lowlands?

The discovery of new species of balloon bush crickets in the Eastern Arc Mountains provided new insights into the former extent of the East African montane forest. Mount Kilimanjaro formed only one to two million years ago, a very young geological age compared with the Eastern Arcs, which are estimated to 100 MYA. If the Eastern Arcs are the evolu-

tionary cradle of *Aerotegmina* and species like *A. kilimandjarica* evolved more recently, as suggested by molecular genetic studies, this supports the hypothesis of a cooler and more humid climate in East Africa one to two million years ago, with a closed area of forest resembling today's montane forest connecting the lowlands between all mountains of the region and connecting the populations of *Aerotegmina*. Molecular genetic studies of the bush crickets supported this hypothesis, which could help explain why the Eastern Arcs are so particularly rich in biodiversity and endemism.

In spite of warnings by scientists already 20 years ago, only few patches of coastal forests are left along the Kenyan and Tanzanian coast, and unabated clearing and destruction is ongoing. Not only will amazingly beautiful habitats and curious plant and animal species be lost, but also a unique archive for the understanding of biogeographical patterns and a resource to learn about the vegetational history of the climatic past of the African continent.

Figure 1 Montane forest on Mount Kilimanjaro. *Aerotegmina kilimandjarica* is found from the submontane to the upper montane forest zone. It is restricted by night frosts. Photo: Andreas and Claudia Hemp.

191

< **Figure 2** Noisy, but not easy to spot: a male Kilimanjaro balloon bush cricket (*Aerotegmina kili-mandjarica*) as usually found in high trees camouflaged under broad leaves. Photo: Andreas and Claudia Hemp.

< **Figure 3** The rare giant balloon bush cricket *Aerotegmina megalop-tera,* much of whose habitat in the Kazimzumbwi Forest near Dar es Salaam had to give way to the growing city of Kisarawe. The males produce click-like alarm sounds that can be heard over long distances in the nocturnal forest and may attract females even in low population densities. Photo: Andreas and Claudia Hemp.

Figure 4 The wings of female balloon bush crickets, as this Kilimandjaro (A) and giant (B) balloon bush crickets, are not inflated as are those of the males. Photo: Andreas and Claudia Hemp.

by Daniela Matenaar

When walking through the veld in southern Africa, I often wondered whether it was a twig that I had kicked or a pebble that remarkably started rolling by itself and crossed my path. Most of the time, I was surprised because it was actually something alive, and as soon as I wanted to reach for it, it was gone. Let me introduce you to the world of perfect camouflage: the lentulids. The family Lentulidae belongs to the superfamily Acridoidea and is distantly related to the grasshoppers of the family Acrididae, which are unfailingly found each summer in gardens in central Europe. The lentulids live elsewhere – in Africa, in the vast sub-Saharan reaches, where they nowadays inhabit almost every conceivable habitat, from the grueling Namibian deserts to the rich, tropical rain forests of eastern Africa. Their origin and main center of diversity, however, lies in what is today southern Africa: the lentulids evolved in the Fynbos biome, characterized by unique, shrub-like vegetation (fig. 1); this biome is regulated by fire, and consequently all living things within it have adapted their life cycles to fire as well. From the Fynbos, the lentulids spread northeastward. In this process, they adapted perfectly to the respective environmental conditions.

Considering their place of origin, was it useful to have wings in a very windy habitat along the ocean while being of very delicate composure? Surely not, as the energetic costs for flight were too high and having wings could prove deadly indeed when living really close to the shore. Moreover, the unique and vast plant diversity and structure only asked for conquering every niche, to adapt, hide, and merge with the surroundings to reach optimal cover. That is what the lentulids achieved. They lost their wings completely and are unable to produce or process sounds, at least not in the usual grasshopper way that is so familiar to us; in other words, they are deaf and unable to chirp. Yet they can occur in striking colors and mimic their surroundings impeccably (fig. 2). Most lentulid species are specialized on certain host plants. They sit on the plant, which feeds them; most of them literally spend their whole life on only a few plants. They find shelter at the bottom or deep inside the plant and heat up through basking on top of it. They drink, rest, court, and finally mate on it.

In case they are disturbed or chased (they have been shown to be a reliable food source for baboons), they have striking defense mechanisms: they act as if they were the plant or stone itself. They behave like the pebble you just kicked accidentally, like a dry twig broken off, or like a seed falling or flipping through the wind to the ground (fig. 3). They may also use the wind on purpose – for locomotion. That is what we observed for the slender restio grasshopper B*etiscoides meridionalis:* when you are a very small, delicate individual with no wings to even reach the very close plant right next to you without wasting too much energy, what do you do? Just jump up and wait for the wind to do the rest, and in a blink of an eye, you are blown away for a couple of meters. If the landing spot does not match the individual's needs, it clambers or

Figure 1 (a) Coastal Fynbos in the buffer zone of Kogelberg Nature Reserve with Proteaceae and different Restionaceae species. Lentulid species of Betiscoides, Devylderia and Gymnidium inhabit this biotope. (b) Habitats of Betiscoides muris and species of Devylderia and Karruia in Groot Winterhoek Wilderness Area with shrub-like Fynbos vegetation. (c) Wetland-habitat of Betiscoides nova at Jonaskop with typical assemblage of different Restionaceae species and spatially altering and temporal flooding. (d) Landscape at Swartberg Nature Reserve after a Fnybos veld fire. All sites in Western Cape, South Africa. Photos: Daniela Matenaar.

jumps in a very controlled manner until it finds shelter again. Yet lentulids are surely capable of holding on tight because some of them – for instance, all species of *Betiscoides* – are adapted to wetland habitats; they are not the best of swimmers and easily drown.

The mating behavior in lentulids is as fascinating as their camouflage. Grasshopper courtship display without the familiar magical sounds? That is hard to imagine. Usually, females get aroused quite quickly when a male starts singing in front of them. With lentulids, it is different. Little is known about their mating behavior, but what has been found is that they prefer to mate at night and that courtship and mating can take a very long time. The male gently approaches the female, by touching her head with his antennae and her back with the feet of his fore and jumping legs. If the female accepts, the male will sit and hold onto her back; the repeated touching continues, sometimes for hours, until the actual mating happens. We filmed a male being rejected and kicked off after nearly 10 hours of intense courtship.

As much as lentulids are adapted to their host plants or surroundings, they also depend on

the perfect soil when it comes to completing their life cycle. It is crucial for a species born in an environment adapted to and dependent on repeated fire events to survive these challenging and abruptly changing conditions. Females leave their host plants to lay their eggs deep into the soils, preferably with appropriate humidity and vegetation cover in order to provide the best possible conditions for their offspring. Observers often reported very vividly how grasshoppers would literally jump for their lives when veld fires occurred. Winged species stand a chance, but for the lentulids, only the wind can help. Some of the just-hatched nymphs face one of the greatest challenges right at the start of their lives: they find themselves on almost bare soil or on a small plant, barely enough to feed them. What do they do – give up and die? No, they just keep moving, until they find their sheltered zone.

Due to their very heterogeneous habitat and their limited mobility, species diversity is extremely high within the lentulids. Every mountaintop in the Fynbos biome is assumed to host a different species. This circumstance speaks for itself, and you can easily guess that due to various reasons (e.g., climate change, habitat destruction and fragmentation), all lentulid species live under imminent and actual threat. Still, so many aspects of their species diversity and ecology still need to be studied, and humankind is urgently obliged to safeguard their source of life and do anything to protect these remarkable creatures.

WHERE LENTULIDS LIVE

Southern African members of the Lentulidae family inhabit some very specific habitats.

Fynbos: Fynbos is a biome containing shrub- and heath-like vegetation, which is adapted to and depends on recurring fire events. It is situated in the southwestern part of South Africa. Fynbos covers a large part of the Capensis or Cape Floristic Region, which is one of the six global floral kingdoms and is considered as one of ~30 biodiversity hot spots on Earth due to its enormous plant diversity and high level of endemism.

Veld: A wide, open, and rural area in southern Africa dominated by shrub or grassy vegetation with only a few small trees.

< Figure 2 (a) Male of *Betiscoides* sp., perfectly camouflaged on its restio host plant in Jonkershoek Nature Reserve. (b) Male of *B. muris* on the run in Groot Winterhoek Wilderness Area, Western Cape, South Africa. (c) Nymph of *Karruia* sp. in Groot Winterhoek Wilderness Area, Western Cape, South Africa. Photos: Daniela Matenaar.

Figure 3 Female individual of *Lithidiopsis* sp. in Namib Greens, Namibia, camouflaged as a pebble. Photo: Daniela Matenaar.

5.5 AMERICAS

5.5.1 *SKY ISLANDS: HOT SPOTS OF ENDEMIC GRASSHOPPER DIVERSITY IN THE AMERICAS*

by L. Lacey Knowles and JoVonn G. Hill

Sky Islands are high-elevation habitats (relative to the surrounding area) that are isolated from each other by the different habitats of lower elevations and are the home of many grasshopper species across the Americas. This geographic setting itself is conducive to species divergence. However, the relative degree of isolation of grasshopper species varies depending upon their habitat preference (e.g., montane specialists versus open forest dwellers), as well as the altitude, geologic history, and environmental conditions of the sky islands. Consequently, sky islands shape the species diversity and distributions of grasshoppers. Most grasshopper diversity across the sky islands of North America occurs in the genus *Melanoplus* (Acrididae: Melanoplinae). Here we provide a survey of some sky-island grasshoppers rather than an exhaustive coverage of sky-island grasshopper diversity.

DISTRIBUTIONS AND SPECIES DIVERSITY OF DIFFERENT SKY ISLANDS

In the Americas, the hyperdiverse genus *Melanoplus* stands out with an astounding diversity of over 364 species. Interestingly, the highest concentration of endemic diversity is found in the sky islands, contrasting with the lower surrounding elevations.

Different sky islands are inhabited by a region-specific assemblage of grasshopper taxa. For example, the grasshopper fauna of the Rocky Mountain sky islands is distinct from and distantly related to that of the Cascade or Sierra Nevada sky islands. There is also taxonomic turnover across the Rocky Mountains. In the central Rocky Mountains of Colorado, for example, you will most likely collect taxa

from the Dodgei species group (fig. 1), but you will not find them in the northern Rocky Mountains sky islands of Montana; there, you will find species of the Montanus or Indigens (fig. 2A–B) groups. Within any one region, there are multiple grasshopper species, each of which is closely related to the others. These regionally circumscribed distributions of sky-island grasshopper species differ from the broad distributions of many non-sky-island taxa whose distributions may span nearly coast to coast in the Americas (e.g., *Melanoplus sanguinipes*, *M. bivitattus*, and *M. femurrubrum*).

Taxonomic composition and regional endemism are not the only ways the sky island grasshopper faunas differ. Interestingly, the number of sky-island species varies across species groups and, consequently, among geographic regions. Across the Americas, the highest concentration of sky-island species diversity is in the West, although there are diversity hot spots in the East, especially in the Appalachian Mountains and in the geologically young, landlocked archipelago of sand ridges in Florida that form an island landscape of habitats isolated by the intervening habitat differences of the low-lying areas. Among the diverse sky islands of the West, grasshopper sky-island diversity varies across subregions; for example, the Cascade Mountains host

Figure 1 Species distributions across Colorado, Utah, and New Mexico, and outlines of the male genitalic shapes of species in the Dodgei group. Figure by Lacey Knowles.

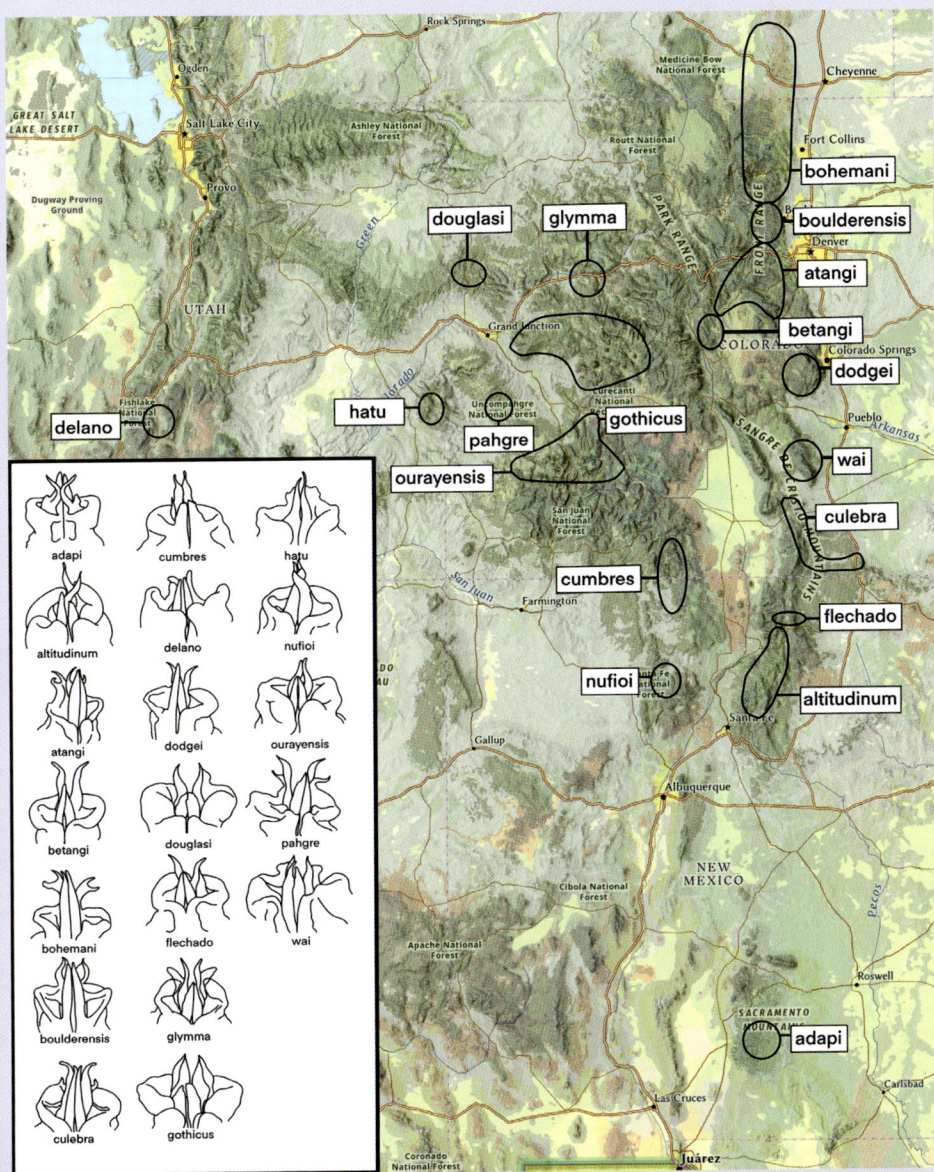

relatively few taxa compared with other mountain systems. Moreover, species diversity can also differ between species groups with geographically overlapping distributions. For example, in the northern Rocky Mountains, the Indigens and Montanus species groups contain 10 and 32 species, respectively, raising intriguing questions about the causes of the different species' diversities.

The factors contributing to species-diversity differences are not known. However, there are some potential candidates related to both abiotic and biotic factors. Among the abiotic factors, differences in the degree of isolation of populations, the number of sky islands in each region, and/or the historical stability of the sky islands (e.g., the extent to which Pleistocene glaciations caused distributional shifts) might contribute to regional differences in species diversity. Biotic factors that may contribute to regional differences in diversity include local adaptation to environmental gradients and/or sexual selection associated with reproductive interactions (e.g., differences in genitalia may be subject to cryptic female choice). For example, the degree of ecological restriction to sky islands differs. In the northern Rocky Mountains, *Melanoplus* species of the Indigens group (fig. 2A) are ecologically restricted to montane meadows (fig. 2B), whereas in the Appalachians, species of the Viridipes group are often abundant on southern Appalachian rocky bald outcrops, but they can also be found in adjacent woodlands (fig. 2E–F).

An examination of the distinguishing traits among the grasshopper species suggests a role for sexual selection, with species differing in the structure of male genitalia. Sexually selected characters are also known to undergo rapid differentiation. Rapid divergence capable of conferring reproductive isolation is required to prevent incipient species divergences from being lost to homogenizing gene flow, especially in regions where temporary connections form among sky islands because of climate-induced distributional shifts during glacial periods. Rapid divergence, as indicated by genetic analyses, suggest that both the abiotic and biotic factors may interact to promote divergence. For example, the founding of isolated mountaintops may have induced genetic differentiation via genetic drift; this partitioning of ancestral genetic variation across mountaintops may likewise have provided opportunities for sexually selected differences to accumulate among the isolated mountaintops. Phenological overlap may also constrain the diversity of species in some sky islands. For example, higher elevation sites (e.g., those above 2,500 m) are highly seasonal, often with condensed growing seasons compared to lower-elevation sites. As such, these sites also experience the strongest warming conditions and have more species with a higher average level of phenological overlap compared with lower-elevation sites.

This brief survey of sky island grasshoppers just scratches the surface of a fascinating group of taxa that accumulated its high diversity over a geologically recent period of about 4 million years, which contrasts with the ancient origin of Orthoptera and divergence times of 50 to 60 and more MYA in other species. If the recency of this radiation of species diversity is confirmed by further analyses of genomic data, it will indicate extremely rapid speciation. Along with a recent taxonomic revision, future phylogenetic and genetic study, and characterization of morphological divergence, the sky island *Melanoplus* species can offer great insights into the evolutionary processes underlying their diversity differences across regions and species groups.

Figure 2 Select sky-island grasshoppers and their habitats. (A–B) *Melanoplus indigens* and alpine grassland near Double Springs Pass in Custer County, Idaho. (C–D) *M. magdalenae* and alpine grassland on South Baldy in the Magdalina Mountains, Socorro County, New Mexico. (E–F) *M. deceptus* and southern Appalachian grassy bald (Gregory's Bald) in Great Smoky Mountains National Park, Blount County, Tennessee. Photos: Lacey Knowles and JoVonn Hill.

5.5.2 BIG AND SHOWY, BUT POORLY KNOWN: MEXICAN ORTHOPTERA

by Paolo Fontana and Ricardo Mariño-Pérez

Mexico is located in a privileged region on the American continent: the northern part of the country is influenced by the Holarctic biogeographic region, while the southern area is influenced by the Neotropical region, resulting in a fascinating diversity of flora and fauna.

A preliminary, but recently updated list of the taxa of Orthoptera present in Mexico was assembled by Fontana and colleagues from the data compilation in the Orthoptera Species File Online. This website today lists 1,054 species from 22 families for Mexico. The most species-rich families are the Acrididae with 399 species, followed by Tettigoniidae (257), Oecanthidae (72), Phalangopsidae (71), Romaleidae (37), Pyrgomorphidae (33), Anostostomatidae (26), Stenopelmatidae (26), and Trigonidiidae (21). Two families endemic to Mexico are the Xyronotidae (4 spp.) and Tanaoceridae (1 sp.).

There are many studies on the diversity of Orthoptera at specific sites throughout Mexico, but few studies analyze the diversity and biogeography at the level of wider regions or the entire country. Two monographs by Daniel Otte provide a very detailed and still very current vision of the Gomphocerinae and the Oedipodinae subfamilies of Acrididae for North America, including Mexico. For other Orthoptera families, complete monographs are lacking, and identification requires a complicated review of the literature and collections.

Mexican Orthoptera include typical tropical genera with a wide distribution in Central and South America. Among the Ensifera, we can mention, for example, the widespread Tettigoniidae *Stilpnochlora, Lichenomorphus, Markia, Cocconotus, Copiphora*, and the Romaleidae *Tropidacris, Titanacris*, and *Chromacris*. There are many long-winged Phaneropterinae species (called popularly *esperanzas*). From a biogeographic point of view, a group of short- to medium-winged Phaneropterinae genera is of particular interest; like other flightless groups, it is very species-rich (high speciation rates due to scarce mobility). Formerly subsumed under the genus *Dichopetala*, it was split into 8 genera and 32 species recently. Another genus of Phaneropterinae endemic to Mexico is the genus *Arachnitus*, which comprises only one squamipterous and one apterous species. Further phaneropterine species with reduced wings and apparently limited distribution are the two brachypterous species *Insara* – namely, *I. oaxacae* and *I. acutitegmina*, recently described for the state of Oaxaca. Another example, this one in the subfamily Conocephalinae, is the species *Brachycaulopsis jovelensis*, so far the only species of this genus and known only from the federal state of Chiapas. The subfamily Listroscelidinae comprises nine species of the genus *Neobarretia*, all of which occur in Mexico, some also in the southern USA. These species are characterized by a conspicuous deimatic display that make them

Figure 1 *Neobarretia spinosa* has a deimatic display. The species of the *Neobarretia* genus have a generally green color, but when disturbed, they exhibit the hind wings, which are characterized by a bright- and dark-spotted pattern. At the same time, they raise their front legs, which are armed with large dark-colored spines. Photo: Paolo Fontana.

Figure 2 The more than 20 Mexican species of the genus *Stenopelmatus* are known by the popular name *cara de niño*, which means "baby-faced" (due to their conspicuous, rounded, and very smooth head). Photo: Paolo Fontana.

Figure 3 The well-camouflaged *Xyronotus aztecus* is a representative of one of the two families completely endemic to Mexico. Photo: Paolo Fontana.

among the most characteristic insects of the arid to desertic habitats of northern Mexico (fig. 1).

In Mexico, there are many species of crickets, especially of the families Gryllidae, Phalangopsidae, Trigonidiidae, Oecanthidae, and Mogoplistidae, while there are few known species of Gryllotalpidae and Myrmecophilidae. This group of Orthoptera is less studied than the Tettigoniidae and Acrididae and may certainly still yield many discoveries. The Rhaphidophoridae, Anostostomatidae, and Stenopelmatidae are also well represented (fig. 2), while there are few known species of Gryllacrididae.

The Caelifera (grasshoppers and locusts) are the most represented and probably best known Orthoptera in Mexico. The groups of which we have a wider knowledge, thanks to Daniel Otte's monographs, are the Gomphocerinae and the Oedipodinae. The group of Melanoplinae in Mexico is very complex (over 160 species; figs. 3, 4), with many species of the genus *Melanoplus* and of the endemic genera *Pedies*, *Oaxaca*, *Netrosoma*, *Perixerus*, and *Phaedrotettix*. The lubbers (Romaleidae) are represented by the genera Tropidacris, Titanacris, and Chromacris. Remarkably, the typically inconspicuous Ripipterygidae, the mud crickets, are also represented by colorful species (fig. 5).

Of great biogeographical interest are a small group of species of the Tanaoceridae and Xyronotidae families. *Tanaocerus rugosus* is a typical species of Baja California Norte. The only other species of the genus is present in California (USA). The genus *Xyronotus* has only three species, all endemic to Mexico: *Xyronotus aztecus* (fig. 6), *Xyronotus cohni*, and *Xyronotus hubbelli*. The second endemic genus is monotypic: *Axyronotus cantralli*.

The Mexican Pyrgomorphidae are likewise of great biogeographic, but also anthropological and folkloric interest. This family is represented in Mexico by approximately 30 species of the subfamily Orthacridinae, with the genera *Calamacris*, *Ichthiacris*, and *Sphenacris*, and of the subfamily Pyrgomorphinae, with the genera *Ichthyotettix*, *Piscacris*, *Prosphena*, *Pyrgotettix*, *Sphenotettix*, and *Sphenarium*. This last genus was the subject of a recent revision that led to the description of 8 new species, bringing the number of species in the genus to 17.

The large number of new species and new genera described in recent years in different habitats and biogeographic areas highlights how the study of the Orthoptera of Mexico promises many new discoveries and how the general picture of this complex and rich fauna is still not completely defined.

Figure 4 A recently described genus of Melanoplini from the state of Oaxaca, *Liladownsia*, has aroused particular interest both for its showy, colorful appearance and its considerable size, which make it almost inexplicable that it had not been found and described earlier. Another reason for the popularity of this discovery was its dedication to the very famous Mexican singer-songwriter and Grammy Award-winner Ana Lila Downs Sánchez, whose stage name is Lila Downs. This taxon was dedicated to her for several reasons, including that she was born in the vicinity of the type locality, that she incorporates indigenous tongues, including the Zapotec language spoken in the type locality, into her music, and that she very often wears colorful traditional Mexican costumes during her concerts. The species name of the only known species *L. fraile* has been given one of the names by which it is called by the people of the place where the species lives, *fraile*; this species was unknown to science, but obviously not to those who cross the wonderful habitats around the town of San José del Pacífico every day. Photo: Paolo Fontana.

Figure 5 *Ripipteryx tricolor* is a conspicuous member of the Ripipterygidae, the pygmy mole crickets. Photo: Paolo Fontana.

Figure 6 *Aztecacris laevis* is one of the many colorful representatives of Melanoplinae. Photo: Paolo Fontana.

5.5.3 SOUTH AMERICA, A CRADLE OF ENDEMIC GRASSHOPPER RICHNESS

by María Marta Cigliano and Martina E. Pocco

In South America, as on other continents of the world, Acrididae are among the most species-rich families of Orthoptera, including 12 subfamilies and approximately 1,153 species. Some of these subfamilies are found across most of the world, while others are restricted to South America. A recent molecular study suggested that South America represents the cradle of the Acrididae, and many basal groups in this family are found almost exclusively on that continent. These groups include the Marelliinae, Pauliniinae, Ommatolampidinae, Leptysminae, and Rhytidochrotinae. While Rhytidochrotinae mostly occur in the forests of Colombia, the other subfamilies are widely distributed in the Amazon basin and northern South America, and some also spread into Central America. The grasshoppers in these groups are quite different from the usual ones. Two specific species, *Marellia remipes* and *Paulinia acuminata* (fig. 1), have a very peculiar way of living. They spend their whole lives on big leaves that float on the water; they eat and also lay their eggs on these leaves. As a specific adaptation, their back legs are flat and wide, helping them swim underwater.

Another interesting species, *Stenopola puncticeps* (fig. 2), belongs to a small subfamily, the Leptysminae, with 77 different species. They are known for their specific body features and are highly adapted to semi-aquatic habitats, like marshes or swampy areas. They like to eat grasses, sedges, or other broad-leaved monocots that grow in these places. *Cylindrotettix dorsalis* (fig. 3) also belongs to Leptysminae, and just like many other grasshoppers that eat grass, these insects have an elongate body shape. While most grasshoppers lay eggs underground, the females of leptysmines bite holes in a plant stem and then lay their eggs inside the plant.

Opaonella tenuis belongs to a small group of 46 species, the Rhytidochrotinae, that mainly live in montane forests from northern Brazil to Costa Rica, with the highest diversity in Colombia. Most members

Figure 1 *Paulinia acuminata* (Pauliniinae). Photo: Martina E. Pocco.

> **Figure 2** *Stenopola puncticeps puncticeps* (Leptysminae). Photo: María Marta Cigliano.

> **Figure 3** *Cylindrotettix dorsalis* (Leptysminae). Photo: María Marta Cigliano.

lack wings as well as auditory organs (tympana) and possess a short thorax. Many species are brilliantly colored. Remarkably, the absence of tympana and the lack of species-specific genitalia suggest an alternative to acoustic and genital courtship in this group. The presence of prominent eyes, vibrant body coloration, and complex color patterns strongly implies that visual signaling may serve as the primary mechanism for mate recognition. Also, unlike most grasshoppers, they like to eat ferns instead of grass.

Psiloscirtus bolivianus belongs to the Ommatolampidinae, which is one of the largest subfamilies of grasshoppers in South America, including over 290 species. While some of them can be found as far north as the southern USA, the majority are concentrated in northern South America and the Amazon basin. The Ommatolampidinae group is incredibly diverse in both appearance and habitat pref-

erences. For instance, there are cryptic and geophilous species like *Vilerna rugulosa*; others inhabit the canopy, while others again, like *Abracris flavolineata*, have a more typical grasshopper appearance. Because of this diversity, there is also a wide range of feeding habits and ways of laying eggs among ommatolampidine grasshoppers.

Oedipodinae (band-winged grasshoppers), Gomphocerinae (slant-faced grasshoppers), and Acridinae are three closely related groups of acridids, represented by over 2,500 species worldwide. Their members are often found in grasslands, and they typically produce sounds by rubbing their wings and legs or by snapping body parts. There is some confusion among scientists about exactly how to classify these closely related groups; at the current state of classification, none of the three groups is probably monophyletic – that is, distinct based on evolutionary

Figure 4 *Hyalopteryx rufipennis* (Acridinae). Photo: María Marta Cigliano.

relationships – which is why we treat them together here. In South America, Acridinae include a small group called Hyalopterigini (17 valid species), the wings of which species are typically obliquely truncated at the apex, as in *Hyalopterix rufipennis* (fig. 4). Gomphocerinae are represented by around 90 South American species. One particular gomphocerine species, *Rhammatocerus schistocercoides* (fig. 5), found in Brazil, Colombia, and Venezuela, stands out for its gregarious behavior. Members of Oedipodinae are more commonly found in warm parts and mostly dry and open habitats and are known for their nearly perfect camouflage, matching the underground they sit on (homochromy). They are not as common in South America (fewer than 10 species) as in North America and the Old World. Only one genus, *Trimerotropis*, is widespread across arid regions of the USA to Argentina and Chile, whereas the other four genera (*Lactista*, *Heliastus*, *Sphingonotus*, and *Diraneura*) are more localized.

Copiocerinae, Proctolabinae, and Melanoplinae (spur-throated grasshoppers) are another trio of closely related groups. The colorful *Adimantus ornatissimus* is among the 91 species of Copiocerinae. This group of grasshoppers is found from Mexico to South America, mostly distributed in northern South America and Amazonia, and they are known to specialize in feeding on rain-forest palms, which is quite unusual for grasshoppers.

The beautiful species of *Poecilocloeus* (fig. 6) are part of the Proctolabinae. This group includes 215 species found across an area stretching from southern Mexico to northeastern Argentina and southern Brazil. Proctolabinae typically inhabit humid forests up to approximately 2,500 m above sea level. Some of them live in large trees, while others, particularly those with short wings (brachypterous species), are found on small trees and shrubs. Many of these insects specialize in eating plants from the Solanaceae and Asteraceae families in the Neotropics. They often lay their eggs inside or on the leaves of their host plants.

Melanoplinae are the largest subfamily of Acrididae in the Americas, with around 1,217 species. These grasshoppers are found from Alaska down to Patagonia. Most are represented in the Northern Hemisphere, but there are about 270 species that live in South America. Among them, the group Jivarini (*Maeacris aptera*), which lives in the Andes and differs in many ways from the rest of the subfamily, appears to be the earliest branch of the group. Melanopline grasshoppers are spread across the continent, except for the Amazon basin, and can be found in various habitats, from sea level to elevations of

Figure 5 *Rhammatocerus schistocercoides* (Gomphocerinae). Photo: María Marta Cigliano.

5,000 m in the Andes. There is a lot of variation between species, especially in the genitalia of males, and sometimes even in females. These grasshoppers do not have any special courtship rituals before mating; instead, males usually approach females abruptly to start mating. Most species live in grasslands and eat a mix of plants, including dicotyledons and grasses, sometimes competing with grazing livestock. Certain species, like those from the genus *Dichroplus*, including *D. maculipennis*, are economically important in Argentine grasslands.

The South American locust *Schistocerca cancellata* belongs to the Cyrtacanthacridinae subfamily, a cosmopolitan group with around 171 species, including some of the most economically important locust species in the world, represented by only two genera with fewer than 20 species in South America (*Halmenus* in Galapagos and *Schistocerca*). *S. cancellata* has long been a significant agricultural pest in Argentina and neighboring countries. It is the only truly swarming locust in southern South America, with a vast invasion area spanning nearly 4,000,000 km^2 across central and northern Argentina, Uruguay, southern Brazil, Paraguay, southeast Bolivia, and central and northern Chile. From the 1870s to 1954, Argentina experienced severe plagues, with 15 out of 22 provinces affected. Similar occurrences were observed in neighboring countries during the same time. Major plagues ceased from 1954 to 2015, due to extensive pesticide campaigns, but a significant outbreak occurred in late 2015 in northern Argentina, followed by expansion into neighboring countries

Figure 6 *Poecilocloeus* sp. (Proctolabinae). Photo: Martina E. Pocco.

over the next few years, the worst in over 60 years. Management of this species has begun to implement some of the world's best practices, but many aspects need to be investigated and then implemented in ways relevant to the political and management systems prevalent in the affected countries.

Since the 19th century, numerous entomologists have contributed to the knowledge of the South American Acrididae, but mostly during the 20th century, several taxonomists worked exclusively on this fauna, contributing to our understanding of its diversity. Most of the region has been studied, but there are some countries of South America that still need to be explored, like Venezuela, the Guianas, Surinam, Bolivia, Chile, most of Brazil (except southern Brazil), and Peru.

The South American Acrididae have suffered taxonomic impediments, as the number of orthopteran taxonomists has been decreasing over the last 30 years. However, this fauna represents a wonderful system to address all sorts of interesting evolutionary questions. The incorporation of genomic scale data to resolve relationships at various levels, the application of novel imaging technologies that allow examination of morphological features in unprecedented detail, and the permanent development of the Orthoptera Species File, the taxonomic database that serves as a backbone for the advancement of orthopteran systematics – all may help to attract students to South American grasshopper systematics.

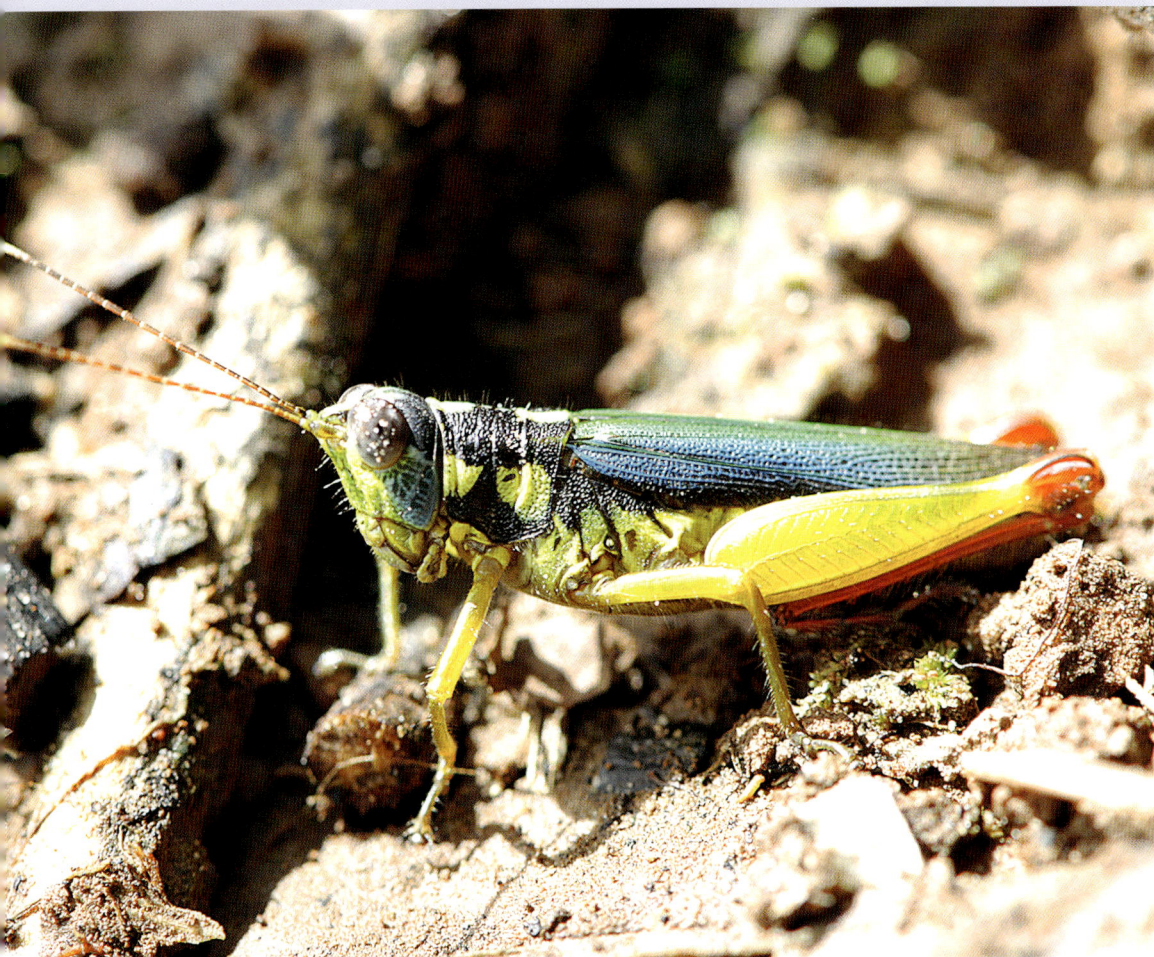

5.5.4   COLORFUL LUBBERS: THE DIVERSITY OF AMERICAN ROMALEIDAE

by Martina E. Pocco and María Marta Cigliano

Commonly known as lubber grasshoppers, Romaleidae is one of the families of Acridoidea endemic to the Americas. Represented by 483 species widely distributed in South and Central America, with a few members extending to the south of North America, it shows a high morphological diversity. In general, lubbers are characterized by their sluggish behavior (leading to the name) and a robust, medium-sized to large body. The family is divided into two big groups, Romaleinae and Bactrophorinae, which differ in morphology and ecology.

The more than 200 species of Bactrophorinae are mostly found in the north of South America and Central America. Most members of this distinctive group live in tropical rain forests. Many are arboreal, living in the canopy, while others can be found in shrubs or herbaceous habitats in clearings. Most bactrophorines are medium-sized insects, with prominent eyes, short or no wings, powerful hind legs, an elongated second segment of the hind tarsus, and prominent sensory hairs covering the lower surface of the abdomen and legs. A common example is *Helicopacris modesta* (fig.1) of Bolivia, which can be found clinging to the trunk or branches of trees. Its ventrally concave thorax is likely an adaptation to this habit, which is also characteristic for other arboreal Bactrophorinae, such as members of the genus *Richnoderma*, which are wingless and normally found resting on medium-sized twigs of trees and shrubs.

The Romaleinae include the largest and most colorful representatives within Neotropical Acridoidea. The 269 species constitute one of the most heterogeneous groups of Acridoidea in shape and habits, from robust and dorsoventrally flattened (e.g., in geophilous members, such as *Phrynotettix*, adapted to desert conditions) to cylindrical and elongated (e.g., in members associated with canopy palm trees, such as *Legua*). Except for the small inhabitants of the crowns of the Amazon rain forest, Romaleinae are generally medium-sized to large grasshoppers – sometimes very large, such as *Tropidacris* (the largest known grasshoppers) and *Titanacris*. Females of *Tropidacris collaris*; fig. 2) can reach about 13 cm, and males a bit more than 10 cm. There is even a record of a female of *T. cristata* with a length of 14.5 cm and a wingspan of 25 cm. In flight, these large specimens look amazingly like small birds. The variation in morphology is seen not only in the general shape of the body, but also in the shape of the head and pronotum. Several species exhibit high pronotal crests, as in *Prionolopha serrata*. The majority of Romaleinae species have fully developed wings, often with colorful hind wings, and sometimes contrasting color patterns. Some species display sexual dimorphism, such as *Staleochlora ronderosi* (fig. 3), with males macropterous and females brachypterous. One of the defining characters for Romaleinae is the presence (with few exceptions) of a tegmino-alar stridulatory apparatus. Unlike many Acridoidea, which produce sound by rubbing the hind legs against the wings, Romaleinae do this by the movement of both wings, rubbing the serrated, parallel, and arcuate veinlets located in the hind wings against a strongly raised vein on the undersurface of the tegmina.

Equally diverse is the variety of habitats the species occupy. Romaleinae are widespread in subtropical areas with mixed forest and savannah vegetation, where many species are common at forest edges, while others are restricted to the Amazon rain forest. They are mostly polyphagous, although some groups have a strong preference for certain plants. *Chromacris* species tend to prefer Solanaceae, while others are adapted to feeding on grasses.

Figure 1 *Helicopacris modesta* (Ophthalmolampini, Bactrophorinae), male, Santa Cruz, Bolivia. Photo: Martina E. Pocco

Figure 2 *Tropidacris collaris* (Tropidacrini, Romaleinae), female, Bahía, Brazil. Photo: María Marta Cigliano.

Lubber grasshoppers are known not only for their sluggishness, but especially for their conspicuous colors. Many species of Romaleinae display striking, aposematic coloration (yellow, red, or orange with a black background) in combination with gregarious behavior at immature stages. The conspicuous coloration sometimes persists in the adult – for example, in *Zoniopoda tarsata* (fig. 4) or in the North American *Romalea microptera* – while in other species there is a change to a cryptic coloration in the adult stage, as in *Tropidacris* and *Chromacris* species. Most aposematic species, such as *R. microptera* and *R. eques*, are capable of consuming poisonous plants, sequestering their toxins, and exuding distasteful chemicals through the spiracles as a defensive mechanism. The nymphs of *T. collaris* display strong gregarious habits and live in well-defined groups, on the ground or in low levels of vegetation. Their coloration is shiny black, with red and cream stripes. On the contrary, the adults live solitarily and are cryptically colored, except for the startling bluish color of their hind wings. This flash coloration appears when grasshoppers spread their colorful wings during flight. Adults live in higher levels of vegetation, in trees or bushes, with a feeding preference for the foliage of trees and shrubs with hard leaves. Likewise, *C. speciosa* nymphs display an aposematic coloration (black with pink or red markings) and exhibit a strong gregarious behavior, normally occurring in clustered form on the leaves or stems of plants. There is a gradual change to a sedentary behavior as they become adults, along with a change into a more cryptic coloration, with a conspicuous orange or yellow-and-black color pattern in their hind wings. This species shows a strong preference for dicots, particularly Solanaceae, which contain toxic compounds and are used by several aposematic grasshoppers as a chemical defense. *Z. tarsata* is another colorful species in which the conspicuous color remains in adults. They prefer herbaceous vegetation and shrubs in low, humid places, near water, but they can also invade higher ground with dense and high vegetation

cover. They are frequently observed perching on plants of the Asteraceae family. This species has a gregarious tendency or is locally abundant in limited areas with adequate vegetation. The gregarious behavior of the nymphs (fig. 4) of this species is not as strong as in the previous ones, but they are commonly found feeding and perching in groups that are moderate in number.

Several Romaleinae exhibit intraspecific variation in body coloration. There are cases in which this variation can be explained by the distribution, such as the South American *Z. tarsata*, which exhibits a color gradient, with lighter and paler colors in the south of its range. *C. speciosa* also exhibits geographical variation of body color, but also within populations. In other cases, the variation is mostly determined by ecological conditions, such as *T. collaris*, with green individuals found in humid and forested areas and brown individuals in dry or semi-arid areas. In other Romaleinae, the variation in color seems to be related neither to geographic nor to ecological factors, as in *Diponthus argentinus*, which exhibits at least four apparently randomly distributed discrete color morphs, all of them with a pattern of contrasting color bands, and even a high degree of variability within each color morph.

The great morphological diversity found in Romaleidae reveals the variety of ecological niches they inhabit. Their conspicuousness in terms of morphology and coloration, and their success in a wide range of habitats make them one of the most interesting grasshopper groups from an evolutionary perspective. However, the classification and phylogenetic relationships within Romaleidae, in particular within Romaleinae, are not yet completely understood.

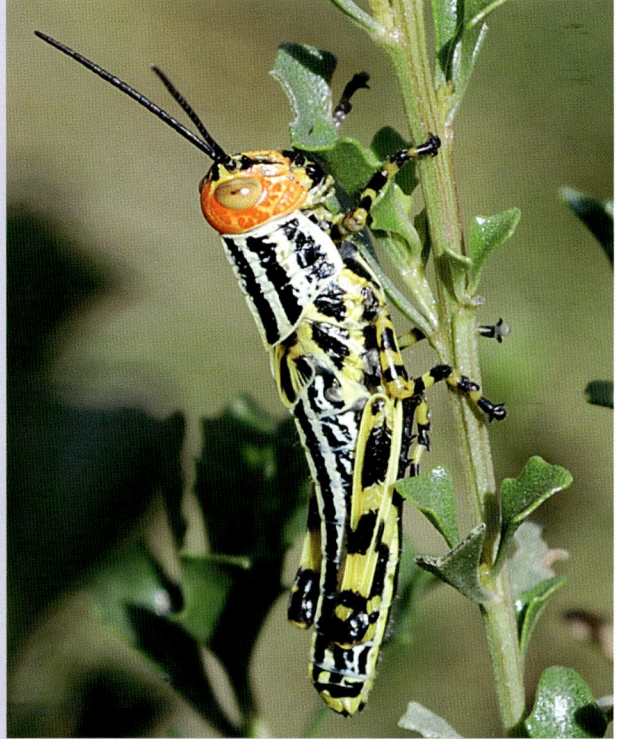

Figure 4 *Zoniopoda tarsata* (Romaleini, Romaleinae). (left) male, Córdoba, Argentina; (right) nymph, Córdoba, Argentina. Photos: Martina E. Pocco.

< Figure 3 *Staleochlora ronderosi* (Elaeochlorini, Romaleinae), female, Formosa, Argentina. Photo: Martina E. Pocco.

5.5.5 THE ENDEMIC JUMPING STICKS FROM MESO- AND SOUTH AMERICA

by Oscar J. Cadena-Castañeda, Alba Bentos-Pereira, and Juan Manuel Cardona-Granda

The proscopiids are commonly known as stick grasshoppers or horse-headed grasshoppers in English, and as *saltamontes palo* ("stick grasshoppers") or *falsos bichos palo* ("false stick animals/stick insects") in Spanish. They are distinguished from other orthopterans by their appearance, which is superficially similar to phasmids, or stick insects. However, phasmids have elongated antennae, a prognathous and more or less quadrangular head, and all legs of similar shape. Proscopiids, as members of Caelifera, have antennae shorter than the femur of the first legs, a hypognathous head with an elongated fastigium, and the third pair of legs developed for jumping. Unlike many Caelifera, they lack abdominal tympanic organs, and most species lack wings. There is a marked sexual dimorphism, with males shorter and more colorful than females.

This group of curious insects is distributed in South America, with a couple of species extending as far as Panama and Costa Rica in Central America. Currently, stick grasshoppers are grouped into three subfamilies: Hybusinae (1 genus and 4 species), only found in central Chile; Xeniinae (3 genera and 16 species), distributed between Chile, Argentina, Paraguay, and Bolivia; and Proscopiinae (2 tribes, 25 genera, and 198 species), ranging from Costa Rica to Argentina and Chile. Six genera are not grouped into any subfamily. Overall, 36 genera and over 230 valid species are known for the entire family. Studies on proscopiids are not as abundant as in other taxa of Orthoptera; hence, their taxonomy remains confusing. The boundaries between most taxa are diffuse, except for groups that have been defined more recently or undergone recent revisions. In all other respects, there is still a lot of work to be done, and verifying the validity of the subfamilies is urgent.

Apparently, the center of the radiation of stick grasshoppers was in a part of southern America, probably around the Patagonian and pre-Andean regions of Argentina and Bolivia, where the most basal genera are still found. The tribe Tetanorhynchini – with the group *Bolidorhynchus-Microcoema*, the group *Anchotatus-Anchocoema*, and the genus *Astroma*, all belonging to the subfamily Proscopiinae – retain the most ancient characters, such as the presence of wings at least in some genera, and an elongated but simple fastigium without significant modifications (figs. 1, 2). They share the region with a high species diversity of genera belonging to the subfamily Hybusiinae: *Hybusa*, *Epigrypa*, and others with uncertain status, such as *Epsigrypa* and *Carphoproscopia*.

Many stick grasshoppers are generalists and occupy a wide variety of habitats. Some prefer open areas, such as grasslands or tropical and subtropical savannahs, where they can be found on grasses or shrubs. Typically, the bodies of these species are particularly elongate, and their heads have an extremely long fastigium, sometimes with an X-section. In the Tetanorhynchini, the subgenital plates are also very elongate, reaching almost the same size as the fastigium. These structures facilitate strategies of blending in among the dominant grass tufts. This group has its center of diversity in the Mato Grosso region and São Paulo, Paraná, and Santa Catarina in Brazil, as well as parts of Paraguay.

Many Proscopiini are adapted to an arboreal lifestyle. Their distribution is vast, from the northern limit of the range of Tetanorhynchini to Costa Rica. While some genera (like *Apioscelis*) are also found in secondary forest formations, other genera are largely restricted to primary forests, as exemplified by *Proscopia*. Some members of this genus have reached gigantism. Compared with their grassland-dwelling

Figure 1 *Cephalocoema* sp., male on top of the female (Proscopiinae: Tetanorhynchini), Natural Park El Impenetrable, Argentina. Photo: Holger Braun.

relatives, their heads are shorter and more compact, and the fastigium and subgenital plate are less elongate. Different levels of the forests – understory, mid-level, canopy – have their specialized sets of species. Especially the canopy specialists, such as the two Central American species in the genus *Pseudoproscopia*, are poorly collected and studied. However, there are also grassland dwellers within Proscopiini, such as *Prosarthria* of the Caribbean region, northern Colombia, and Venezuela. Convergently to Tetanorhynchini, they have evolved a more streamlined morphology.

A quite different group is composed of *Corynorhynchus* and *Nodutus*, found in the Atlantic Forest, along the coastal strip of Brazil. Both genera are very similar in their external and internal morphology, with large, X-shaped fastigia, inflated nota, and remarkable homogeneity in many characters. They represent an evolutionary lineage independent from Proscopiini, with whom they apparently share the arboreal lifestyle.

As mentioned previously, Proscopiidae are mostly generalists, preferring angiosperms, and when they specialize, like the genus *Astroma*, they also do so on angiosperms (*Larrea* spp.). In terms of economic importance, few have been recorded as pests, with species of *Stiphra* and *Tetanorhynchus* (fig. 2) rarely affecting cashew (Anacardiaceae) and eucalyptus (Myrtaceae). Stick grasshoppers are understudied insects, and further research and more researchers interested in this curious group are needed. Additionally, many areas in South America and Central America are unexplored or lack records, leaving several distributional gaps that could harbor genera and species still unknown to science. Furthermore, many species have only been described based on one sex, and no new data have been obtained since their original description, which could lead to the same species being described twice. Little to nothing is known about their behavior and survival strategies, which likely vary among species or taxonomic groups. This suggests another line of study for this group of orthopterans.

5.5.6 THE NEOTROPICAL MONKEY GRASSHOPPERS: SOME OF THE MOST COLORFUL INSECTS IN THE WORLD

by Oscar J. Cadena-Castañeda and Juan Manuel Cardona-Granda

Eumastacidae, commonly called monkey hoppers, are striking insects due to their coloration and the airplane-shaped position of the third pair of legs. Worldwide, they comprise a notable diversity of 255 species. In the Neotropical region, they are distributed from northern Argentina to Honduras. In Guatemala and farther north, they are substituted by the Episactidae, a closely related family with similar external morphology, with which they overlap from Costa Rica to Mexico. Currently, Neotropical monkey grasshoppers are grouped into 5 subfamilies: Eumastacinae (17 genera and 81 species; fig. 1), Paramastacinae (*Paramastax* is the only genus, with 15 species; fig. 2), Pseudomastacinae (*Pseudomastax* is the only genus, with 23 species; fig. 3), Parepisactinae (2 genera and 5 species; fig. 4), and Temnomastacinae with 2 tribes, Eumastacopini (previously treated as an independent subfamily with 8 genera and 33 species) and Temnomastacini (2 genera and 13 species). In the Americas outside the Neotropics, there are 2 additional subfamilies: Masynteinae (*Masyntes* is the only genus with 6 species) endemic to Cuba, and Morseinae (4 genera and 15 species), with a curious disjunct distribution in the eastern United States and the California peninsula in Mexico, as well as in South America with *Daguerreacris tandiliae*, known from the Buenos Aires province in Argentina.

The monkey grasshoppers vary in shapes and colors, but many species sport conspicuous and bright colors of blue, green, yellow, and metallic red, which give it the common name *saltamontes payaso* ("clown hoppers") in Latin America. The eumastacid species with vibrant colors are grouped in the subfamilies Eumastacinae, Paramastacinae, and Temnomastacinae; in the forests and jungles where they live, they stand out among the vegetation for their peculiar coloration and abundance. The other subfamilies have less striking color patterns: the Pseudomastacinae are mainly black or brown with some whitish spots, with females usually more cryptic than males, whereas the Parepisactinae are olive green or light brown and easily go unnoticed on the fern vegetation they live on.

The taxa mentioned here have wings developed to variable stages, with winged, brachypterous (short wings), subapterous (minute wings), and apterous (without wings) forms. All of them, however, are difficult to capture due to their very long and energetic jumps, which are also the origin of their English common name monkey grasshoppers. The eumastacids lack both a tympanic organ and stridulatory structures that allow them to emit sounds, as is common in many other orthopteran groups. However, it has recently been discovered that they communicate with vibratory signals, generated by vibrations of their bodies on a substrate near another individual, which perceives the presence of the emitter. Additionally, the large eyes suggest that vision plays an important role in their communication.

These slender, colorful grasshoppers are recorded from sea level to 3,000 m, with centers of diversity in the Andes and the Amazon. The countries that host the most taxa are Colombia and Brazil; the Colombian Andean region is the major hot spot of genera and species of Eumastacinae, Parepisactinae, and Paramastacinae. The Colombian Andes are divided into three cordilleras (mountain ranges)

< **Figure 2** *Tetanorhynchus* cf. *carbonelli* (Proscopiinae: Tetanorhynchini), Natural Park Iguazú, Argentina. Photo: Holger Braun.

known as the eastern, central, and western. The Eumastacinae and some members of the subfamily Temnomastacinae (the genus *Malenamastax*) have diversified mainly on the eastern mountain range, and the Paramastacinae are only found in the western and central ranges. Some genera, such as *Maripa* (Temnomastacinae), reach gallery forests in the savannahs north of the Amazon. Farther north, in the Venezuelan Andes, which emerge from the Colombian eastern mountain range, a similar pattern occurs with some species of Eumastacinae and Temnomastacinae. In the middle and lowlands (toward the Amazon foothills) of the Andes of Ecuador, Peru, and Bolivia, the Pseudomastacinae are the most widely distributed taxon, followed by the Eumastacinae and Paramastacinae.

The Amazon region is most diverse in Temnomastacinae (Eumastacopini), followed by the Eumastacinae (almost exclusively *Eumastax* species), and the non-Andean *Pseudomastax* species (Pseudomastacinae). Toward the south, outside the Amazon, in biomes such as the Cerrado, Pantanal, and Caatinga, the predominant group is the Temnomastacinae (Temnomastacini). The Parepisactinae have a peculiar disjunct distribution; four species of *Parepisactus* (fig. 4) are found in the Andes of Colombia and Ecuador, whereas *Chapadamastax diamantina* inhabits completely different habitats in the Brazilian state of Bahia. Toward the Chocó biogeographic region, between Colombia and Ecuador, and toward Mesoamerica reaching Honduras, *Homeomastax* (fig. 1; the Mesoamerican counterpart of the

Figure 1 *Homeomastax dereixi* male (Eumastacinae). Photo: Oscar Cadena-Castañeda & Juan Manuel Cardona-Granda.

Amazonian genus *Eumastax*) predominates, while *Helicomastax* is only known from Panama.

These beautiful and particular insects are abundant, but locally not very diverse: typically, a single or at most three species are found at any specific locality at which several species of Acrididae or other groups may occur. Neotropical eumastacids inhabit humid places and dense vegetation, mainly preferring pteridophytes (ferns) and vascular plants of the Melastomataceae, Malvaceae (*Sida* spp.), and Rosaceae (*Rubus* spp.) families, as well as some Mimosaceae and sometimes Euphorbiaceae (*Acalypha diversifolia* Jacq.). Only few species are found on grasses in open areas. However, it is clear that our current knowledge on the group is limited, and additional studies are needed to understand the diversity and biology of monkey grasshoppers in the Neotropics; many areas remain unexplored, and there are distributional gaps that may house genera and species still unknown to science. Little or nothing is known about their behavior and survival strategies, which likely vary between species or taxonomic groups.

Figure 2 *Paramastax rosenbergi* male (Paramastacinae) Photo: Oscar Cadena-Castañeda & Juan Manuel Cardona-Granda.

Figure 3 *Pseudomastax personata* male (Pseudomastacinae). Photo: Oscar Cadena-Castañeda & Juan Manuel Cardona-Granda.

Figure 4 *Parepisactus norcentralis*
male (Parepisactinae). Photo: Oscar
Cadena-Castañeda & Juan Manuel
Cardona-Granda.

6 RESEARCH AND RESOURCES

6.1 DIVERSITY IN BOXES: NATURAL HISTORY MUSEUMS AS ARCHIVES AND RESEARCH PLATFORMS OF ORTHOPTERA

by Luc Willemse

Across the globe, there are hundreds, possibly even thousands of entomological collections containing Orthoptera. This chapter introduces Orthoptera collections by looking briefly at what lies behind us, zooming in on the present, and concluding with a peek into the future. The term *collection* is used throughout the text for a coniguous collection of boxes (or vials) with Orthoptera specimens; the term *repository* is used for the site (building, museum, or institute) in which collections are kept.

ORTHOPTERA COLLECTIONS: THE PAST

The origin of natural-history collections can be traced back to the cabinets that were assembled by wealthy nobles and royal families as early as the European Renaissance of the 16th and 17th centuries. Such cabinets included exotic specimens, along with other types of objects that were collected out of curiosity and admiration and served to impress and astonish visitors, rather than as a base for scientific research. These cabinets were assembled haphazardly by buying objects on sale, by exchange, or on request, from a variety of sources with no or hardly any information on their origin. Overall practices for storing objects differed significantly from the protocols that museums use today. Many of these early cabinets suffered from mold and damage and were short-lived. Starting around the turn of the 19th century, when cabinet ownership fell out of favor, natural-history objects were auctioned off and widely dispersed among obscure collectors, family members, and institutions of higher learning. As such, a great many type specimens from the first known cabinets of natural history are irretrievably lost or presumed destroyed. Still, it was the contents of these very cabinets that often formed the basis for new natural-history collections that functioned as museums. The first museum with zoological specimens, which opened its doors to the public in 1683, was the Ashmolean Museum in Oxford, England. Next, in London, the British Museum (which in 1881 gave rise to today's Natural History Museum) was created by an Act of Parliament in 1753; it opened in 1759. Later, more natural-history museums were erected across Europe, the United States, and Australia: some examples are in Vienna, Austria (1765), Charleston, South Carolina, USA (1773), Florence, Italy (1775), Karlsruhe, Germany (1785), Paris, France (1793), Berlin, Germany (1810), Prague, Czech Republic (1818), Leiden, The Netherlands (1820), Sydney, Australia (1827), Saint Petersburg, Russian Federation (1832), and Brussels, Belgium (1846).

In addition to the onset of museum collections from the mid-1700s onward, another important development in the mid-18th century was the introduction of binomial nomenclature by Carl Linnaeus, the founder of modern biological taxonomy. In his monumental work *Systema Naturae*, he not only introduced binomial nomenclature, but also the classification of the plant and animal kingdom, a system still used up to the present day to classify life on this planet. The 10th edition of *Systema Naturae*, published in 1758 with 4,400 species of animals, is used as the official starting point of zoological nomenclature. Many of the specimens described there were from Linnaeus's personal collection.

Between 1850, when most of the larger museums of natural history in the western world had already been established, and today lies an interval of 175 years, during which many more national, regional, and local museums of natural history were established across the globe. Many of these contain Orthoptera. As a result of collecting trips by internal staff or affiliated citizen scientists and donations of private collections, or both, collections expanded over time. Every museum and every collection has its own story, with noteworthy events, decisive dates, and people who significantly contributed one way or the other to the collection as collector, curator, or taxonomist.

ORTHOPTERA COLLECTIONS: THE PRESENT

Today, repositories housing Orthoptera collections are found all over the world. They vary in size from small, with only a handful or tens of boxes, to very large, with thousands of boxes. Besides their unique history, they differ from each other in taxonomic and geographic composition, the funding they may or may not receive, their accessibility to visitors, and their national and international exposure. Below, we focus on two important aspects of Orthoptera collections. First, we will look at the whereabouts of Orthoptera collections, and second, we zoom in on some features of these collections that determine their usefulness for research.

LOCATIONS OF ORTHOPTERA COLLECTIONS AND THEIR MATERIAL

For anyone studying grasshoppers, the need may arise, sooner or later, to study additional material or gather more data. To gain access to material and data, one needs to know the whereabouts of repositories with Orthoptera collections. Before the digital era, information about Orthoptera collections in museums or in private ownership was scattered. Between 1935 and 1937, Horn and Kahle published a book containing an alphabetical list of entomologists, with details of their collections. A more general overview of insect and spider collections for all countries of the world was published in 1993 by R. H. Arnett et al. (see number 3 in the Websites section below); it lists some 1,300 repositories, only 45 of which specifically mention the presence of the order Orthoptera. Besides these published overviews, information about the locations and/or contents and composition of Orthoptera collections is scattered across numerous papers, each of which provides various levels of details as to the content of the collection. Some of these papers may be biographies of orthopterists – for instance, Antonio Galvagni or Boza Pokopac; others are lists of materials collected during an expedition – for instance, the third Archbold expedition – and the history of a museum, like the one in Vienna. In the digital era, repositories, each at their own pace, started digitizing their collections. Nevertheless, information about collections remains scattered, summarized in dashboards, registries, or descriptive text containing coarse information only, without providing much detail. Examples of websites that provide information about orthopteroid collections and repositories are:

- GRSciColl (Websites section 4): The Global Registry of Scientific Collections, the most elaborate registry about natural-history collections that is currently available; a search for Orthoptera yielded 27 hits.
- Insect and Spider Collections of the World (3).
- Museum Collections Phasmodea (5): Provides a list of museums with Phasmatodea collections.

Most websites providing information about Orthoptera in collections are dedicated to information at the specimen level. Besides numerous institutional and regional portals, the most important websites and data gatherers are (as of November 2024):
- GBIF (Websites section 6): The Global Biodiversity Information Facility. Provides detailed information for close to a million Orthoptera specimens kept in repositories worldwide.
- iDigBio (7): Integrated Digitized Biocollections. A national resource of the USA for advancing digitization of biodiversity collections; provides detailed information for close to 700,000 Orthoptera specimens mainly kept at and originating from the Americas.
- OSF (8): The Orthoptera Species File. A taxonomic database of the world's Orthoptera (grasshoppers, locusts, katydids, crickets, and related insects), both living and fossil, with full taxonomic and synonymic information for all taxa.
- ecdysis (9): A portal for live-managing arthropod occurrence data that is focused on the Americas.
- The Barcode of Life Data Systems (10): An online workbench that aids collection, management, analysis, and use of DNA barcodes. Currently, BOLDSYSTEMS (11) stores barcode data for some 83,000 specimens linked to about 3,500 species of Orthoptera.

WEBSITES
> The Linnean Collections:
  https://linnean-online.org/insects.html
> The Archive of the Natuurkundige Commissie:
  https://dh.brill.com/nco/
> The Insect and Spider Collections of the World:
  http://hbs.bpbmwebdata.org/codens/
> Global Registry of Scientific Collections:
  https://scientific-collections.gbif.org/
> Phasmodea:
  https://www.phasmatodea.com/museum-collections
> Global Biodiversity Information Facility: occurrence data, including collection specimen:
  https://www.gbif.org/
> Integrated Digitized Biocollections:
  https://www.idigbio.org/
> Orthoptera Species File:
  https://orthoptera.speciesfile.org/
> Ecdysis: A Portal for Live-data Arthropod Collections:
  https://ecdysis.org/index.php
> The Barcode of Life Data Systems:
  https://www.ibol.org/phase1/bold/
> The BOLD systems:
  https://v3.boldsystems.org/
> Distributed System of Scientific Collections (DiSSCo):
  https://www.dissco.eu/

There is no website or portal that allows a full global search for repositories with Orthoptera collections, let alone a search for detailed information about their contents. To enable this, wide awareness of the need for such a website, followed by international commitment and collaboration, is needed. The usefulness of such a website would be increased substantially if it also allowed queries about taxa and their geographic origin. While this would be a tremendous help for anyone interested in finding repositories with specific taxa in their collection, the downside is that it requires all repositories to create digital indexes of their collections. A published example of a digital species index is the one carried out by the Academy of Natural Sciences of Drexel University (ANSP) in Philadelphia.

FEATURES OF ORTHOPTERA COLLECTIONS

The composition and size of Orthoptera collections are only two of several features that determine the overall usefulness of a collection and its strength as a research tool. Other features that affect usefulness are:

Exposure, publicity, and reputation: If a collection and/or its composition is unknown to the outside world, the chances of it being consulted and used are slim. Lack of publicity may apply to private collections if the owner has not (yet) published or to museum collections that have not been actively curated for long periods of time, particularly those that have no or only a few type specimens. Obviously, an important aspect contributing to exposure is making information about a collection and its taxa and specimens available online.

Accessibility: The building and rooms where the collections are being kept should be accessible so that the collection can be studied or consulted. Housing a collection in a building closed to visitors, even if it is kept in perfect condition, obviously does not facilitate its use.

Findability: Being able to find (all the) specimens and taxa in a collection one is interested in may seem trivial but, for various reasons, this proves quite challenging. Reasons for suboptimal or poor findability are backlogs in keeping the classification and nomenclature in a collection up to date and dealing with unidentified material.

Logistics: The infrastructure that allows visitors to reach the repository, the availability of a nearby guesthouse or accommodation, and the availability of equipment and laboratory facilities are factors that may be taken into account.

Completeness: Another important feature is the "degree of completeness." When Orthoptera collections contain only very few or only the commonest species, then the need to visit the repository will be low.

Reliability of identifications: Who has been responsible for the identifications in the collection? Identifications in natural-history collections have been and are being conducted by multiple people with varying backgrounds and taxonomic expertise, ranging from students to world specialists.

All these features determine, to some extent, the quality and accessibility of Orthoptera collections as research tools. Lack of human resources, funding, and/or taxonomic expertise pose challenges that each repository has to deal with in its own way to optimize features and create optimal circumstances to exploit natural-history collections.

ORTHOPTERA COLLECTIONS: THE FUTURE

Based on current developments in techniques, legislation, taxonomic expertise, collection digitization, and other areas, outlines can be drawn about changes that may take place in Orthoptera collections in the coming decades.

— Findability of repositories, taxa, and specimens: As a result of ongoing digitization and increased international collaboration (e.g., the DiSSCo program in Europe (Websites section 12), a visualization tool could be developed that allows prospective users to easily and quickly find the locations of Orthoptera collections, not only institutional ones, but also private ones. Results could be visualized as an interactive map, like Google Maps for finding restaurants or hotels, with features like opening hours, logistics, contacts, or even information about taxa and their geographic origin.

— Taxonomic expertise: In Europe, taxonomy as a research discipline in museums and universities is under pressure. A recent "red list assessment" of taxonomic expertise in Europe showed that taxonomic expertise is no longer available for every group in every country, including Orthoptera. It is not clear if this trend in Europe is a global trend, whether it will continue, and what its effect will be in the coming decades. On the other hand, an analysis of author and co-author names from Orthoptera taxa from 1758 up to and including 2020 showed that the number of unique personal names linked as an author or co-author to Orthoptera taxa has been continuously increasing, starting with only 6 in the

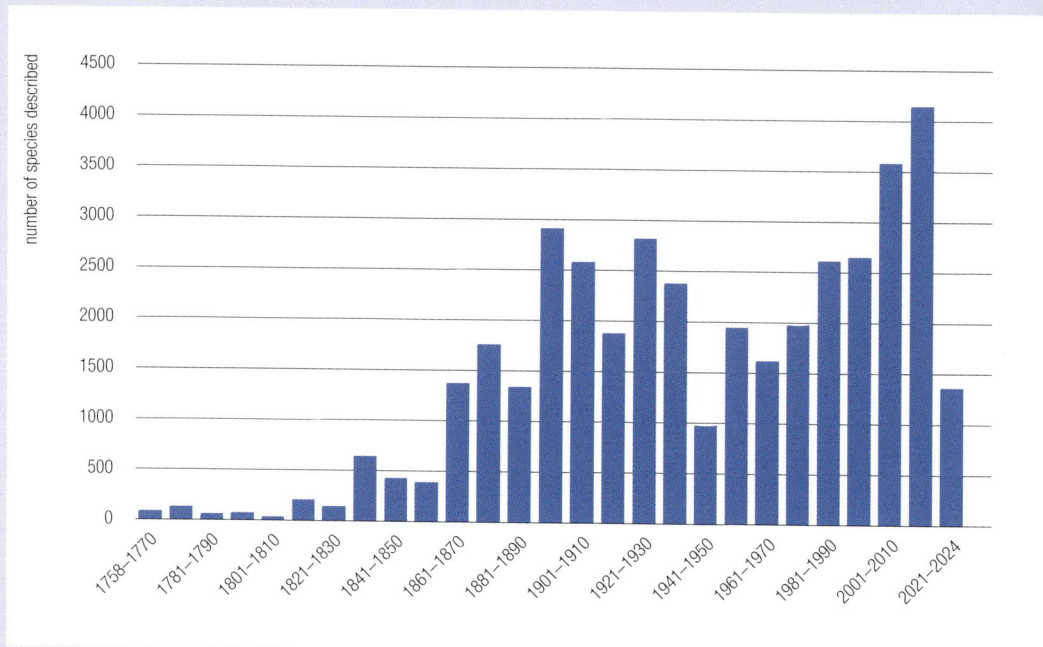

period 1758–1770 and becoming 452 for the period 2011–2020.

— Acquisition and distribution of new material: In recent years, more effort has been put worldwide into regulating access and benefit sharing of biodiversity data and specimens (Nagoya Protocol), and there has been a growing awareness on the part of countries in hot-spot areas of their role in nature conservation. As a result, collecting activities are becoming more and more regulated, and material more often stays in the country of origin. Over the coming years, this may result in a decrease of material being collected – a shift in acquisition of collected material from repositories in the industrialized countries to those in developing countries – but it may also lead to increased collaboration.

— New species: Based on queries using Taxonworks in combination with the OSF dataset, an analysis was made to illustrate the number of species that have been described per decade since 1758 (fig.1). As much as we like to, it is

impossible, based on this kind of information, to predict the trend in the number of species being described in the coming years and decades or to make any substantiated estimate about the overall total number of species of Orthoptera in the world.

— Development of new techniques: Various technical disciplines have shown rapid developments in recent years and are expected to continue doing so. This applies to molecular techniques (DNA), artificial intelligence (AI), imaging (e.g., 3-D imaging [fig.2] and CT), and mass digitization. Future developments of each of these techniques may have profound effects on the accessibility and use of natural-history collections.

CONCLUDING REMARKS

While collating and analyzing information and data for this section, two challenges linked to Orthoptera collections surfaced that are presented here to conclude this chapter.

The distribution of collections in the world: For all the described reasons, Orthoptera collections are not distributed in a uniform way across the globe. Most collections, including those with types, are situated predominantly in the industrialized countries in the Northern Hemisphere. However, if collections were distributed according to the richness of the local Orthoptera fauna, most collections would be situated in the tropical hot-spot areas in the Southern Hemisphere in America, Africa, and southeast Asia. Orthoptera collections are still lacking in large parts of Africa, southeast Asia, and, to a smaller extent, South America. If taxonomy and taxonomists want to make meaningful contributions toward nature conservation, the establishment and expansion of collections in these regions is of utmost importance, as the presence of a collection forms the starting point for education.

The absence of national biodiversity centers: The current distribution of repositories with type specimens across the globe shows a large spread across countries but also within countries. Most countries lack a national center of biodiversity,

so type specimens have been deposited in many repositories. An inventory showed that there are 7 countries with more than 10 repositories holding primary type specimens. Likewise, out of the roughly 300 repositories worldwide with primary type specimens of Orthoptera, almost half (145) have primary types for fewer than 5 taxa. Although there are arguments against centralizing collections in general and type specimens in particular, it may be worthwhile, in some cases and some countries, to consider options to do exactly that. Developing national strategies for natural-history collections, including the storage of type specimens, may help in making well-considered choices.

< **Figure 1** Number of Orthoptera species described per decade since 1758. Data from Orthoptera Species File.

Figure 2 A 3-D scan of *Scelimena celebica* (Tetrigidae). Scan: J.-H. Pamin.

HOW IT ALL STARTED:
THE LINNAEAN ORTHOPTERA COLLECTION

by Luc Willemse

The insect collection that Carl Linnaeus (1707–1778) assembled (fig. 1) contained, among others, the species described in his book *Systema Naturae*. The collection not only included specimens Linnaeus himself collected in Sweden, but also material sent to him by his "apostles" (students) from all corners of the globe – for instance, by Carl Peter Thunberg (1743–1828) from South Africa and Japan. After Linnaeus died, his son Carl Linnaeus the Younger. (1741–1783) took it upon himself to look after his father's botanical and zoological collection, which was kept under poor conditions and suffered from mold and insect and rodent pests. Linnaeus the Younger died within five years after his father's death; shortly thereafter, in 1784, Linnaeus's widow sold the entire collection of books, manuscripts, and specimens to Sir James Edward Smith, who took the collection to England. In the following years until his death in 1828, Smith expanded the Linnaeus collection with specimens collected personally and those he re-

Within the image:
Hemiptera Box 22
LOCUSTA Continued
morbillosus
38
LSL INS 8969
38. morbillosus
LSL INS 8970

ceived from a wide network of friends. The Linnaean collection kept by Smith was already being studied by his contemporaries, like Johan Christian Fabricius and William Kirby. After Smith died, the collection was bought by the Linnaean Society, founded by James Edward Smith in 1788. Today the Linnaeus collection is kept by the Linnaean Society at Burlington House, Piccadilly, in London. The insect collection holds 9,000 specimens, 3,200 of which belong to the original collection of Linnaeus. Because so much attention was focused on Linnaeus's herbarium in references to the Linnaean Society's holdings, some entomologists were unaware of the fact that Linnaeus's insect collection has also survived to the present day. A detailed overview of the orthopteroid insects described by Linnaeus and notes on the specimens in the Linnaeus collection has been presented by Marshall (1983).

Figure 1 Specimens from the Linnaean collection. Photo: The Linnean Society of London.

A DYNASTY OF ORTHOPTERISTS: THE NATURALIS ORTHOPTERA COLLECTION

by Luc Willemse

Naturalis, short for Naturalis Biodiversity Center (fig. 1), was formed in 2010 following the merger of the Zoölogisch Museum Amsterdam (ZMA), the Nationaal Herbarium Nederland, and the National Natural History Museum (aka Naturalis) in Leiden. The latter, in turn, was erected in 1995 as successor of the Rijksmuseum van Natuurlijke Historie (RMNH), created by royal decree in 1820. The history of Naturalis and its precursors dates back more than 200 years. Regarding its Orthoptera collection, four periods can be distinguished:

— 1820–1850: At the time of its foundation, the RMNH orthopteroid collection contained orthopteroid specimens that had been held in old Dutch cabinets, including those used by Caspar Stoll for his magnum opus (Stoll 1787, 1813). In this period, the collection expanded through material collected in Indonesia (former Dutch East Indies) as part of the exploration of the flora and fauna that took place in the context of the Natuurkundige Commissie voor Nederlandsch-Indië. In the same period, the RMNH also received material from other parts of the world, like Japan, South Africa (Cape of Good Hope), North Africa, Europe (Italy, Dalmatia), Colombia, USA (Tennessee), and Australia and Tasmania. This material was used by W. de Haan (1801–1855) to publish his *Bijdragen tot de kennis der Orthoptera* (Contributions to the knowledge of Orthoptera; 1842–1844), in which 135 new taxa of Orthoptera were described.

— 1850–1960: After 1850, the orthopteroid collection of the RMNH expanded further with material collected during expeditions of Dutch entomologists and curators from RMNH. Most of these expeditions went to Indonesia, but they also traveled to other parts of the world, like Surinam and the Karakorum mountains in Kashmir. Material collected in Indonesia, particularly on the Caelifera, was studied by C.J.M. Willemse (1888–1962), who published numerous papers describing ca. 500 new taxa.

— 1960–2010: After the independence of Indonesia, the frequency of expeditions there dwindled, and shipments with freshly collected orthopteroids for RMNH quickly dried up. Between 1960 and 2010, the Orthoptera collection of Naturalis continued to expand significantly, but this time the expansion took place at the ZMA via material collected by its curators and affiliated citizen scientists during their holidays in Europe and beyond. In this way, the Orthoptera collection of the ZMA expanded enormously with material from all parts of the world, but especially the Palearctic region.

— 2010–today: After the merger in 2010, two more important Dutch Orthoptera collections were added: that of C.J.M. Willemse from the Natuurhistorisch Museum Maastricht, where it had been kept since 1963, and the collection of F.M.H. Willemse from his private address at Eygelshoven.

— In 2024, the Orthoptera collection of Naturalis held close to 400,000 specimens, including 4,800 species and primary types for some 500 taxa.

Figure 1 Four generations of orthopterists of the Willemse family
in the Naturalis collection. From left
to right: Joost, Luc, Fer, and Cornelis
(Cees). Foto: Naturalis

ARTIFICIAL INTELLIGENCE, THE FUTURE OF ORTHOPTERA IDENTIFICATION?

by Luc Willemse

Most collections have considerable numbers of un-identified specimens due to researchers' lack of time or in-house taxonomic expertise. These specimens may be stored as singletons or as groups in boxes, either preliminarily sorted by higher taxonomic groupings (order, family) or by geographic region, or they may be completely mixed. To reduce the amount of unidentified material, repositories, especially the larger ones, have started in recent years to digitize their collections at the drawer level. In insect collections, for instance, whole drawers are being imaged and published online, to be browsed through by the entomological community (fig. 1). In this setting, machine-learning applications can follow the image-capturing step by splitting the image into segments, each of which features an individual specimen. The algorithms then assess the identity of the specimens, beginning at higher taxonomic

Identification of taxa

Figure 1 (left) Algorithms recognize and number individual specimens in drawer BE.2286032 of unsorted items. (right) AI algorithms identify specimens to a higher taxon (line color) and accuracy (line type). Photo: Oxford University Museum of Natural History (drawers are mockups).

levels. Subsequently, collection staff can add this information to the corresponding specimen on the image, either physically, using identification labels; digitally, in the object level registration; and/or via efficiently sorting and integrating these specimens into the taxonomic group in the collection. As a result, searching among unidentified material in collections becomes much more efficient, as specimens have already been grouped.

6.2 ORTHOPTERA SPECIES FILE (OSF): THE TAXONOMIC DATABASE OF THE WORLD'S ORTHOPTERA

by María Marta Cigliano and Holger Braun

The study of crickets, katydids, grasshoppers, and related insects requires determining which organisms belong to which species. The discipline of defining biological species and naming them is called taxonomy. Species names are fundamental for categorizing these units of biodiversity and are the basis of most other biological disciplines. Knowing the name of a species unlocks everything that has been learned about its biology, distribution, and relevance to scientists. Thus, the primary mission of the Orthoptera Species File (OSF) is to provide a freely accessible and constantly updated taxonomic database of the world's Orthoptera, ensuring that accurate information is readily available to researchers around the globe.

OSF was originally based on the Species File Software developed by David Eades. David, who also supported the Species File Group at the Illinois Natural History Survey, introduced a significant innovation in this new platform: an editing infrastructure facilitating online data addition and modification. The editing functionalities related to taxon names and nomenclature incorporated all the requirements of the International Code of Zoological Nomenclature. The content for the database was sourced from the works compiled by Daniel Otte from the Academy of Natural Sciences of Philadelphia, published as the Orthoptera Species File book series from 1994 to 2000 (fig. 1). Daniel Otte and Piotr Naskrecki collaborated between 1997 and 2002 to create an initial

Figure 1 Orthoptera Species File book series from 1994 to 2000.

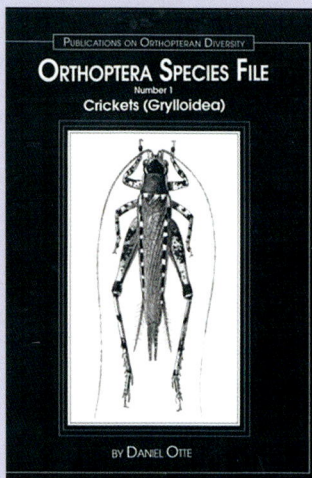

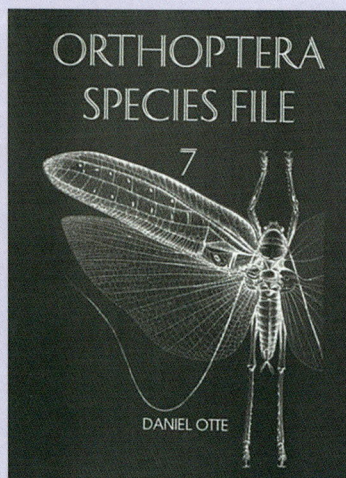

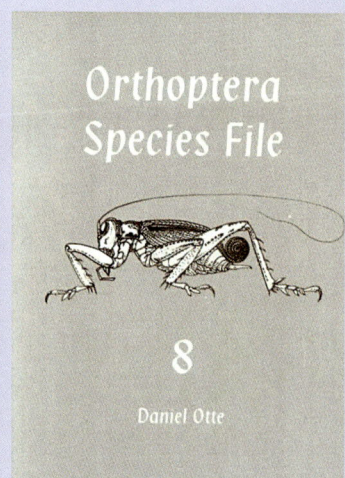

online version of the Orthoptera Species File, which included photographs and sound recordings.

Currently, the OSF has full taxonomic and synonymic information for around 30,000 valid species (both extant and fossil), with complete references, images, sound recordings, and a large set of specimen records. Every year 300 to 500 new species are added (between 2010 and 2022 averaging 367 species, including 10 new fossil species). At the same time, new synonymies and nomenclatural changes are incorporated. The information is obtained by constant monitoring of the Zoological Record database, *Zootaxa* and other taxonomic journals, the *Journal of Orthoptera Research*, Google Scholar, and ResearchGate, and new papers shared on social media. Often papers and information on incorrect or incomplete data are directly received from the authors via email or other media.

The Orthopterists' Society, in cooperation with the Illinois Natural History Survey, provides grants for work supporting the OSF. Since 2013, this program has facilitated 90 projects of researchers from over 30 countries. It has become an instrumental force in expanding the repository. The grants usually provide photo series of type specimens, images of live insects with geo-referenced localities, and sound recordings.

Through the collaboration with the Orthopterists' Society, the OSF operates within a governance structure that ensures both community engagement and the sustained management of its

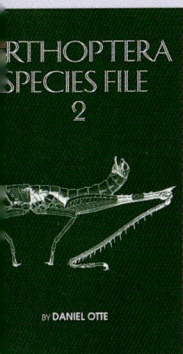

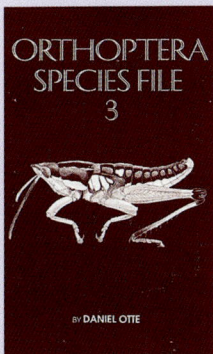

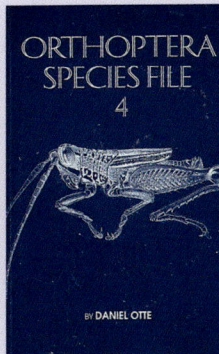

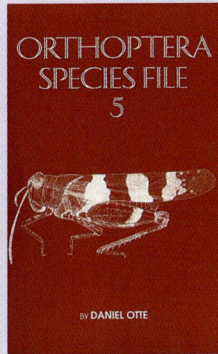

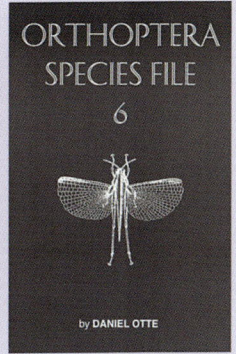

ORTHOPTERA SPECIES FILE 2

BY DANIEL OTTE

ORTHOPTERA SPECIES FILE 3

BY DANIEL OTTE

ORTHOPTERA SPECIES FILE 4

BY DANIEL OTTE

ORTHOPTERA SPECIES FILE 5

BY DANIEL OTTE

ORTHOPTERA SPECIES FILE 6

by DANIEL OTTE

extensive database. This tripartite system comprises the Editorial Board, the Governance Committee, and the Data Management Team. The Editorial Board, at the Museo de La Plata in Argentina, oversees the maintenance and accuracy of the OSF. The classification and nomenclature of the database reflect the most recent published information. The Governance Committee evaluates grant proposals and advises on conflicting classifications. The Data Management Team, comprising the Species File Group based at the Illinois Natural History Survey and the Museo de La Plata, maintains the database and develops the TaxonWorks software.

In September 2023, OSF migrated to the TaxonWorks platform, and the old OSF has been archived for reference (http://orthoptera.archive.speciesfile.org/. The Species File Group has created two products: TaxonWorks, which serves as the new software infrastructure for data curation and data filtering, and TaxonPages, which displays information to the public.

The new public interface (accessible at https://orthoptera.speciesfile.org/), currently provides an overview of the most relevant information. It comprises various panels, including the taxonomic history of names, references, distribution maps, and nomenclature statistics. Species pages also include an image gallery, information about type specimens, and links to other websites. Sound recordings are currently accessible through links to the old OSF. Additional tabs displaying the complete information of the database, as well as different search and filter options, will be progressively developed according to the needs of the orthopterist community.

The transition to TaxonWorks is expected to enhance the OSF in several ways:

— Improved accessibility: The redesigned interface allows for easy access to information on both desktop and mobile devices.
— Advanced search and analysis: TaxonWorks provides powerful tools for researchers to explore Orthoptera taxonomy and distribution through various filters and data-combination options, facilitating in-depth analysis.
— Modern data management: This transition ensures that the OSF stays up-to-date with contemporary standards in the digital landscape, supporting efficient data-management practices and maintaining relevance in the field of cyber-taxonomy.

In parallel with the technological evolution, the OSF maintains its reputation as a reliable resource for Orthoptera taxonomy. The commitment to accuracy and authoritativeness remains as its central objective.

In summary, the migration to TaxonWorks is a groundbreaking moment in the evolution of the OSF. The accessibility of the public version from mobile devices, as well as the analytical capabilities and collaborative potential of the real database curated within the new infrastructure, position the OSF at the forefront of digital taxonomy to ensure its continued relevance and reliability in the pursuit of Orthoptera research.

6.3 DNA BARCODING: A GENETIC TOOL FOR CATALOGING THE DIVERSITY OF ORTHOPTERA

by Oliver Hawlitschek, Martin Husemann, Mattia Ragazzini, and Lara-Sophie Dey

Orthoptera may be less species-rich than the mega-diverse orders of insects, but 30,000 known species is still a huge number. Consequently, identification of these species is a difficult task. There are few specialists worldwide who have a good knowledge of all taxonomic groups of Orthoptera and of the global orthopteran fauna. Identifying a grasshopper, cricket, or katydid from a poorly studied group or an under-studied region may be difficult even for these experts: sometimes, historical literature in a variety of languages needs to be consulted, and the individual in question must be compared with specimens deposited in museums worldwide (see chapter 6.1, Diversity in boxes). Fortunately, the Orthoptera Species File (see chapter 6.2, Orthoptera Species File) provides images of many species. Yet, some taxa can only be identified after dissection and inspection of their internal anatomy (mostly male genitalia). However, for some taxa, only one sex is known to science.

These problems are not unique to Orthoptera. Fortunately, the advancement of molecular genetics has provided tools that have revolutionized the task of species identification. The most prominent of these is called DNA barcoding: As with barcodes in a supermarket, this method is based on the idea that any animal (or other organism) can be identified through a specific barcode and a scanner. The technique is more complicated in real life than in this allegory, but not by much. DNA barcoding requires a molecular genetic laboratory and some knowledge of the method, but it does not a priori require the rare expert knowledge necessary to identify many insect species using traditional means. More recent technological advances have sent costs plummeting and have even produced sequencing machines that can be connected to a laptop and are small enough to be carried anywhere.

As implied by the name of the method, any organism's barcode is hidden in its DNA. The DNA codes all hereditary information of an organism in four different bases, identified by letters of the genetic alphabet (A, T, G, C), which can be read by a sequencing machine. An organism's entire genome is typically very large, consisting of hundreds of millions of bases or more. The genomes of Orthoptera are especially large, measuring up to 22 billion bases (or base pairs, counting both strands of DNA), almost an order of magnitude larger than the human genome (3.1 billion base pairs). Sequencing such a large genome remains very costly, and assembling it is complicated. This is why, as early as in 2003, the cytochrome C oxidase I gene, variously abbreviated as COI, CO1, or COX1, was established as the standard barcoding marker in animals. COI is only 645 base pairs long and has been shown to allow the identification to species level of high proportions of species of almost all animal groups.

While generating a barcode is rather simple, its identification requires comparison to a reference database, in which barcodes of previously identified specimens have been deposited. Only with a database that is complete for a region and well-curated will DNA barcoding show its full powers. This, again, depends on having taxonomists who are capable of establishing such a dataset. Since 2003, large-scale barcoding projects have been established around the world, either by independent research teams or in consortia. The iBOL (International Barcode of Life) Consortium was launched in 2010, aiming to establish a DNA barcode reference library for all multicellular life. Its latest project, BIOSCAN, utilizes the most recent advances in DNA sequencing technology to monitor biodiversity through the bar-

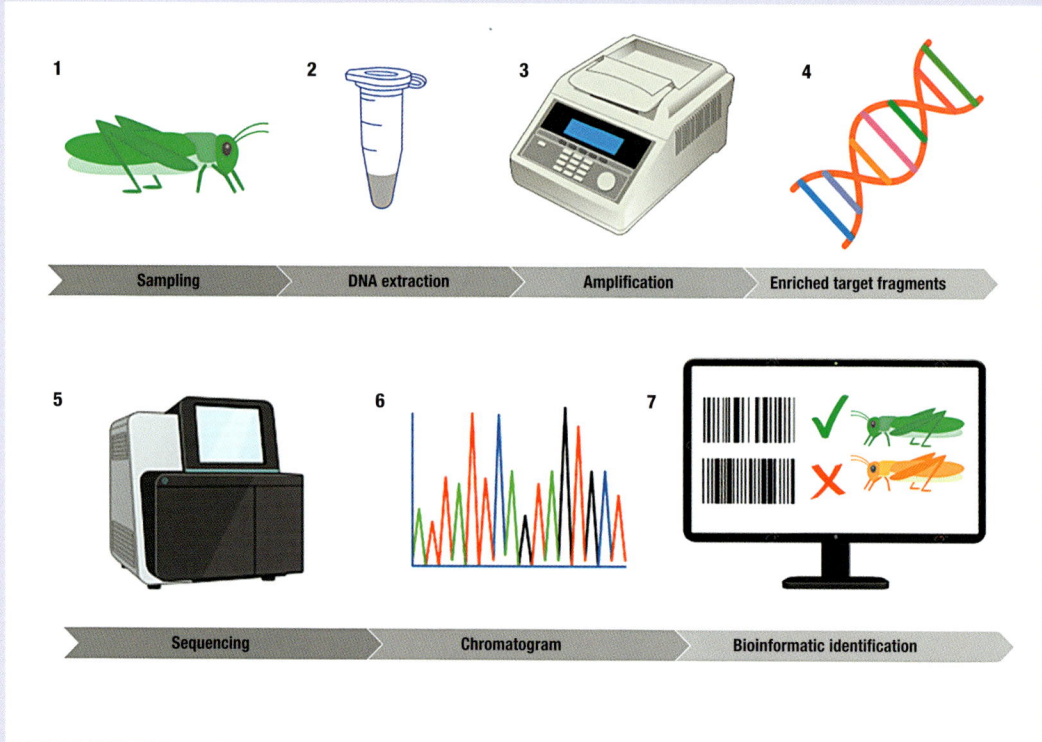

1	2	3	4
Sampling	DNA extraction	Amplification	Enriched target fragments

5	6	7
Sequencing	Chromatogram	Bioinformatic identification

code approach. So far, these various projects have accumulated more than 22 million barcode sequences across all taxa stored in a common repository – BOLD, the Barcode of Life Data Systems based in Guelph, Canada.

The most basic application of barcoding is helping taxonomists with species identification, sparing them the laborious tasks of sorting through samples and manually clustering individuals and, thus, making their work more efficient. Beyond this, barcoding has introduced whole new avenues for taxonomic research and species discovery. In many barcoding studies outside the well-studied geographical regions and taxonomic groups, 20% of barcodes or more cannot be identified at the species level because they belong to species for which no reference barcodes exist. Studies in tropical regions, and even in temperate ones, frequently discover tens or hundreds of species previously unknown to science. Through various clustering and delimitation algorithms, DNA barcoding facilitates the grouping of individuals into molecular Operational Taxonomic Units (mOTUs), potentially representing cryptic species that can later, after morphological analyses, be formally described.

No rule in biology is without exception, and so DNA barcoding also has its limits. Of all animals, and compared with other insect orders, Orthoptera are among the groups that are posing most challenges in this field. One reason is that their notoriously large genomes contain many pseudogenes – that is, non-functional copies of genes that may be mistakenly targeted and lead to errors in the sequencing of their barcodes. Moreover, many species of Orthoptera are known to hybridize. As the COI marker is mitochondrial and therefore inherited only through the maternal line, it may be "transferred" across species boundaries through

mOTU

Notostaurus anatolicus

Dociostaurus hispanicus

Dociostaurus brevicollis

Dociostaurus maroccanus

Dociostaurus jagoi

Dociostaurus genei

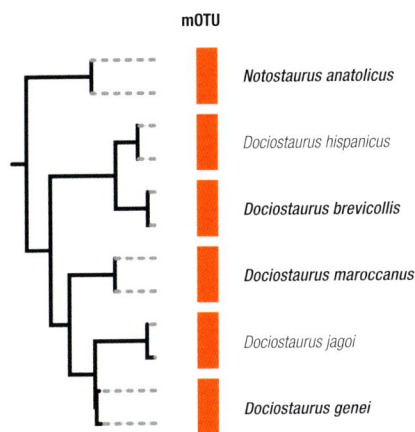

cross-breeding without showing up in the morphology of the specimen. Finally, species in many complexes of Orthoptera, especially grasshoppers, are of such an evolutionarily young age that there was not enough time for their COI genes to diversify and differentiate through natural mutation. These species simply cannot be distinguished using DNA barcoding, requiring a combination of deeper DNA sequencing approaches integrated with morphology and bioacoustics.

As in many fields of science, new methods answer some questions and open many others at the same time. DNA barcoding supports the work of taxonomic experts, but it certainly does not replace them. So far, only 4,429 species of Orthoptera have been referenced in the BOLD database, representing 15% of the total known species diversity. Taxonomic expertise is needed more than ever to fill these gaps.

< **Figure 1** A simplified DNA barcoding workflow from specimen via DNA extraction, amplification, sequencing, and databasing barcodes.

Figure 2 DNA barcoding is very helpful in groups of morphologically similar species. These members of the genus *Dociostaurus* look very much alike, but their DNA barcodes form clusters (or molecular operational taxonomical units, mOTUs) that correspond to species and allow the identification of individuals using their barcodes.

Figure > Human activity is rapidly destroying the last remaining untouched natural habitats, not only in Northern Madagascar as shown on this photo, but around the world. Orthopterans dependent on these habitats will disappear along with them, some even before they are scientifically recorded. Photo: Oliver Hawlitschek.

7 CONSERVATION

7.1 CONSERVING ORTHOPTERA DIVERSITY: RESARCH AND MANAGEMENT

by Axel Hochkirch

Similar to other animal, plant, and fungal species, many Orthoptera species are in peril. The most important resource providing information about the conservation status of Orthoptera, as well as their threats and required conservation action, is the International Union for Conservation of Nature (IUCN) Red List of Threatened Species. Red List assessments follow strict quantitative criteria against which each individual species is assessed. Depending on the status, it can be assigned to one of eight Red List categories: Least Concern (LC, no strong extinction risk), Near Threatened (NT, close to the thresholds of VU, EN, or CR), Vulnerable (VU, high extinction risk), Endangered (EN, very high extinction risk), Critically Endangered (CR, extremely high extinction risk), Extinct in the Wild (EW, survives only in captivity or outside its native range), Extinct (EX, no doubt that it is no longer extant), and Data Deficient (DD, inadequate information to assess extinction risk). About one-third of the ca. 1,500 Orthoptera species that have been assessed so far for the IUCN Red List of Threatened Species are threatened with extinction or even extinct. The drivers of these extinctions are manyfold, but most are related to the increasing human demand for land and resources. In order to halt the decline of Orthoptera, we have to understand the requirements of threatened species as well as their major threats. Unfortunately, data on many Orthoptera species are still scarce, and many species are only known from the type material, while detailed information on their distribution, population trends, habitat, and threats is lacking. Fieldwork is therefore required to close such knowledge gaps and develop useful conservation measures for threatened Orthoptera species.

Understanding the major threats to Orthoptera is key to implementing meaningful conservation action. Numerous threats are impacting Orthoptera, depending on their habitat requirements and the human impacts on the habitats. Species dwelling in grasslands are often threatened by changes in agricultural land use, including habitat conversion to farmland, intensification of agricultural land-use practices (larger fields, larger and heavier machines, fertilization, pesticides, higher livestock densities, more frequent mowing), but also by abandonment and subsequent shrub encroachment. Forest species are often threatened by deforestation, plantations of monocultures (and introduction of non-native trees), or industrial forestry, but also by increasing frequencies of wildfires (fig. 1). Threats specific to wetland species are drainage and increasing frequencies of droughts from climate change. Species in natural floodplains often suffer from damming and disruption of the natural flooding regime. Overgrazing is the main threat to most dry-land species. Cave Orthoptera can be threatened by mining and touristic exploration of caves, while coastal species are often threatened by touristic development and the rising sea level. On oceanic islands, many Orthoptera are threatened by invasive species, including plants that may alter the habitat structure and microclimate, predators and pathogens that may increase mortality, and herbivores that may compete with Orthoptera for food. In mountains, species may be threatened by abandonment of traditional grazing regimes, but also by climate change and winter-sport development. Some general threats affecting various habitats are urbanization, road construction, and industrial development. The effects of pollution on Orthoptera are poorly understood, but, particularly, pesticides may affect many species. Furthermore, the increasing nitrogen deposition from intensive

Figure 1 The Gomera stick grasshopper *Acrostira bellamyi* (Pamphagidae) is a Critically Endangered endemic of the island of La Gomera in the Canaries Archipelago, where it inhabits laurel forests (threatened by conversion to pine cultures) and *Euphorbia* shrublands (threatened by grazing and wildfires). Photo: Axel Hochkirch.

agriculture alters the vegetation structure of habitats, which may become unsuitable for species that require open habitats.

A comprehensive threat analysis is the basis for conservation planning. Some threats may have more immediate effects on a population than others. For example, the main threat to coastal endemic grasshoppers in the Mediterranean is touristic development, particularly the destruction of habitats for touristic infrastructure. However, in the future, other threats that alter the vegetation structure – such as the rising sea level caused by climate change or the introduction of invasive alien plants – may become more important (fig. 2). While potential future threats must always be taken into consideration, it is important to mitigate the current major threats first. Habitat loss and deterioration are most often reported as threats to Orthoptera, but it must be considered that these are also easier to recognize compared with many other threats (e.g., introduction of a pathogen). Therefore, threat identification may become difficult if no obvious changes to the habitat are visible.

Strategic conservation planning is a tool to ensure that the most suitable conservation measures are initiated. There is ample guidance for conservation planning available, such as the IUCN Guidelines for Species Conservation Planning. Strategic conservation planning is evidence-based, considers knowledge gaps, and includes the input of all stakeholders in a participatory process, so that the implementation of the plan will be supported by all relevant players. At the beginning, a status review is compiled as the basis of the action plan. This should include all known information on the species (published or unpublished), including sections on taxonomy, present and historic distribution, species biology and ecology, conservation context, and threats and their drivers. Afterward, a vision is drafted, describing the desired future state of the species in the long term in an inspirational way. Goals specify the vision with a clear time frame and concrete numbers. In a stakeholder workshop, the vision and goals are thoroughly discussed, and a threat analysis, based upon the status review, is undertaken, in which the importance of each threat is evaluated. Objectives are positive statements of what is to be done to contribute to meeting a goal and the framework on which actions will be developed. For each action, one or two results are defined that shall meet the SMART criteria, where SMART means "specific, measurable, achievable, realistic, and time-bound." Actions are usually defined in the following terms: What is to be done? When is it to be done? Where is it to be done? Who will do it? Who is responsible and accountable for it being done? What resources are needed? What shows that it has been done? The close and precise relationship between actions and their results is a key aspect of an effective conservation action plan.

A typical basic structure of the goals is (1) research to close knowledge gaps relevant to the conservation of the species, (2) habitat management to reverse the negative population trend (by mitigating threats and/or establishing tailored in situ habitat or species management), and (3) raising awareness for the species and its threats among the most important stakeholders. Obviously, the second goal is the most important to preserve species from extinction. The actions related to management are, therefore, at the core of any conservation project and the key element of any funding proposal.

As habitat loss and deterioration are the main threats to Orthoptera, conservation action usually must focus on the restoration or management of habitats. It is important to understand the habitat requirements of Orthoptera species and the effects of any management changes. Monitoring the response of the target species (or species group) to any conservation action allows adaptation of management, if required. For example, the field cricket *Gryllus campestris* was listed as Critically Endangered on the regional Red List for Lower Saxony, Germany. At the beginning of the 1990s, this species had a single population left in the western part of Lower Saxony, located at the edge of a peat bog. A survey in 1991 showed that only 32 singing males were remaining. This population was restricted to a small heathland fragment, which was partly over-

grown by pine forests *Pinus sylvestris*. This habitat deterioration was identified as the major threat to the species. The county owned the site and agreed to log down the pines and remove the topsoil to promote restoration of heathland vegetation. In the next year, the first young heather plants *Calluna vulgaris* sprouted, and the field cricket population started to expand to the restored parts of the site. A year later, some singing males were heard on a grassland patch 500 m north, which was subsequently managed by occasional sheep grazing to improve habitat conditions. After two more years, a subpopulation established on this grassland, and the species spread farther to the north. Dispersal of the population was promoted by creating wider heathland corridors along roads. After ten years, the population reached a new maximum of 949 singing males, and a translocation project was started, because it was believed that the peat-bog areas directly adjacent to the population provided no suitable habitat. The population nevertheless continued to disperse and even crossed the peat bog and reached a new maximum of nearly 2,500 singing males in 2006.

Figure 2 The Endangered *Amphinotus nymphula* (Tetrigidae) is restricted to a few fragmented populations in the cloud forests of the Seychelles Islands, which are continuously declining in extent and quality due to invasive species and potentially global warming. Photo: Axel Hochkirch.

Translocation or reintroduction projects are still rare for Orthoptera. Such projects are often more complex than one initially would expect, and many translocation projects have failed. A key to the success of such projects is to first restore the habitat and mitigate the threats at the introduction site before starting a translocation or reintroduction. Captive breeding can help to facilitate such projects, but breeding can also be challenging without proper knowledge of the climatic requirements and development times of the eggs (fig. 3). Mortality of the first nymphal instars is often high, and a couple of years may be required until breeding protocols are optimized in a way that allows reintroductions. In particular, pathogens may hamper introductions from captive-bred stocks. Therefore, translocations from wild populations are usually preferred, if the population size of the potential source population is large enough to allow removal of individuals. The population dynamics of such a source population also needs to be monitored after the translocation.

The beauty and diversity of Orthoptera are amazing, and it will be important to initiate and implement conservation measures to preserve them. While scientists can help to close knowledge gaps and assess the conservation status of species, the implementation of conservation action is usually conducted by other parties, such as conservation authorities, NGOs, landowners, and tenants (fig. 4). In the majority of cases, the public is not aware of the occurrence of a threatened species in a certain area or the necessary steps to preserve it. Collaboration with stakeholders is, therefore, crucial in order to preserve the incredible diversity of Orthoptera on our planet.

Figure 4 The Adriatic marbled bush cricket *Zeuneriana marmorata* (Tettigoniidae), an endemic of the Gulf of Venice, is listed as Endangered. Its wetland habitats are threatened by draining, conversion of wetlands into crop fields, and pesticide application for mosquito control. Photo: Axel Hochkirch.

< Figure 3 The Crau Plain grasshopper *Prionotropis rhodanica* (Pamphagidae) is a Critically Endangered species endemic to a small area of 12 to 16 km^2 of the Crau Steppe in southern France, with a total population of likely fewer than 5,000 adult individuals. It is one of the few orthopteran species for which a dedicated captive-breeding program has been put in place. Photo: Axel Hochkirch.

7.2 THE RESPONSES OF CENTRAL EUROPEAN ORTHOPTERA TO CLIMATE CHANGE

by Thomas Fartmann, Felix Helbing, Franz Löffler, and Dominik Poniatowski

Climate change, particularly since the late 1980s, has been responsible for a rise in temperatures and a higher frequency of extreme weather events. Annual precipitation in temperate Europe has only slightly changed. Nevertheless, summer droughts have become more frequent, mainly caused by precipitation shifts from summer to winter and increased evapotranspiration due to higher summer temperatures. As a result, global warming already has had a major impact on insects, and its effects are likely to increase in the future. Both gradual climatic changes and the increasing frequency of extreme events, such as droughts, fires, and floods, contribute to this. However, the effects of the latter are much more serious, particularly in fragmented habitats. Extreme events are highly unpredictable in both space and time, which makes it almost impossible for species to adapt to them. In general, the response to climate change is often rather complex and may even differ among related species and due to the landscape configuration.

Figure 1 Female of the bog bush cricket *Metrioptera brachyptera*. Photo: Dominik Poniatowski.

Both the bog bush cricket *Metrioptera brachyptera* (fig. 1) and Roesel's bush cricket *Roeseliana roeselii* (fig. 2) are normally short-winged and flightless. However, the habitats of the two closely related bush-cricket species are quite different. *M. brachyptera* is a habitat specialist that mainly occurs in calcareous grasslands and mires in low population densities. By contrast, *R. roeselii* is a habitat generalist that inhabits different types of open grassy habitats; moreover, its abundance is usually much higher than that of its sibling species. In both bush crickets, density stress in the early nymphal instars leads to the release of a stress hormone, which triggers the development of long-winged (macropterous) morphs that are capable of flight. Here, in a literal sense, stress leads to wings.

Over the last 20 years, springs were often disproportionally warm and dry, which increased the survival of nymphs and fostered their abundance, leading to increases in population densities (fig. 3).

Figure 2 Long-winged female of Roesel's bush cricket *Roeseliana roeselii*. Most individuals have short wings and are flightless, like the bog bush cricket (fig. 1). Photo: Dominik Poniatowski.

Climate change
↓
Warm spring
↓
Low nymphal mortality
↓
Higher population density
↙ ↘

Habitat specialist (normally low density):	Habitat generalist (normally high density):
M. brachyptera	*R. roselii*
↓	↓
No density stress	Density stress
↓	↓
Low proportion of macropters	High proportion of macropters
↓	↓
Population losses	Range expansion

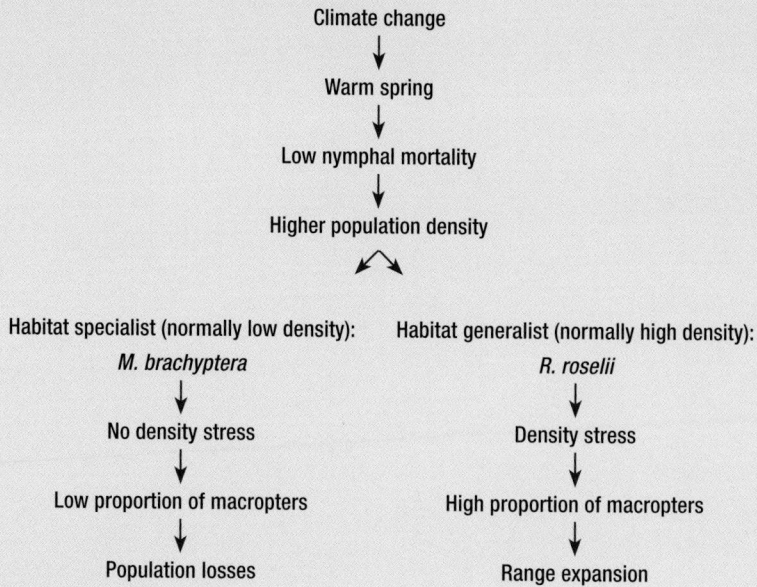

However, the further effects on the two species were found to be very different. Since the natural abundance of *M. brachyptera* is usually low, reduced nymphal mortality due to favorable climatic conditions has mostly not been sufficient to cause density stress and to produce long-winged individuals. In the rare case of macropters emerging, the high degree of habitat fragmentation in central Europe strongly limits the dispersal of *M. brachyptera* and its ability to track climate change. Additionally, the eggs of the species are sensitive to droughts, which are expected to become more frequent with global warming, threatening the populations of *M. brachyptera*. By contrast, due to the naturally higher abundance of *R. roeselii*, the warmer and drier springs during the last two decades have regularly led to density stress among young nymphs of this species. Nowadays, macropters of *R. roeselii* can be observed frequently, and proportions of up to 20% long-winged individuals within populations are quite common. These morphs exhibit a higher mobility than the short-winged ones because they are capable of flight. Due to the high availability of suitable habitats, this species can keep pace with global warming and has recently expanded its range in western, central, and northern Europe.

Overall, many thermophilous and mobile habitat generalists have benefited from climate change in temperate Europe. For example, about one-third of the 82 native grasshopper species in Germany expanded their ranges due to global warming during the last three decades (figs. 2, 4). By contrast, the dispersal of many habitat specialists was largely constrained by limited habitat availability. In particular, hygrophilous habitat specialists with low mobility and cold-adapted species that are typical of higher elevations have suffered from climate change. However, the abundance of grasshoppers even collapsed in dry and mesic habitats due to large-scale desiccation of aboveground vegetation during recent droughts. In the long term, climate change increases the risk of biotic homogenization of grasshoppers in temperate Europe, favoring widespread habitat generalists at the expense of many sedentary habitat specialists.

< Figure 3 Effects of climate change on two related bush-cricket species during post-embryonic development: the habitat specialist bog bush cricket *Metrioptera brachyptera* (fig. 1) and the habitat generalist Roesel's bush cricket *Roeseliana roeselii* (fig. 2). Figure by Thomas Fartmann.

Figure 4 Female of the slender green-winged grasshopper *Aiolopus thalassinus*. Photo: Marcel Kettermann.

7.3 ALPINE GRASSHOPPERS OF THE MEDITER-RANEAN: ISOLATED REFUGES SHRINKING FROM GLOBAL WARMING

by Joaquín Ortego

Anthropogenic greenhouse gas emissions are warming the world at a pace unprecedented in at least the last 2,000 years, which has already increased Earth's global surface temperature by around 1.1 °C (ca. 1.59 °C on land) compared with the average in the period 1850–1900. Alpine ecosystems are expected to be particularly vulnerable to the impacts of climate warming, and for this reason, they have been targeted as ideal indicators for the detection of early-warning signals of the ecological impacts of rising temperatures. Mediterranean sky-island archipelagos – clusters of mountaintops surrounded by lowlands characterized by a warmer climate – represent the distributional margin for many cold-adapted species and often sustain the southernmost living representatives at different taxonomic ranks. Beyond representing the distributional limits of northern European species, Mediterranean sky islands also harbor high levels of local microendemism, with many taxa of great conservation concern exclusively distributed in a single or a few neighboring mountaintops. This is the case for a high number of narrow endemic species of alpine grasshoppers that often form extremely small and fragmented populations. The high vulnerability of alpine Orthoptera from the Mediterranean region to climate warming in comparison with their boreal-temperate counterparts, together with their short generation times and fast responses to environmental disturbances, make them an ideal study system to evaluate and anticipate the impacts of human-induced climate change.

ALPINE GRASSHOPPERS OF THE TRIBE PODISMINI

Some of the most fascinating alpine grasshoppers inhabiting Mediterranean sky islands belong to the tribe Podismini (Orthoptera, Acrididae, Melanoplinae; fig. 1). Whereas some genera (e.g., *Cophopodisma*, *Chortopodisma*, and *Epipodisma*) are either monotypic or represented by a single species in the Mediterranean region and adjacent mountain ranges (Alps, Pyrenees), other genera such as *Podisma*, *Italopodisma*, *Oropodisma*, and *Peripodisma* have likely

Figure 1 Historical and contemporary elevational distribution of *Podisma carpetana carpetana* in Sierra de Guadarrama, central Spain. Panel (A) shows elevational ranges occupied by the species according to historical records (1887–1974, in dark and light blue) and recent resurveys (2017, in red). Historical elevational distribution is based on records from the literature and the 410 specimens deposited at the National Museum of Natural Sciences (MNCN-CSIC, Madrid, Spain). Note that blue/red areas were depicted based on the minimum elevation at which the species was recorded, and thus they are likely to considerably overestimate the total area actually occupied by the species at each time period due to heterogeneity in microclimatic conditions and strong preferences for specific microhabitats. The arrow indicates the upslope distributional shift experienced by the species between historical and contemporary records. Panel (B) shows the small area currently occupied by the species in Sierra de Guadarrama according to the prospective surveys performed in 2017. Dot size is proportional to the number of recorded specimens (range = 1–16 individuals) at each point where the species was detected. Panel (C) shows the typical habitat currently occupied by the species. Images: Joaquín Ortego.

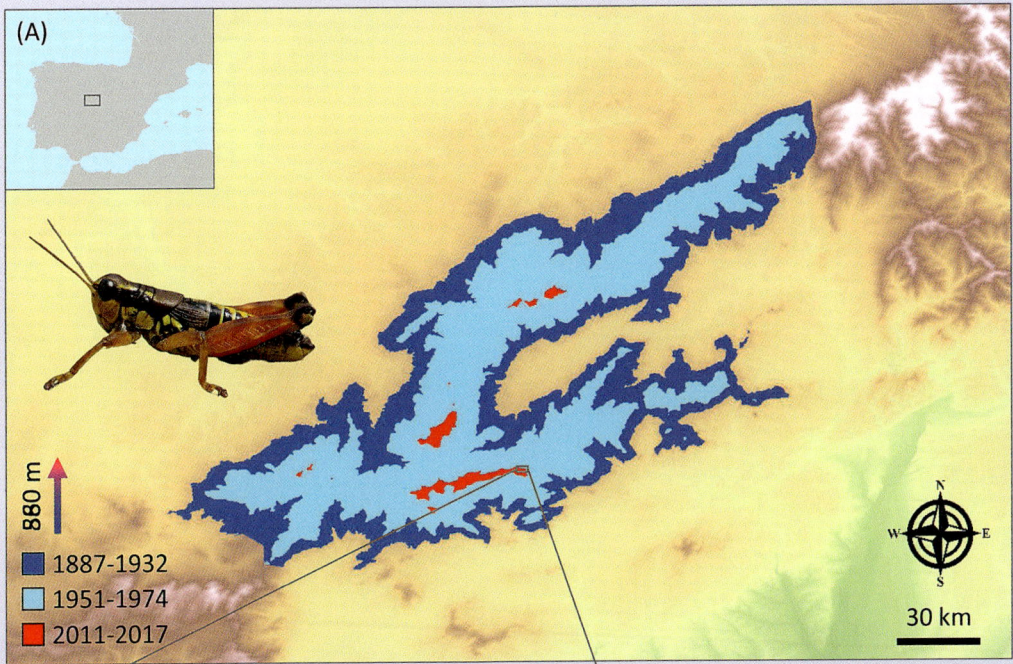

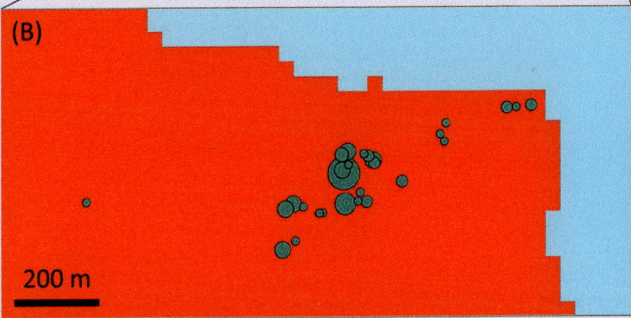

radiated in situ in different sky-island archipelagoes from the northern Mediterranean peninsulas. This phenomenon has led to the formation of a considerable number of narrow endemic species. Most taxa within these radiations show allopatric distributions, and empirical evidence suggests that they originated during the Pleistocene (<2.6 MYA), pointing to diversification in geographical isolation as a consequence of range fragmentation fueled by Quaternary climatic oscillations. Most species are flightless and have very limited dispersal capacities, which has likely contributed to the marked genetic fragmentation of their contemporary populations. Small distributions and declining and fragmented populations have led to the inclusion of most Mediterranean Podismini species in the IUCN Red List of Threatened Species with the categories Critically Endangered, Endangered, Vulnerable, or Near Threatened. Comparisons of historical records with recent surveys performed by our research team for the genera *Podisma* in the Iberian Peninsula and *Oropodisma* in the Balkans suggest that several populations have experienced marked shifts toward higher elevations and severe range contractions during the past 50 to 100 years, which points to the impacts of human-induced climate warming.

THE CASE OF *PODISMA CARPETANA* IN CENTRAL IBERIA

Recent field surveys and comparisons with historical records indicate that the populations of the taxon *Podisma carpetana carpetana* in the Sierra de Guadarrama (Spain) have experienced marked declines and upslope distributional shifts. The proximity of these populations to the city of Madrid, where the prominent orthopterist Ignacio Bolívar and colleagues were based, has resulted in a high number of historical records for the species in the area (e.g., >400 specimens deposited at the National Museum of Natural Sciences, MNCN-CSIC). The species remained unrecorded during ca. 40 years: the last historical specimen was collected in 1974, and the species was not rediscovered until 2013, when the rangers of Sierra de Guadarrama National Park found a

small population. Our intensive surveys in the area between 2011 and 2017 confirmed that the species has disappeared from most places where it was recorded between 1887 and 1974, with the only remaining population located in an isolated habitat patch at the highest elevational ranges (>2,100 m) of Sierra de Guadarrama. This corresponds to an upslope shift of >800 m with respect to the lowest elevation records (1,240–1,500 m) for the species between 1907 and 1961 (fig.1A,B). A prospective survey in the area where the species is still present today revealed that it exclusively occupies a very small habitat patch (<0.1 km²) covered with dwarf junipers (*Juniperus communis*) and brooms (*Cytisus oromediterraneus*) interspersed with bare soil and stony grasslands (fig.1C). This elevation displacement and contraction of populations is probably a consequence of climate warming. However, the role of other impacts on the progressive decline of the species in the area cannot be ruled out, including massive recreational activities due to the proximity

Figure 2 Historical and contemporary elevational distribution of four taxa from the genus *Oropodisma* endemic to the Peloponnese Peninsula, southern Greece. Panel (A) shows the geographical location of the studied populations presented in bottom panels: (B) *O. erymanthosi* from Mount Erymanthos, (C) *O. kyllinii* from Mount Kyllini, (D) *O. chelmosi* from Mount Mainalon, (E) *O. chelmosi* from Mount Parnon, and (F) *O. taygetosi* from Mount Taygetos. Bottom panels show the elevational ranges occupied by each species in different mountain ranges according to historical records (1970–1978, in blue) and recent resurveys (2022–2023, in red). Note that blue/red areas were depicted based on the minimum elevation at which the species was recorded, and, thus they are likely to considerably overestimate the total area actually occupied by the species at each time period due to heterogeneity in microclimatic conditions and strong preferences for specific microhabitats. Arrows on the lower right corners of panels B–F indicate the upslope distributional shift experienced by the species between historical and contemporary records. Inset picture in panel A shows a female of *O. chelmosi*. Drawing: Marina Trillo.

of Sierra de Guadarrama to the capital, overgrazing of alpine habitats, and the creation of the ski resort of Puerto de Navacerrada in 1940 in an area where the species was previously widespread. Estimates of genetic diversity also suggest a pessimistic prospect for the relict population of *P. carpetana* from Sierra de Guadarrama, which shows the lowest levels of genetic diversity among all populations of the species in the Iberian Peninsula. The comparatively low levels of genetic diversity in the remnant population from Sierra de Guadarrama is expected to reduce its adaptive potential to face the impacts of ongoing environmental change, which can ultimately lead to population extinction over short time scales.

THE CASE OF *OROPODISMA* IN THE PELOPONNESE REGION

The marked range contraction reported for *P. carpetana* in southern Iberia is comparable to the upslope distributional shifts experienced by some populations

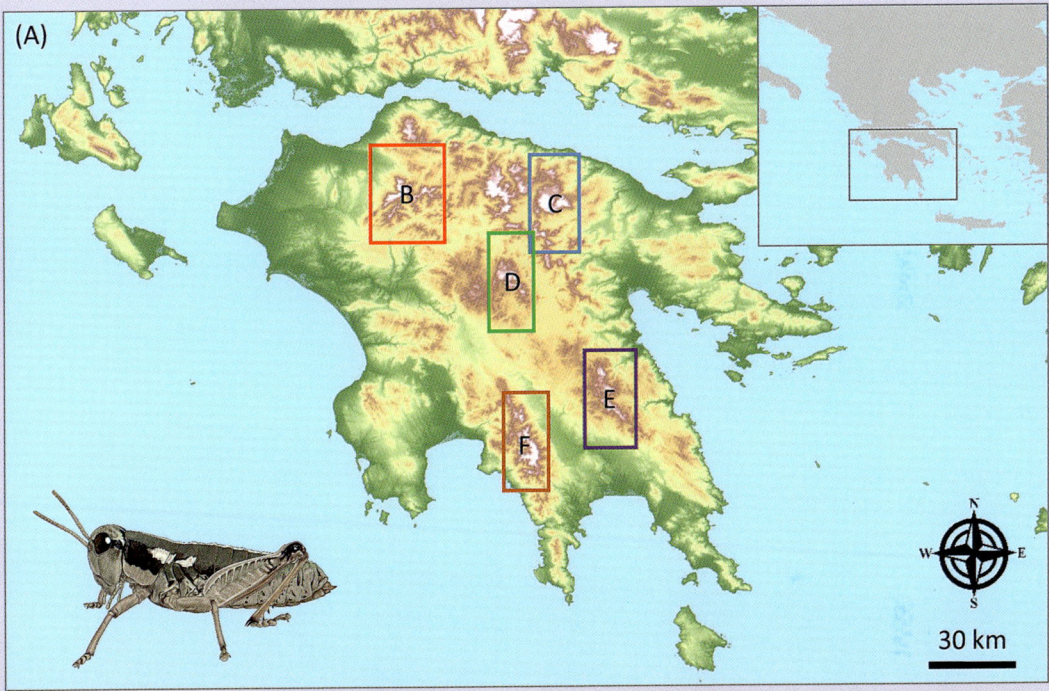

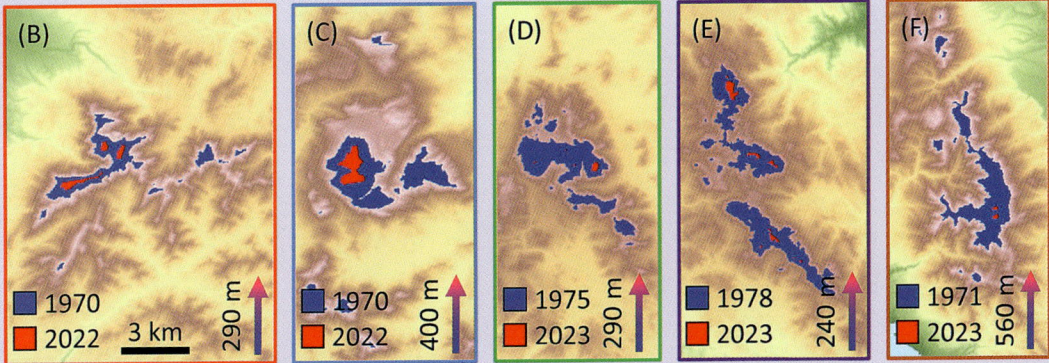

(A)

(B) 1970 2022 | 3 km | 290 m

(C) 1970 2022 | 400 m

(D) 1975 2023 | 290 m

(E) 1978 2023 | 240 m

(F) 1971 2023 | 560 m

261

from the genus *Oropodisma* in the Balkan Peninsula. In 2021 and 2023, we performed extensive surveys and specimen sampling of most known populations of the genus *Oropodisma* for detailed taxonomic, demographic, and conservation genomic studies. In the course of the fieldwork, we noted the considerable population contractions experienced by taxa distributed at the southernmost latitudes, especially those inhabiting the mountain ranges from the Peloponnese Peninsula (Greece). According to our surveys, the single known populations for the taxa *O. erymanthosi*, *O. kyllinii*, and *O. taygetosi* and the populations of *O. chelmosi* from Mount Parnon and Mount Mainalon have experienced elevational shifts of 240–560 m with respect to the lowest elevation at which the species were recorded in the same localities during the early 1970s (fig. 2). Although populations of *O. erymanthosi* and *O. kyllinii* still sustained reasonably abundant populations, only a very few individuals of *O. taygetosi* could be detected after intensive prospecting in Mount Taygetos, Mount Parnon, and Mount Mainalon. These data suggest that most narrow endemic taxa of the genus from the Peloponnese Peninsula are probably at the verge of extinction and that their IUCN Red List conservation statuses should be probably elevated one or more steps in future species reassessments (e.g., from Vulnerable to Critically Endangered in the case of *O. taygetosi*).

CONCLUSIONS

Collectively, these preliminary results point to severe population contractions in some Podismini grasshoppers from Mediterranean sky-island archipelagos, which can be most likely attributed to rising temperatures and progressive deterioration of suitable environmental conditions for those alpine species distributed at the southernmost latitudes of their respective genera. Baseline monitoring schemes and assessments of historical population trends of alpine Orthoptera is urgent to update the conservation status of a high number of species with extreme vulnerability to climate warming.

Figure 1 The speckled buzzing grasshopper in its habitat in Sweden. Photo: Martin Husemann.

7.4 THE SPECKLED BUZZING GRASSHOPPER – THREATENED AND EXTINCT DESPITE SUITABLE CLIMATE

by Lara-Sophie Dey and Martin Husemann

The speckled buzzing grasshopper *Bryodemella tuberculata* is a specialist of open soil habitats with very sparse vegetation and is one of the rarest grasshopper species of central Europe (fig.1). While it was widespread in many parts of Europe until the early 20th century and remains so in central Asia, only few European relic populations are known from the Alps and isolated sites of northern Europe.

Why has the species declined so dramatically in Europe over the past hundred years? Is it, perhaps, a victim of human-induced global warming, and does that mean that more populations will be extinct due to rising temperatures in the future? We used mathematical modeling to compare the climatic factors affecting extant versus extinct localities with present-day data and future climate models. The results clearly indicate the opposite of what we thought: this grasshopper should actually benefit from the warming climate rather than being negatively affected. A comparison of the climate of all sites shows that the locations from which *B. tuberculata* has already disappeared would still be climatically suitable (fig.2).

So why did they disappear? We were curious and did some legwork, visiting some former habitats of the speckled buzzing grasshopper and exploring its history. One of these sites is the Lüneburg Heide, one of Germany's largest heathland areas. B. tuberculata was last recorded here around 1910. At this time, the heathland was being transformed into traditional agricultural land because the products of local sheep farming, wool and honey, could no longer compete with the low prices of imported goods. Shepherds abandoned their pastures, heathland was transformed into cropland, and the speckled buzzing grasshopper quickly disappeared after the destruction of its habitats. Although the farmers turned back to livestock grazing after a few decades and the heathlands spread again, the grasshopper did not return. We found the same story all over Europe: heathlands were converted into agricultural lands, and gravel banks were destroyed when rivers were straightened for navigation purposes and later restored, but even after extensive habitat restoration efforts, B. tuberculata never returned to any of these sites. Most likely, the remaining populations are too far apart. Reintroduction may be the only solution, but it has not yet been attempted.

The situation of the remaining European populations of the speckled buzzing grasshopper is precarious, as exemplified by the Isar River in the German part of the Alps. One of the few remaining near-natural rivers of the region, with a braided structure and extensive gravel beds, the Isar is the habitat of many locally rare species, such as the rattle grasshopper Psophus stridulus, Tetrix tuerki, Chorthippus pullus, and many other threatened animals and plants. While the area represents a protected natural habitat, the deduction of water for hydroelectric power generation has resulted in a decline of floods that kept the gravel banks free from vegetation. The speckled buzzing grasshopper is declining, along with its habitat. Measures to protect this species would also benefit many of the other organisms restricted to this regionally unique site. B. tuberculata, as a large, conspicuous and attractive grasshopper, could and should therefore act as a flagship species for the protection of its habitats.

Figure 2 Global distribution and modeled climate suitability of the speckled buzzing grasshopper. White dots represent extinct locations; black dots represent current occurrences. Background colors represent modeled habitat suitability according to climate parameters: yellow = high suitability, blue = low suitability. Bryodemella tuberculata disappeared from many sites despite suitable climate conditions. Graphics modified from Dey et al. 2021.

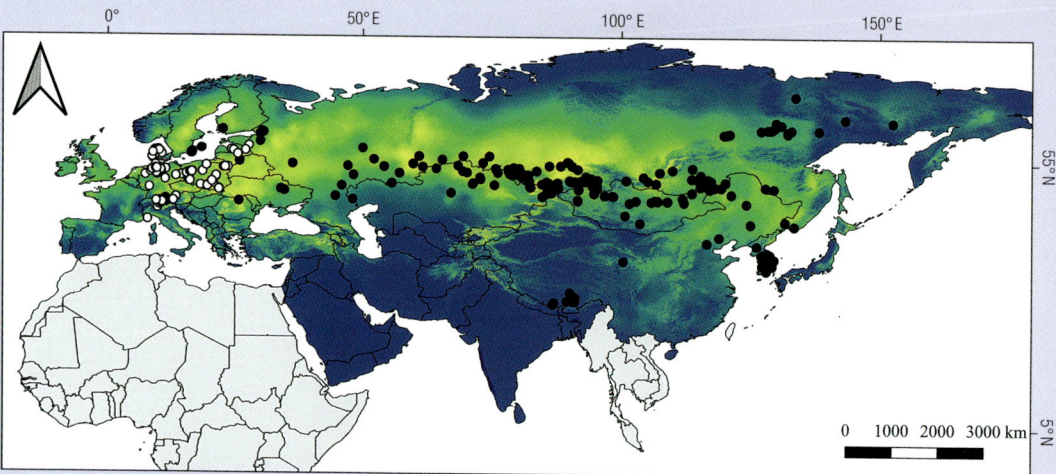

7.5 THE ROCKY MOUNTAIN LOCUST: FROM MAGNIFICENT PROFUSION TO MYSTERIOUS EXTINCTION

by Jeffrey A. Lockwood

In the 19th century, swarms of the Rocky Mountain locust *Melanoplus spretus* swept across North America, consuming a hundred thousand pounds of vegetation a day. At their peak, the locusts outweighed the bison population of the continent. During the outbreak of 1874–1877, the locust inflicted a staggering $200 million in damage to agriculture (equivalent to $116 billion today).

Western American pioneers were subsistence farmers, needing their crops for survival. To forestall starvation, they took diverse approaches to suppress the locust. To destroy locust eggs, farmers flooded, plowed, and harrowed the soil. Once eggs hatched, hand-dug ditches made effective pitfall traps for nymphs. Various machines were invented, including horse-drawn implements to crush, vacuum, and smother young locusts.

Through extensive explorations, scientists discovered the locust's "Permanent Zone." In the fertile, montane river valleys of the Rockies, the insect could always be found, and from these habitats swarms originated during outbreaks. An immense swarm in 1875 covered 198,000 square miles with an estimated 3.5 trillion locusts. But just 27 years later, this insect was extinct. What could possibly account for the disappearance of an organism that had spread from California to Missouri and from Texas to Canada?

Some scientists argued that the species actually persisted in the solitary form, as a common rangeland grasshopper. Researchers tried to induce a phase transformation using extant species, but

Figure 1 Left: Specimen of the Rocky Mountain locust (visible above the pen cap) melting out of the ice on Knife Point Glacier, Wyoming. Right: Compared to a one cent-coin. Photos: Jeffrey A. Lockwood.

Figure 2 Scene from *Locust: The Opera*, in which the soprano playing the role of the Rocky Mountain locust interacts with the tenor playing the role of a scientist. Photo: Jeffrey A. Lockwood; the photo was taken by Lockwood during a public performance of the opera, which he produced

failed to reincarnate the locust. Recent anatomical and molecular studies show that *M. spretus* was a unique species.

Other entomologists suggested that ecological disturbances generated outbreaks, but locust specimens from Rocky Mountain glaciers were radiocarbon-dated to 700 years before the present (fig. 1). Furthermore, anthropologists found evidence of locusts being consumed by Paleoindians 5,000 years ago. Scientists also speculated that other large-scale alterations (climate change, reduction in bison, and extirpation of Native Americans) could account for the extinction, but empirical evidence was lacking.

The mystery was solved by understanding ecological bottlenecks. During recession periods, as in the late 1800s, the locust was restricted to the Permanent Zone. At this time, the mountain valleys were being converted to agriculture to support the mining industries. The sanctuaries of the locust were plowed, irrigated, grazed, harrowed, flooded, and trampled just when the insect was most vulnerable. *M. spretus* succumbed to unwitting habitat destruction.

The Rocky Mountain locust presents a morally complex story that is captured in *Locust: The Opera* (https://youtu.be/L_4xzj7gAjA; fig. 3). The devastation by these locusts caused terrible suffering, so extinction was arguably a good thing. However, something of great value was lost along with this humbling, remarkable species.

7.6 BALANCING BIODIVERSITY: HABITAT MANAGEMENT AND GRASSHOPPER RESILIENCE IN INDIA'S PROTECTED AREAS

by Dhaneesh Bhaskar and P. S. Easa

India is a mega-diverse country with many endemic and threatened species protected and managed by law in various protected areas, such as national parks and tiger reserves. Habitat-management practices in India's protected areas are mostly tailored to flagship species, particularly large, threatened, and charismatic mammals, like tigers and elephants. This is largely a consequence of colonial-era forest laws and priorities: many of today's protected areas were hunting or game reserves during colonial rule, when the understanding of nature conservation and habitat management was still limited. This resulted in biased approaches that neglected the largest parts of biodiversity.

Such is the case of Eravikulam National Park, situated in the Western Ghats biodiversity hot spot of southern India. In the national park, the habitat-management strategy is centered around a threatened endemic ungulate species, the Nilgiri tahr *Nilgiritragus hylocrius* the world's largest viable population inhabits the park's high-elevation grasslands. The management strategy of the park focuses on prescribed burning, intentionally setting fire to selected plots of grasslands. Unsurprisingly, this is a legacy from colonial times, when the area was a hunting and game reserve.

Prescribed burning, initially used to attract Nilgiri tahr for hunting, has evolved into a method for controlling the Nilgiri tahr population in the protected habitat. Remarkably, this more than 60-year-old fire-management approach remained unexamined for its impact on other taxa. The first study to assess the effects of prescribed burning in Eravikulam National Park on other organisms was conducted no earlier than the period 2015–2018 (figs. 1, 2). The aim of the study was to compare the approach of large-scale prescribed burning of the park to small-scale fire management in the neighboring Parambikulam Tiger Reserve. Grasshoppers, as the predominant herbivorous insects of open grassland habitats, were selected as an indicator group. The study confirmed that large-scale burning, as conducted in Eravikulam, had much more severe detrimental effects on grasshopper communities than the small-scale fire management practiced in the tiger reserve. In areas with small-scale fires, grasshopper communities and individual species abundance recovered rapidly after burning, with no significant differences in total grasshopper abundance between burnt and control plots from the first post-monsoon season onward. Conversely, Eravikulam National Park, with its extensive burning, experienced an extended period of reduced grasshopper abundance and showed lower overall diversity, probably due to the lasting impact of its fire history. The mosaic pattern of small-scale burning in the tiger reserve facilitated grasshopper recolonization, while the large-scale burning in Eravikulam National Park caused greater distances between unburnt plots, impeding recolonization. Fire intensity and its impact on soil temperature also played a role, potentially exposing grasshopper eggs to higher temperatures. Changes in vegetation, such as the invasion of ferns and the dominance of the shrub *Strobilanthes kunthiana* (Neelakurinji) in frequently burnt grasslands, also influenced grasshopper populations. Flightless grasshopper species like *Zygophlaeoba sinuatocollis* and *Paramastacides ramachendrai* have shown a notable decline in burnt areas of both the

Figure 1 Researchers involved in studying the grassland prescribed burning in Eravikulam National Park. Photo: Dhaneesh Bhaskar.

protected areas; their recolonization rate has also been observed to be at slow pace.

The study highlighted the necessity for adaptive and scientifically informed habitat-management practices within protected areas. Instead of concentrating solely on single keystone species, management strategies should consider the broader ecosystem, including often-overlooked invertebrate communities. Small-scale, mosaic-pattern burning emerged as a more favorable approach, offering benefits for both large herbivores and invertebrate diversity.

> Figure 2 Individuals of *Carliola carinata*, an endemic flightless grasshopper, were observed to change from their normal brown coloration (above) to ashy black in response to fire (left), aiding their camouflage in the burned habitat. Melanism in burned areas is considered a special case of homochromy and has also been observed in African and European orthopterans. Gradual decrease of melanized forms of *C. carinata* was also observed as the habitat recovers from the fire, though dark individuals could still be found one month after the fire (right). The color change is seen as an adaptive response to reduce predation risk. Photos: Dhaneesh Bhaskar.

Figure > Pillar with foliage decoration and small animals, including a bush cricket, from the Canopus of Villa Adriana (Tivoli, Rome); white marble, Hadrian age (1st–2nd century). Photo by Paolo Fontana.

8 CULTURAL ASPECTS

8.1 CRICKET FIGHTING IN CHINA, A TRADITION MORE THAN TWO AND A HALF MILLENNIA OLD

by Wenhui Zhu and Sheng-Quan Xu

Fighting games are an important part of ancient Chinese entertainment and competitive folk customs. Fighting games have been documented since the Spring and Autumn Period and the Warring States Period (roughly, the 7th to the 2nd centuries BC). They were not only a way for the powerful to amuse themselves and boast about their wealth and victories, but were also a form of entertainment for the common people during their daily leisure time and New Year celebrations. There were many types of fighting, from cock fighting, bull fighting, and cricket fighting, to other animal and insect fighting, and even plant fighting – all contributing to rich and colorful ancient folk practices.

Cricket fighting is a traditional folk game in China, dating back over two millennia to the Warring States Period (fig. 1). At that time, cricket fights were a prevalent form of gambling. By the Tang Dynasty, over a thousand years ago, cricket fighting had expanded beyond gamblers and become a popular part of common street culture, alongside activities like fishing, bird keeping, and flower growing. During the Song Dynasty, cricket fighting transitioned from a folk practice into the imperial courts, captivating emperors and officials alike, even leading some to neglect their duties. Scenes of emperors engaging in cricket fights can be witnessed in movies and television shows, such as the movie *The Last Emperor*, which portrays the Qing Dynasty's last emperor, Puyi, tending to crickets.

The main participants in cricket fighting are the crickets known as *ququ* in Chinese, which belong to the several species of the genus *Velarifictorus*. *Velarifictorus* primarily inhabit surface environments, such as grassy areas or bare ground. They consume almost exclusively the vegetation found on the ground, thus limiting their food sources in the wild. Faced with limited food and increasing competition, the only way for them to survive is through competition and combat. After people observed this trait, they invented cricket-fighting activities. Cricket fighting is not a year-round activity due to the short lifespan of the crickets, typically only about six months. People usually commence cricket fights around mid-August. Typically, they take place in ceramic or porcelain cricket containers (fig. 2). When two male crickets confront each other, an intense battle ensues. Initially, crickets vigorously flap their wings, both to boost their own spirits and to intimidate their opponent. Subsequently, they engage in combat, rolling their antennae and rotating their bodies to strike, bite, or grasp their opponent. After several rounds, the defeated cricket appears despondent, while the victorious one exudes pride. Cricket fights are unpredictable, each cricket exhibiting its unique personality, making the outcome uncertain until the final moments. This unpredictability is the charm of cricket fighting and the reason why people always have been drawn to it.

Due to the different climate zones across China, the maturation time of crickets varies greatly. In mid-June, large crickets in the provinces of Guangdong and Guangxi begin to mature. Their chirping frequency is slow, but their sounds are rough and loud. Shandong Province is one of the most famous cricket-producing areas in China. It has two major markets (fig. 3), Ningyang and Ningjin. Every August, cricket season is opened by Ningyang locals. Shandong crickets are renowned for their aggressive nature and prowess in fighting. As cricket fighting gained popularity, the demand for "high quality" crickets rose. After cricket fighting became an enter-

Figure 1 Cricket fighting. Photo: Zhuqing He.

Figure 2 Ceramic cricket container. Photo: Hao Tang.

tainment for the royal nobility, Shandong crickets became an exceedingly rare royal tribute. The cricket fights of Ningyang were not only recognized in 2009 as a provincial-level intangible cultural heritage project, but were also designated as a national geographical indication product – that is, a product of certified origin and quality – showcasing their significance in both cultural and economic spheres.

With changing times and technological advancements, the traditional practice of cricket fighting is gradually fading away. The rapid urbanization and emergence of new forms of entertainment have led to the gradual decline of this traditional culture. Nonetheless, some enthusiasts of traditional culture are trying to protect and preserve this ancient pastime, hoping to continue the legacy of cricket fighting through various means.

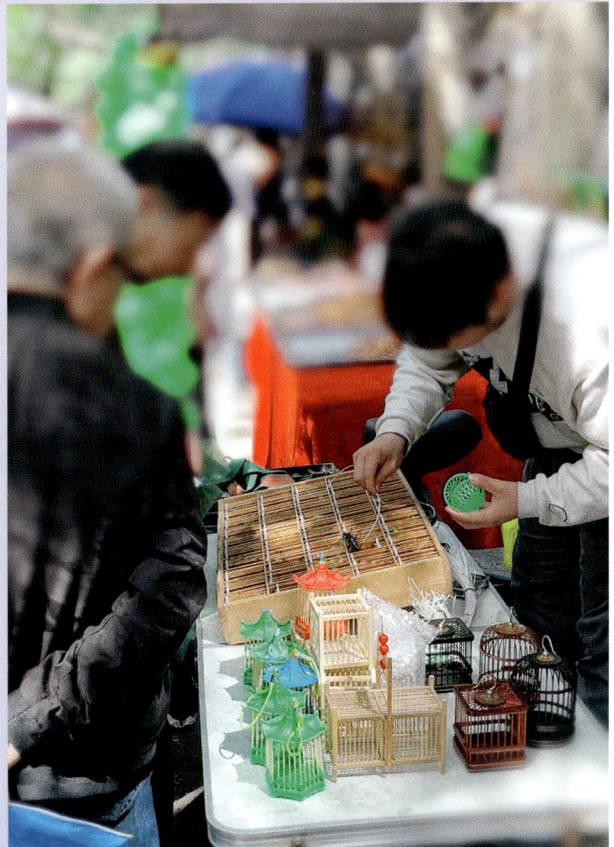

Figure 3 Cricket trade at the market. Photo: Wenhui Zhu.

8.2 THE GIANT HOODED KATYDID, A PET LIKE NO OTHER

by Frank Glaw

Giant hooded katydids *Siliquofera grandis* are among the world's largest members of Orthoptera. Females can reach lengths of up to 13 cm, weights of more than 30 g, and wingspans of up to 25 cm (fig. 1). Their natural range extends from New Guinea to northern Australia and may include other parts of southeast Asia. Scientifically described 170 years ago, living individuals of this species were imported to Europe for the first time only a few years ago, and the species' biology remains poorly known. What is known is that *S. grandis* are nocturnal canopy-dwellers of tropical rain forests. They move slowly and seem to rely on their camouflage when threatened, but are also capable of jumping, flying, and defending themselves with a strong bite. Any songs audible to humans are not known, and intraspecific communication is assumed to rely on vibrations.

Recent attempts of breeding giant hooded katydids in captivity have been successful; keeping them is easy and inexpensive. Throughout her life, a female is capable of laying up to 370 eggs in humid soil with her ovipositor. The nymphs hatch after about three months, molt for the last time after another three months, and may live for more than a year as adults. The katydids even take food from a keeper's hand. When eating sweet apples, the saliva can be seen dripping from their mouths.

S. grandis is so big that many structures that require a stereomicroscope to examine in other insects can be seen with the naked eye (fig. 2). This includes the "ears," the tympana, situated on the fore femora, as in other ensiferans. The sole of every foot carries eight adhesive pads with little holes, allowing giant hooded katydids to climb vertical glass panes with ease. When climbing, these holes secrete a thin layer of fluid that supports adhesion to the surface.

A surprising amount of these insects' time is dedicated to individual grooming. The katydids can be observed cleaning their adhesive pads, their antennae, and their mouthparts with much dedication. Their repertoire includes many types of behavior that may seem surprising. Nymphs use their hind limbs to throw fecal pellets as far away as they can, which may be several meters. The presumed reason for this behavior is that their visual camouflage is excellent, but the smell of the feces may attract predators. Removing the feces as far away as possible may mislead predators.

Their good temper and slow, steady movements make giant hooded katydids the ideal creatures to keep as pets, allowing people to observe their interesting behaviors and get to know them as individuals who have their own personalities. If presented correctly, they are perfectly suited for allowing children to experience insects, to indulge their fascination with the "creepy-crawlies," and to feel less anxiety around them. Therefore, they are ideal outreach animals for schools and universities and may help in the training of future entomologists and conservationists.

CARE OF *SILIQUOFERA GRANDIS*

Cage: Glass cage, minimum size 40 cm x 40 cm x 60 cm, with good ventilation and 10 cm of humid soil as a substrate.

Temperature: Room temperature, no heating necessary.

Food: Brambles, hazel, also apples, carrots, salad, and dandelion.

Reproduction: Eggs are laid into soil and hatch after two and a half to three months.

Comments: Mist plants daily so the katydids can drink the water.

Figure 1 Giant hooded katydid walking on glass. Photo: Frank Glaw and Timon Glaw.

Figure 2 (left) Ventral side of a tarsus, with eight adhesive pads and claws. (right) Tympana, (the "ears") of *Siliquofera grandis*. Photo: Frank Glaw and Timon Glaw.

275

8.3 ON THE HUNT FOR THE BEST PHOTO: ORTHOPTERA ECOTOURISM IN SOUTHEASTERN EUROPE

by Christan Roesti

While ecotourism has experienced a real boom in recent years, Orthoptera have rarely been at the center of touristic attention. Yet many orthopteran species exhibit extremely interesting behavior, and many rare species make exciting discoveries for naturalists. I spent days and weeks in search of these species to observe and photograph them. Grasshoppers and crickets offer fantastic photo opportunities that can be perfectly staged with a good eye and a little patience. The best time to photograph them is mostly in the early morning, when the light is soft and diurnal grasshoppers are often basking in exposed spots. The exciting singing and mating behavior of many species allows for impressive video recordings. When the light conditions for photographing deteriorate toward midday, it's time for a break – for orthopterans and photographers alike.

Over time, I developed the desire to show the areas I visited and the species I observed to other people interested in natural history. While many companies organize wildlife or ornithological tours on a large scale, grasshopper tourism is not yet really a business. Only some small companies have specialized in organizing nature and photo tours with a focus on grasshoppers and other insects. One of these companies is Orthoptera.ch GmbH from Switzerland (www.orthoptera.ch). Since 2012, Orthoptera.ch has organized at least one nature trip focusing on grasshoppers or other insects each year. These excursions are aimed at nature enthusiasts who enjoy observing and photographing or who want to pass on their acquired knowledge to others. This market is very limited. It is particularly important for us to work with local tour guides in order to involve the local population in the planning. In Bulgaria and Romania, for example, there are the well-known orthopterists Dragan Chobanov (see chapter 5.3.1, The Mediterranean Hot Spot I: The Balkano-Anatolian region of the East) and Ionut Iorgu.

The southern countries of Europe are the most popular destinations for insect tours. The region is home to a high diversity of Orthoptera, with many locally endemic species. Even species completely new to science may still be found (see box in chapter 5.3.3: Naturalistic serendipity). The Balkan Peninsula, where the photos shown here were taken, has proved to be very rewarding (figs. 1–5). People are very open to strangers and have always welcomed them warmly, and practically all habitats are freely accessible to tourists; areas away from the roads are very rarely fenced in. The mountainous climate is very pleasant in summer, and it rarely rains, providing ideal conditions for grasshopper watching. Most of the mountain ranges in the southern Balkans are home to at least one endemic grasshopper species (see chapter 5.3.1). Due to the different altitudes from the sea to the mountaintops, interesting locust species are continuously active from May to September, and the same is true for many remarkable species from other animal groups.

The accompanying pictures are meant to show why a visit to the Balkan Peninsula will be a fascinating experience for those interested in Orthoptera. An entomological trip can also be combined with a beach vacation, as recreational locations and observation areas for grasshoppers are often close to each other.

Figure 1 A group of ecotourists lying or kneeling in the dry grass and intently keeping their gazes on the ground: a typical view of an orthopterological excursion. Clusters like this often mean that somebody has found a particularly interesting specimen, and everyone will help other group members take the best possible photographs. Photo: Christian Roesti.

Figure 2 The arid karst habitats of the Peloponnese Peninsula in the south of Greece are home to a multitude of specialized species of Orthoptera. Photo: Christian Roesti.

> **Figure 3** The beautiful caeliferan *Ramburiella turcomana* is an inhabitant of ancient cultured land, as shown in **Figure 2** Photo: Christian Roesti.

> **Figure 4** There are some species with whom every entomologist must fall in love. For me, one of them is the stone grasshopper *Glyphanus obtusus*, which occurs in very localized populations on the Peloponnese Peninsula and has tiny wings. This photo shows an adult male. Photo: Christian Roesti.

> **Figure 5** *Saga natoliae* is one of Europe's largest orthopterans. It alone is a good reason for a trip to Greece, where it shares its habitat with *S. hellenica* and *S. campbelli*. Photo: Christian Roesti.

8.4 GRASSHOPPERS AS TRADITIONAL ROYAL FOOD AND MODERN PROTEIN SOURCE IN MADAGASCAR

by Sylvain Hugel and Brian L. Fisher

The first written record of the tradition of eating insects on Madagascar, whose people have a long history in entomophagy, describes the culinary use of Orthoptera. This account comes from D'Azvedo, captain of an expedition with Jesuit priests, who notes in a letter dated May 23, 1617, that the inhabitants of west-central Madagascar were extremely fond of eating locusts. Locusts were noted by many subsequent authors as an important food for people, as well as feed for their livestock. Most early records indicated that locusts were first boiled in water and then fried in fat for immediate consumption. To prepare this food for long-term storage, people dried the insects in the sun, then crushed them into a powder. Legs and wings were often winnowed out in the process (fig. 1). Apparently, all levels of Malagasy society enjoyed snacking on Orthoptera, as these insects were one of the delicacies served at the table of Queens Ranavalona I, II, and III. Queen Ranavalona II even had aides specifically dedicated to collecting grasshoppers for her table. Historical accounts viewed the gathering of locusts as an important food source for the lean months. Chase Osborne, author and explorer, even stated in 1924 that "thanks to locusts and other edible insects, there is no famine on Madagascar, and that because of edible insects, there is little worry over the question of food".

Accounts in historical literature emphasize the consumption of locusts during periodical swarms, as these are extremely impressive events. However, entomophagy probably occurred throughout the year, as is the case today. The tradition of eating insects is still deeply rooted in Madagascar and involves locusts of both solitary and gregarious phases (*Locusta migratoria* and *Nomadacris septemfasciata*), along with many other orthopteran species. Our recent work recorded over 100 insect species consumed on the island. More than a third of

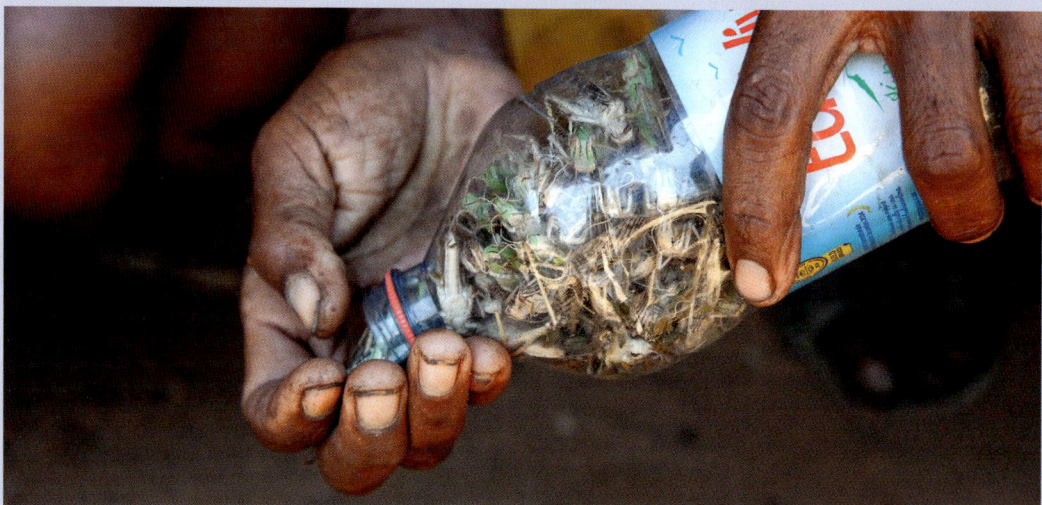

Figure 2 Legs and wings of locusts are frequently winnowed before frying them in oil. Photo: Brian Fisher.

these species belong to Orthoptera. Among these, about two-thirds are grasshoppers and locusts (Caelifera) and one-third are crickets and katydids (Ensifera). The most common vernacular name for edible Orthoptera is *valala*, though this name is often restricted to acridoids.

Our survey indicated that the acridoid grasshoppers *Paracinema tricolor*, *Oxya hyla*, and *Locusta migratoria* were the most commonly consumed orthopterans (fig. 2). We recorded the consumption of these species in approximately half of the ethnic groups in Madagascar. Apart from periods of locust invasions, *P. tricolor* and *O. hyla* are the most abundant grasshoppers in the rice paddies of Madagascar. The widespread consumption of these species can be explained by their opportunistic gathering during the

< **Figure 1** *Locusta migratoria* collected from the wild in western Madagascar. Photo: Brian Fisher.

day in and around crops and fields; these opportunistic collections are largely made by children.

Our survey also indicated that a few Orthoptera species have been recognized as inedible for various reasons (fig. 3). For example, mole crickets (Gryllotalpidae), while eaten in other parts of the world, are considered inedible based on a widespread cultural belief. These insects are called *zazavery*, meaning "lost child," and are considered bad omens.

Although selling orthopterans for food is rare nowadays, *Brachytrupes membranaceus colosseus*, a large species locally known as *sahobaka*, is intentionally collected for marketing purposes in western Madagascar (fig. 4). In the hours before nightfall, foragers follow the loud call of these large crickets to locate the entrance to their sandy burrow.

Although the consumption of Orthoptera may be fading in large cities, these insects are still widely consumed across all Madagascar and play an important role in the island's fragile food security, particularly for rural communities.

Figure 3 The colorful pyrgomorphid *Phymateus saxosus* is one of the few orthopterans that is not eaten. It is called the dog cricket (*vala-lan'alika*) in Malagasy and has a repulsive odor: a Malagasy proverb says that "even dogs won't eat it." This species feeds on various plant species, including Apocynaceae milkweed, and is presumed to be poisonous by Malagasy people. Photo: Sylvain Hugel.

Figure 4 *Brachytrupes membranaceus colosseus* ready to be sold on the market in western Madagascar. Photo courtesy of Faneva Rajemison.

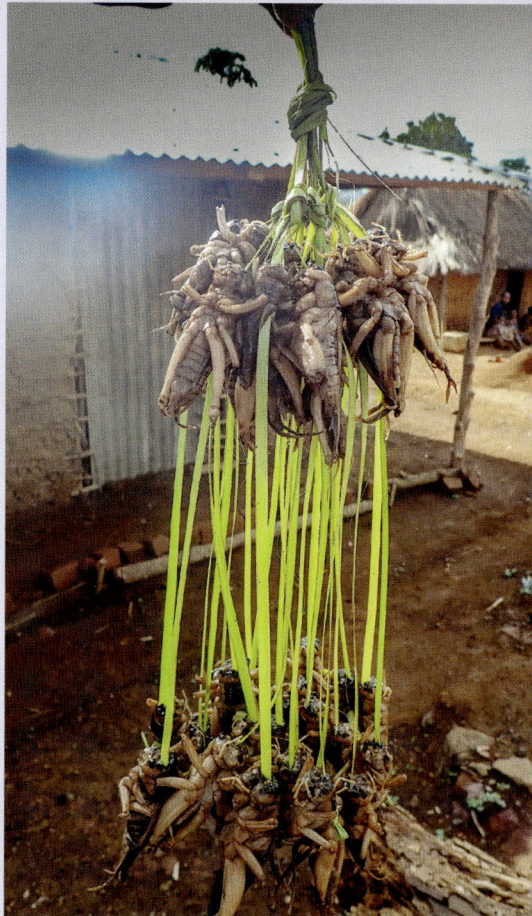

8.5   SALT, LIME, AND *CHAPULINES*: THE REVIVAL OF AN ANCIENT MEXICAN CULINARY TRADITION

by Ricardo Mariño-Pérez

The contemporary human population of Mexico is a combination of a variety of indigenous ethnic groups, Europeans, and people from the Caribbean (of both indigenous and African descent). This has caused a combination of "oriental" and "occidental" aspects in Mexican culture. For instance, entomophagy – eating insects – would be considered a taboo on the "occidental" or European side, while it is a century-old tradition in the "non-occidental" (oriental and Mesoamerican) part of the culture. For historical reasons, Mexico is, broadly speaking, divided into two main cultural areas. The north of Mexico is considered more "occidental," the south more "non-occidental"; consequently, entomophagy is today much more widespread in the south. Although Mexico City is not exactly in the geographical center of Mexico, it acts as a subtle division between these two parts of the country.

Despite this divide, grasshoppers are called *chapulines* throughout all parts of Mexico, instead of the general Spanish term *saltamontes* (literally, "hill jumpers"). The term *chapulin* comes from the Nahuatl language spoken by the Aztecs: *chapa(nia)*, meaning "bounce," and *ulli*, meaning "rubber," so a *chapulín* is an "insect that jumps like a rubber ball." The first written reference to this word appeared in *The Florentine Codex* (AD 1540–1585) bilingual book written in Nahuatl and Old Spanish, authored by the Franciscan monk Fray Bernardino de Sahagún (1577). (See the box "*Chapulines* in *The Florentine Codex*.")

The recent re-awakening of interest in elements of pre-Hispanic (pre-Columbian) gastronomy has led to an increase in popularity of the consumption of insects, in particular *chapulines*, throughout the entire territory of Mexico. Nevertheless, the division between the northern and southern parts of the country is still clearly visible. One of the most common species, the corn grasshopper *Sphenarium purpurascens* (Pyrgomorphidae), is considered a plague in the federal states of Queretaro and Guanajuato, northwest of Mexico City, whereas it is considered a delicacy in the southeastern states of Tlaxcala, Puebla, and Oaxaca. Corn grasshoppers are harvested in the northwestern states, but exclusively shipped to the southern part of Mexico.

Harvesting for personal consumption is mostly done non-professionally and/or for sale, primarily at local fresh markets. People collect nymphs and adults early in the morning, using conical nets and sticks to shake the plants where corn grasshoppers spend the night. In central and southern Mexico, at least 75–100 tons (fresh weight) of the insects are extracted annually. Indirectly, this harvesting method helps to control the populations, because they are also considered a plague (mainly in corn and bean fields), providing additional economic and environmental benefits for the people. The typical way to prepare *chapulines* is to leave them one day without food or to feed them paper in order to clean the gut, then to boil them, sun-dry them, and season them with salt and lemon, black pepper, chili pepper, or garlic juice. Finally, they are fried or grilled.

Because nymphs and adults are different in size and consistency, they are sold separately. Nymphs are softer and are of particular demand, which is why they tend to be more expensive than adults. The typical manner of consuming them is as a snack, with guacamole, in tacos, as an ingredient in sauces, or even on pizzas. Sometimes the *chapulines* are ground, and the powder is combined with salt and other spices to create "*chapulines* salt."

Most popular are the above-mentioned corn grasshopper *chapulines*. In addition, some close relatives in the genus *Sphenarium*, such as *S. mexicanum* and *S. histrio*, are also commonly consumed. The basic reason for their consumption is their abundance; they are not more or less nutritive than other less-abundant grasshoppers. It is quite common to find other species as bycatch in the fresh markets, such as melanoplines of the genera *Melanoplus* and *Akamasacris*. According to Jongema 2017, some 78 species of Orthoptera belonging to 42 genera in 5 families are used as food in Mexico today. Similarly, but less frequently as food for humans, other Orthoptera are also powdered (mainly the abundant locust genus *Schistocerca* and some crickets); this powder is used in flour to prepare tortillas.

Orthoptera consumption as human food has been present in Mexico for centuries. Due to the deep roots of this practice and the high abundance of certain species, coupled with an increasing interest in sustainable and more efficient protein sources other than cattle, they can be expected to remain a central component of Mexican cuisine for many years to come.

Figure 1 Living *chapulines* (*Sphenarium purpurascens*), both nymphs (greenish) and adults (black), in Tecomatlan, Hidalgo, Mexico. Photo: Ricardo Mariño-Pérez.

RECIPE: GRILLED *CHAPULINES*

by Ricardo Mariño-Pérez

Figure 1 Nymphal and adult stages of *Sphenarium* spp., Oaxaca City Fresh Market, Oaxaca, Mexico. Photo: Ricardo Mariño-Pérez.

Ingredients:

2 cups of *chapulines* (fresh or dried; fig. 1)

Juice of 2 limes

Pequin pepper

Salt

Preparation:

Soak the *chapulines* in the lime juice with salt for 20 minutes. Strain the *chapulines* and grill them over a slow fire in a frying pan until they are dried and have a reddish coloration. Serve over lettuce leaves. Sprinkle pequin pepper and salt to taste.

CHAPULINES IN *THE FLORENTINE CODEX*

by Ricardo Mariño-Pérez

There are few surviving written records about pre-Hispanic Mesoamerican culture in general. One of the best-preserved sources, commonly referred to as *The Florentine Codex*, is a part of the *Historia general de las Cosas de Nueva España*, a 16th-century ethnographic research study of Mesoamerica by the Spanish Franciscan friar Bernardino de Sahagún; today, it is held in the Laurentian Library of Florence, Italy.

Book 11 of the codex, *Earthly Things*, describes properties of animals, birds, fish, trees, herbs, flowers, metals, stones, and colors. The book lists several species of grasshoppers (all members of Caelifera) with their Nahuatl names. As in other languages, the species names share a root word, *chapoli*, combined with aspects of morphology, ecology, and importance to humans (e.g., for consumption). For instance, the codex refers to the locust species of the genus *Schistocerca* (fig. 1) as follows: "There are several types of locusts in this land. They are similar to the ones found in Spain. Some are very large and are called *acachapoli*, which means 'arrow grasshopper.' Because, when flying, they are fast and make a noise like an arrow. They are edible."

For species in the genus *Melanoplus*: "Others are called *tiectli chapoli*, are medium size, brownish-reddish and collected during the harvesting of the cornfields. They are edible."

For species in the genus *Sphenarium* (mainly *S. purpurascens*): They were known as *xopanchapoli*, which means "summer grasshoppers" and are commonly called *chapulines*. The local description is as follows: "They are big and thick. They don't fly, only crawl. They eat green beans. Some are black, others brown and others green. They are edible."

Others are called *colacachapoli*; they have a color pattern similar to the birds known as quails. They are edible. Here, the codex is referring to the romaleid species of the genus *Taeniopoda*. As in the previous cases, all the genera mentioned here are generally abundant in central and southern Mexico and relatively easy to find, even for the untrained eye.

The codex also mentions *tlalchapoli* or *ixpopoiochapoli*, which mean "blind grasshoppers." The text says: "They are very abundant and very small. They are crawling on the trails and they don't move even if you step on them. They are edible." Here, the codex is referring to the nymphs of previously mentioned grasshoppers, treating them as different species.

Ensifera, on the other hand, are called *cacatecuilichtli*, because their songs could be interpreted as "chij, chicchi, chi, chi." They are described as edible and to be found in hay. Probably, this account refers to the katydids *Microcentrum* spp. and/or *Neoconocephalus* spp. Curiously, the root *chapoli* is not present and, unlike the aforementioned species, there is no reference to the movement behavior in the name.

Figure 1 *Acachapoli* (*Schistocerca* sp.) from *The Florentine Codex*, volume 11, page 101v. World Digital Library, https://www.loc.gov/item/2021667856/.

Santanderia lita, a member of the monkey grass-hoppers (Eumastacidae), is one of many colorful tropical orthopterans. Photo: Juan Manuel Cardona-Granda and Oscar J. Cadena-Castañeda.

THE AUTHORS

Roberto Battiston
Museo di Archeologia e Scienze Naturali "G. Zannato", Montecchio Maggiore, Italy

Alba Bentos-Pereira
Museo Nacional de Historia Natural, Montevideo, Uruguay

Dhaneesh Bhaskar
Care Earth Trust, Chennai, Tamil Nadu 600061, India

Holger Braun
División Entomología, Museo de La Plata, La Plata, Argentina

Oscar J. Cadena-Castañeda
Universidad Distrital Francisco José de Caldas, Grupo de Investigación en Artrópodos "Kumangui", Bogotá, Colombia
Universidad INCCA de Colombia, Grupo en Biotecnología y Medio Ambiente BIOMA, Bogotá, Colombia

Juan Manuel Cardona-Granda
Associate Director, Nature-Based Solutions, Clear-Blue Markets

Arianne Cease
Global Locust Initiative, School of Sustainability, School of Life Sciences, Arizona State University, Tempe, Arizona, USA

Dragan Chobanov
Institute of Biodiversity and Ecosystem Research, Bulgarian Academy of Sciences, Sofia, Bulgaria

María Marta Cigliano
División Entomología, Museo de La Plata, La Plata, Argentina
Centro de Estudios Parasitológicos y de Vectores (CEPAVE), CONICET–UNLP, La Plata, Argentina

Matthew Connors
James Cook University, Cairns, Queensland, Australia

Laure Desutter-Grandcolas
Institut de systématique, biodiversité et évolution, Muséum National d'Histoire Naturelle, CNRS, SU, EPHE, UA, Paris, France

Lara-Sophie Dey
Senckenberg Deutsches Entomologisches Institut (SDEI), Müncheberg, Germany

Swati Diwakar
Department of Environmental Studies, University of Delhi, Delhi, India

Cheten Dorji
Department of Forest Science, College of Natural Resources, Royal University of Bhutan, Punakha, Bhutan
Wildlife and Ecology, Massey University, Palmerston North, New Zealand

Rachael Y. Dudaniec
School of Natural Sciences, Macquarie University, Sydney, New South Wales, Australia

P. S. Easa
Care Earth Trust, Chennai, Tamil Nadu 600061, India

Thomas Fartmann
Department of Biodiversity and Landscape Ecology, Osnabrück University, Osnabrück, Germany

Brian L. Fisher
Department of Entomology, California Academy of Sciences, San Francisco, California, USA

Paolo Fontana
Fondazione Edmund Mach, Università degli Studi di Padova, Padova, Italy

Frank Glaw
Zoologische Staatssammlung München (ZSM-SNSB), Munich, Germany

Oliver Hawlitschek
Deptarment of Evolutionary Biology and Environmental Studies, Universität Zürich, Switzerland

Danilo Hegg
Wētā Conservation Charitable Trust, Dunedin, New Zealand

Felix Helbing
Department of Biodiversity and Landscape Ecology, Osnabrück University, Osnabrück, Germany

Claudia Hemp
Senckenberg Biodiversität und Klima Forschungs-
zentrum, University of Bayreuth, Bayreuth, Germany

JoVonn G. Hill
Mississippi Entomological Museum, Department of
Entomology and Plant Pathology, Mississippi State
University, Mississippi State, Mississippi, USA

Axel Hochkirch
Musée National d'histoire Naturelle de Luxembourg,
Luxembourg, Luxembourg
Department of Biogeography, Trier University, Trier,
Germany

Terry Francis Houston
Western Australian Museum, Welshpool D.C.,
Western Australia, Australia

Sylvain Hugel
Institut des Neurosciences Cellulaires et Intégra-
tives, Centre National de la Recherche Scientifique,
Université de Strasbourg, Strasbourg, France
Madagascar Biodiversity Center, Antananarivo,
Madagascar

David Hunter
Locust and Grasshopper Control, Canberra, Australia

Martin Husemann
Staatliches Museum für Naturkunde, Karlsruhe,
Germany

Slobodan Ivković
Staatliches Museum für Naturkunde, Karlsruhe,
Germany

Niko Kasalo
Laboratory of Evolutionary Genetics, Ruđer Bošković
Institute, Bijenička cesta 54, HR-10000 Zagreb,
Croatia

L. Lacey Knowles
Department of Ecology and Evolutionary Biology,
Museum of Zoology, University of Michigan, Ann
Arbor, Michigan, USA

Sebastian König
Ecosystem Dynamics and Forest Management
Group, School of Life Sciences, Technical University
of Munich, Freising, Germany
Berchtesgaden National Park, Berchtesgaden,
Germany

Ulrich Kotthoff
Leibniz Institute for the Analysis of Biodiversity
Change, Centre for Biodiversity Monitoring and
Conservation Science, Hamburg, Germany

Santosh Kumar
Cholistan University of Veterinary and Animal
Sciences Bahawalpur, Punjab, Pakistan

Michel Lecoq
French Agricultural Research Centre for Interna-
tional Development (CIRAD), Montpellier, France
CBGP, University of Montpellier, CIRAD, INRAE,
Institut Agro, IRD, Montpellier, France

Chien C. Lee
Institute of Biodiversity and Environmental Conservation, Universiti Malaysia, Sarawak, Kota Samara-
han, Sarawak, Malaysia

Arne W. Lehmann
Stahnsdorf, Germany

Gerlind U. C. Lehmann
Evolutionary Ecology, Humboldt University, Berlin,
Germany
NABU (Nature and Biodiversity Conservation Union),
Berlin, Germany

Jeffrey A. Lockwood
Department of Philosophy and Religious Studies,
University of Wyoming, Laramie, Wyoming, USA

Franz Löffler
Department of Biodiversity and Landscape Ecology,
Osnabrück University, Osnabrück, Germany

Ricardo Mariño-Pérez
Department of Ecology and Evolutionary Biology,
University of Michigan, Ann Arbor, Michigan, USA
Facultad de Ciencias, Universidad Nacional
Autónoma de México, Mexico City, Mexico

Bruno Massa
Dipartimento Scienze Agrarie Alimentari e Forestali,
Università degli Studi di Palermo, Palermo, Italy

Daniela Matenaar
Hessisches Landesmuseum Darmstadt, Darmstadt,
Germany

Fabio Leonardo Meza-Joya
Wildlife and Ecology, Massey University, Palmerston
North, New Zealand

Fernando Montealegre-Zapata
School of Life Sciences, University of Lincoln,
Lincoln, United Kingdom

Mary Morgan-Richards
Wildlife and Ecology, Massey University, Palmerston
North, New Zealand

Joaquín Ortego
Department of Ecology and Evolution, Estación Biológica de Doñana (EBD-CSIC), Seville, Spain

Martina E. Pocco
Centro de Estudios Parasitológicos y de Vectores (CEPAVE), CONICET – UNLP, La Plata, Argentina
División Entomología, Museo de La Plata, FC-NyM-UNLP, La Plata, Argentina

Dominik Poniatowski
Department of Biodiversity and Landscape Ecology, Osnabrück University, Osnabrück, Germany

Mattia Ragazzini
Department of Evolutionary Biology and Environmental Sciences, University of Zurich, Zurich, Switzerland

Mira W. Ries
Global Locust Initiative (GLI), School of Sustainability, School of Life Sciences, Arizona State University, Tempe, Arizona, USA

Christian Roesti
Berne, Switzerland

Ole-Kristian O. Schall
Leibniz Institute for the Analysis of Biodiversity Change, Centre for Taxonomy and Morphology, Hamburg, Germany

Roberto Scherini
Pavia, Italy

Jens Schirmel
iES Landau, Institute for Environmental Sciences, University of Kaiserslautern-Landau (RPTU), Landau, Germany
Eusserthal Ecosystem Research Station, University of Kaiserslautern-Landau (RPTU), Eusserthal, Germany

Nikita Sevastianov
Institute for Information Transmission Problems of the Russian Academy of Sciences (Kharkevich Institute), Moscow, Russia

Kerry L. Shaw
Department of Neurobiology and Behavior, Cornell University, Ithaca, New York, USA

Josip Skejo
Department of Biology, Evolution Lab, University of Zagreb, Zagreb, Croatia

Hojun Song
Department of Entomology, Texas A&M University, College Station, Texas, USA

Thomas Stalling
Inzlingen, Germany

Riffat Sultana
Department of Zoology, University of Sindh, Jamshoro, Sindh, Pakistan

Soňa Svetlíková
Department of Ecology, Comenius University, Bratislava, Slovakia

Ming Kai Tan
Lee Kong Chian Natural History Museum, National University of Singapore, Singapore

Clara Therville
SENS, IRD, CIRAD, University Paul Valery Montpellier 3, University of Montpellier, Montpellier, France

Steven A. Trewick
Wildlife and Ecology, Massey University, Palmerston North, New Zealand

Karim Vahed
Buglife – The Invertebrate Conservation Trust, Peterborough, United Kingdom

Varvara Vedenina
Institute for Information Transmission Problems of the Russian Academy of Sciences (Kharkevich Institute), Moscow, Russia

Luc Willemse
Naturalis Biodiversity Center, Leiden, Netherlands

Sheng-Quan Xu
College of Life Sciences, Shaanxi Normal University, Xi'an, China

Sonu Yadav
Biosecurity and Animal Welfare, Northern Territory Government, Darwin, Australia
Research Institute for the Environment and Livelihoods, Faculty of Science and Technology, Charles Darwin University, Darwin, Northern Territory, Australia

Jeanne Agrippine Yetchom Fondjo
Staatliches Museum für Naturkunde, Karlsruhe, Germany
Zoology Unit, Laboratory of Biology and Physiology of Animal Organisms, Graduate School in Fundamental and Applied Sciences, University of Douala, Douala, Cameroon

Wenhui Zhu
College of Life Sciences, Shaanxi Normal University, Xi'an, China

Many species of Orthoptera show some variability in coloration, and red, violet, or pink forms may occur occasionally. Mostly, these conspicuous colors are a disadvantage compared to the camouflage typically provided by shades of green, brown, or grey. In the pink leaf katydids *(Eulophophyllum lobulatum)* from Borneo, the green morph is more common (top), but the pink morph (bottom) matches the color of equally pink young leaves of some food plants, thus also acting as camouflage. Photos: Jonah Voo.

GLOSSARY

abdomen   The hind body of an insect containing the digestive tract and the sexual and other inner organs.

ala (plural: alae)   The hind wings of Orthoptera.

analogous   Characters of two organisms that are similar in form and/or function but do not share a common evolutionary ancestry.

apomorphy   A character that has newly evolved in the ancestor of a group of organisms and therefore defines the group.

apterous   Without wings; not capable of flight.

bioacoustics   The study of sound production, emission, and reception in biological systems.

brachypterous   With short wings; not capable of flight, but sometimes capable of stridulation.

Caelifera   The group of Orthoptera comprising grass-hoppers.

Cenozoic   The Earth's current geological era, beginning 66 million years ago.

convergent   The independent evolution of similar features in different groups of organisms.

dorsal   The "back" side of an animal, visible from above in Orthoptera.

echeme   A complex structure of orthopteran song, comprising one or several syllables.

endemic   Exclusively occurring in a certain geographic location.

Ensifera   The group of Orthoptera comprising crickets and bush crickets.

entomophagy   Consumption of insects as food by humans.

femur   The upper part of an insect leg, close to the body.

file   A part of the stridulatory – that is, sound-producing – apparatus of many Ensifera.

gregarious   The form of locusts that forms swarms and may cause outbreaks or plagues.

hemimetabolous   The type of life cycle of many insects, including Orthoptera, in which a nymph precedes the imago (adult) stage without a pupa stage.

homologous   Characters of two organisms that have derived from a common evolutionary ancestry.

hot spot   A region with significant diversity.

imago   The adult stage of insects. Imagoes are capable of flight (in species in which adults are) and repro-duction, and do not molt.

invasive   A species that has been introduced by humans outside its natural distribution range and harms its new environment.

locust   A term for grasshoppers that form swarms and may cause outbreaks or plagues.

macropterous   With long wings, usually capable of flight.

mandible   Part of the mouthparts of insects.

Mesozoic   The Earth's geological era from 252 to 66 million years ago; the time of the dinosaurs.

micropterous   With short wings, not capable of flight, but sometimes capable of stridulation.

monophyletic   In taxonomy: a group of taxa comprising a common ancestor and all descendants, and no others.

monotypic   A taxon of any rank comprising a single taxon of the next lowest rank – for example, a genus comprising a single species.

morphology   The study of the form and structure of organisms.

nymph   The immature stage of hemimetabolous insects. Nymphs are flightless and not sexually mature, and they grow via molting.

Orthoptera Species File   An online database of the world's Orthoptera.

oscillogram   A graphical representation of a sound re-cording, showing amplitude over time.

ovipositor   The egg-laying apparatus of female Ensifera.

Paleozoic   The Earth's geological era from 540 to 252 million years ago.

Pangaea   A supercontinent on Earth during the late Paleozoic and early Mesozoic eras.

paraphyletic   In taxonomy: a group of taxa comprising a common ancestor and all descendants, but also non-descendants.

phylogenetics   The study of the evolutionary relationships between organisms using genetics.

pronotum   The exoskeleton part covering the dorsal side of the thorax of insects.

proprioreceptor   A sensory receptor receiving information on self-movement and position.

pulse   The basic element of orthopteran song. Several pulses may form a syllable.

radiation   The evolutionary process in which organisms rapidly diversify from an ancestor into many new forms.

scraper   A part of the stridulatory – that is, sound-producing – apparatus of many Ensifera.

solitarious   The form of locusts that does not form swarms and normally does not cause outbreaks or plagues.

speciation   The evolutionary process generating new species.

spermatophore   A packet or mass of sperm produced by males and externally transferred to females. Ensifera males typically produce spermatophores.

stridulation   Sound production by animals that rub parts of their body – in Orthoptera, typically wings and legs – against each other.

syllable   An intermediate structure of orthopteran song, comprising one or several pulses. Several syllables may form an echeme.

tarsus   The lower part of an insect leg; the "foot."

taxon (plural: taxa)   A taxonomic group of any rank – for example, a subspecies, a species, a genus, a family, a suborder.

taxonomy   The scientific discipline of defining and naming groups of organisms based on characteristics.

tegmen (plural: tegmina) The fore wings of Orthoptera.

thorax   The mid body of an insect, where the legs and wings attach.

tibia   The mid part of an insect leg.

trachea (plural: tracheae) A chitinous tube that is part of the respiratorial system of insects.

tympanum   The hearing organ of Orthoptera. Situated on the front legs in Ensifera and on the sides of the abdomen in Caelifera.

ventral   The "belly" side of an animal, visible from below in Orthoptera.

REFERENCES

Adriaansen, C., Woodman, J. D., Deveson, E. D., and Drake, V. A. (2016). Chapter 4.1: The Australian plague locust: Risk and response. In Shroder, J. F., and Sivanpillai, R., eds. *Biological and environmental hazards, risks, and disasters*, Academic Press.

Adžić, K., Deranja, M., Pavlović, M., Tumbrinck, J., and Skejo, J. (2021). Endangered pygmy grasshoppers (Tetrigidae). In *Imperiled: Enyclopaedia of conservation*. Elsevier. https://doi.org/10.1016/B978-0-12-821139-7.00046-5

Agrawal, A. A., Petschenka, G., Bingham, R. A., Weber, M. G., and Rasmann, S. (2012). Toxic cardenolides: Chemical ecology and coevolution of specialized plant–herbivore interactions. *New Phytologist* 194:28–45.

Akino, T. (2008). Chemical strategies to deal with ants: A review of mimicry, camouflage, propaganda, and phytomimesis by ants (Hymenoptera: Formicidae) and other arthropods. *Myrmecological News* 11:173–181.

Alexander, G., and Hilliard, J. R. (1969). Altitudinal and seasonal distribution of Orthoptera in the Rocky Mountains of northern Colorado. *Ecological Monographs* 39:385–432. https://doi.org/10.2307/1942354

All, J., Groves, C., and Kambesis, P. (2005). *Ghost Cave, Eastern Himalayas, Bhutan*. Hellenic Speleological Society. 14th International Congress of Speleology.

Allegrucci, G., and Sbordoni, V. (2019). Insights into the molecular phylogeny of Rhaphidophoridae, an ancient, worldwide lineage of Orthoptera. *Molecular Phylogenetics and Evolution* 138:126-138. https://doi.org/10.1016/j.ympev.2019.05.032

Amédégnato, C. (1977). Étude des Acridoidea centre et sud américains (Catantopinae sensu lato) : Anatomie des genitalia, classification, répartition, phylogénie. PhD Thesis, Université Pierre et Marie Curie, Paris.

Archbold, R., Rand, A.L., and Brass, L. J. (1942). *Summary of the 1938–1939 New Guinea Expedition*. Results of the Archbold Expeditions. No. 41. *Bulletin of the American Museum of Natural History* 79(3):197–288.

Arnett Jr., R. H., Samuelson, G. A., and Nishida, G. M. (1993). *The insect and spider collections of the world. Flora and Fauna Handbook* 11.

Avdeyev, V. I. (2008). Stages of steppe landscapes formation in Eurasia: General aspects of the problem. (In Russian). *News of the Orenburg State Agrarian University*, pp. 38–42.

Barbolini, N., Woutersen, A., Dupont-Nivet, G., Silvestro, D., Tardif, D., Coster, P.M.C., Meijer, N., Chang, C., Zhang, H.-X., Licht, A., Rydin, C., Koutsodendris, A., Han, F., Rohrmann, A., Liu, X.-J., Zhang, Y., Donnadieu, Y., Fluteau, F., Ladant, J.-B., Hir, G. Le, and Hoorn, C. (2020). Cenozoic evolution of the steppe-desert biome in Central Asia. *Science Advances* 6(41):eabb8227.

Barrientos-Lozano, L. (2011). Dinámica poblacional, biología y ecología de la Langosta Suramericana (*Schistocerca cancellata*, Serville). Instituto Tecnológico de Ciudad Victoria, Tamaulipas, México. Pp. 93–135.

Barrientos-Lozano, L., Rocha-Sanchez, A. Y., Buzzetti, F. M., Méndez-Gomez, B. R., and Horta-Vega, J. V. (2013). *Saltamontes y esperanzas del noreste de México. (Insecta: Orthoptera). Guía ilustrada.* Miguel Ángel Porrúa.

Belovsky, G. E., and Slate, J. E. (2018). Grasshoppers affect grassland ecosystem functioning: Spatial and temporal variation. *Basic and Applied Ecology* 26:24–34.

Benediktov, A. (2005). Vibrational signals in the family Tetrigidae (Orthoptera). *Proceedings of the Russian Entomological Society* 76:131–140.

Bennet-Clark, H. C. (1970). The mechanism and efficiency of sound production in mole crickets. *Journal of Experimental Biology* 52:619–652.

Benton, T. (2012). *Grasshoppers and crickets.* The New Naturalists Library. Collins.

Bentos-Pereira, A. (2003). The tribe Tetanorhynchini, nov. (Orthoptera, Caelifera, Proscopiidae). *Journal of Orthoptera Research* 12:159–171.

– – – . (2006). The tribe Proscopiini, nov. (Orthoptera, Eumastacoidea, Proscopiidae). *Journal of Orthoptera Research* 15(2):143–148.

Bentos-Pereira, A., and Rowell, C.H.F. (1999). The genus *Proscopia* Klug, 1820 (Orthoptera, Caelifera, Eumastacoidea, Proscopiidae) in Central America, with description of a new species. *Revue Suisse de Zoologie* 106:627–641.

Bentos-Pereira, A., Cadena-Castañeda, O. J., and Cardona, J. M. (2015). Familia Proscopiidae. In *Introducción a los saltamontes de Colombia* (Orthoptera: Caelifera: Acrididea: Acridomorpha, Tetrigoidea & Tridactyloidea), pp. 95–141.

Béthoux, O. (2007). Cladotypic taxonomy applied: Titanopterans are orthopterans. *Arthropod Systematics and Phylogenetics* 65:135–156.

Béthoux, O., Nel, A., Lapeyrie, J., Gand, G., and Galtier, J. (2002). *Raphogla rubra* nov. gen. nov. sp., the oldest representative of the clade of modern Ensifera (Orthoptera: Tettigoniidea and Gryllidea). *European Journal of Entomology* 99:111–116.

Bhaskar, D., Easa, P.S., Sreejith, K. A., et al. (2019). Large-scale burning for a threatened ungulate in a biodiversity hotspot is detrimental for grasshoppers (Orthoptera: Caelifera). *Biodiversity and Conservation* 28:3221. https://doi.org/10.1007/s10531-019-01816-6

Bigelow, R. S. (1967). *The grasshoppers (Acrididae) of New Zealand.* University of Canterbury Publications.

Bland, R. G. (1991). Antennal and mouthpart sensilla of Tetrigidae (Orthoptera). *Annals of the Entomological Society of America* 84(2):195–200.

Blankers, T., and Shaw, K. L. (2024). The biogeographic and evolutionary processes shaping within and among population genetic variation. *Molecular Ecology* 33:e17444. https://doi.org/10.1111/mec.17444

Blumer, P., and Diemer, M. (1996). The occurrence and consequences of grasshopper herbivory in an alpine grassland, Swiss Central Alps. *Arctic and Alpine Research* 2:435–440.

Blunt, W., with the assistance of William T. Stearn (1971). *The complete naturalist: A life of Linnaeus.* Collins.

Bolívar, I. (1887). Excursión ortopterológica a Peñalara. *Actas de la Sociedad Española de Historia Natural* 16:4–9.

– – – . (1888). Excursión a San Ildefonso por Peñalara. *Actas de la Sociedad Española de Historia Natural* 17:63–66.

Brader, L., Djibo, H., Faye, F. G., Ghaout, S., Lazar, M., Luzietoso, P. N., and Ould Babah, M. A. (2006). Towards a more effective response to desert locusts and their impacts on food security, livelihoods and poverty. Food and Agriculture Organization of the United Nations.

Bridle, J. R., Saldamando, C. I., Koning, W., and Butlin, R. K. (2006). Assortative preferences and discrimination by females against hybrid male song in the grasshoppers *Chorthippus brunneus* and *Chorthippus jacobsi* (Orthoptera: Acrididae). *Journal of Evolutionary Biology* 19:1248–1256.

Brijlal, R., Rajak, A., and Armstrong, A. J. (2021). Aspects of the life history and ecology of two wingless grasshoppers, *Eremidium armstrongi* and *Eremidium browni* (Lentulidae), at the Doreen Clark Nature Reserve, KwaZulu-Natal, South Africa. *Journal of Orthoptera Research* 30:73–80.

Brongersma, L. D. (1978). Rijksmuseum van Geologie en Mineralogie 1878–1978: Past, present, and future. *Scripta Geologica* 48:37–96.

Buzzetti, F.M.M., Stancher, G., and Marangoni, F. (2021). Sixty years of work on Italy's Orthopteroids biodiversity, the big data of Galvagni collection. *Biodiversity Data Journal* 9:e65953. https://doi.org/10.3897/BDJ.9.e65953

Cade, W. (1975). Acoustically orienting parasitoids: Fly phonotaxis to cricket song. *Science* 190:1312–1313.

Cadena-Castañeda, O. J., and Barrantes, M.A.P. (2013). Estudios sobre el comportamiento de los Ortópteros Neotropicales. *Bioma* 14:53–64.

Cadena-Castañeda, O. J., and Cardona-Granda, J. M. (2015). *Introducción a los Saltamontes de Colombia (Orthoptera: Caelifera, Acridomorpha, Tetrigoidea & Tridactyloidea).* Lulu. s

Cadena-Castañeda, O. J., and Chamé-Vásquez, E. R. (2024). Studies on Eumastacoid grasshoppers: *Paralethus montanus* n. sp. (Orthoptera: Episactidae) from Chiapas, Mexico. *Zootaxa* 5410(1):67–78.

Cadena-Castañeda, O. J., Garay, A., Castañeda, M.D.A., Cardona-Granda, J. M., and García García, A. (2016). Systematics and phylogeny of the genus *Caenomastax* Hebard, 1923 (Orthoptera: Eumastacidae: Eumastacinae). *Zootaxa* 4117(2):241–264.

Camboué, R. P. (1886). Les sauterelles à Madagascar sur le riz malgache. *Bulletin mensuel de la Société nationale d'acclimatation de France* 33:168–172.

Carbonell, C. S. (1977). Origin, evolution, and distribution of the Neotropical acridomorph fauna (Orthoptera): A preliminary hypothesis. *Revista Sociedad Entomologica Argentina* 36:153–175.

Carbonell, C.S., Cigliano, M. M., and Lange, C .E. (2024). Acridomorph (Orthoptera) species from Argentina and Uruguay. Versión II, September 2024. https://biodar.unlp.edu.ar/acridomorph/

Cardona, J. M. (2012). *Grasshoppers of northwest South America: A photo guide. Volume 1: The western fauna.* Blurb.

– – – . (2015). *Grasshoppers of northwest South America: A photo guide. Volume 2: The eastern fauna.* Blurb.

– – – . (2020). *Grasshoppers of northwest South America: A photo guide. Volume 3: The Amazonian fauna.* Blurb.

Cardoso, P., Barton, P.S., Birkhofer, K., Chichorro, F., Deacon, C., Fartmann, T., Fukushima, C.S., Gaigher, R., Habel, J., Hallmann, C. A., Hill, M., Hochkirch, A., Kwak, M.L., Mammola, S., Noriega, J. A., Orfinger, A. B., Pedraza, F., Pryke, J. S., Roque, F. O., Settele, J., Simaika, J. P., Stork, N. E., Suhling, F., Vorster, C., and Samways, M. J. (2020). Scientists' warning to humanity on insect extinctions. *Biological Conservation* 242:108426. https://doi.org/10.1016/j.biocon.2020.108426

Carstens, B. C., and Knowles, L. L. (2007a). Estimating species phylogeny from gene-tree probabilities despite incomplete lineage sorting: An example from *Melanoplus* grasshoppers. *Systematic Biology* 56:400–411.

Carstens, B. C., and Knowles, L. L. (2007b). Shifting distributions and speciation: Species divergence during rapid climate change. *Molecular Ecology* 16:619–627.

Cassar, L. F., Ebejer, M. J., and Massa, B. (2020). Annotated checklist of Orthoptera of the Maltese Islands. *Zootaxa* 4885 (1):107–124.

Cease, A. J. (2024). How nutrients mediate the impacts of global change on locust outbreaks. *Annual Review of Entomology* 69:527–550.

Cease, A.J., Elser, J. J., Fenichel, E.P., Hadrich, J.C., Harrison, J.F., and Robinson. B.E. (2015). Living with locusts: Connecting soil nitrogen, locust outbreaks, livelihoods, and livestock markets. *BioScience* 65: 551–558.

Cease, A.J., Trumper, E. V., Medina, H., Bazán, F.C., Frana, J., Harrison, J., Joaquin, N., Learned, J., Roca, M., Rojas, J.E., Talal, S., and Overson, R.P. (2023). Field bands of marching locust juveniles show carbohydrate, not protein, limitation. *Current Research in Insect Science* 100069.

Centre for Overseas Pest Research (1982). *The locust and grasshoppers agricultural manual.* Centre for Overseas Pest Research.

Cerritos, R., and Cano-Santana, Z. (2008). Harvesting grasshoppers *Sphenarium purpurascens* in Mexico for human consumption. *Crop Protection* 27:473–480.

Chamot-Rooke, N., Rabaute, A., and Kreemer, C. (2005). Western Mediterranean Ridge mud belt correlates with active shear strain at the prism-backstop geological contact. *Geology* 33:861–864.

Channa, S. A., Sultana, R., and Wagan, M. S. (2013). Morphology and burrowing behaviour of *Schizodactylus minor* (Ander, 1938) (Grylloptera: Schizodactylidae: Orthoptera) of Pakistan. *Pakistan Journal of Zoology* 45.

Chapman, R. F., Page, W. W., and McCaffery, A. R. (1986). Bionomics of the variegated grasshopper (*Zonocerus variegatus*) in west and central Africa. *Annual Reviews of Entomology* 31:479–505.

Cheke, R. A. (1990). A migrant pest in the Sahel: The Senegalese grasshopper *Oedaleus senegalensis*. *Philosophical Transactions of the Royal Society of London (B)* 328:539–553.

Chopard, L. (1938). *Faune de de France: La biologie des Orthoptères.* P. Lechevalier.

Chopard, L. (1938). Les Orthoptères désertiques de l'Afrique du Nord. In P. Lechevalier, ed., *La vie dans la région désertique nord-tropicale de l'ancien monde. Mémoires de la Société de Biogéographie* 6:219–230.

Cigliano, M. M., and Lange, C. E. (1998). Orthoptera. In Morrone, J. J., and Coscarón, S., eds. *Biodiversidad de artrópodos Argentinos.* Ediciones Sur.

Cigliano, M. M., and Eades, D. (2010). New technologies challenge the future of taxonomy in Orthoptera. *Journal of Orthoptera Research* 19(1):15–18.

Cigliano, M. M., Pocco, M. E., and Lange, C. E (2014). Acridoideos (Orthoptera) de importancia agroeconómica. *Biodiversidad de Artrópodos Argentinos* vol. 3. Roig-Juñentt S., Claps, L. E., and Morrone, J. J., eds., INSUE.

Cigliano, M. M., Braun, H., Eades, D. C., and Otte, D. (2024). Orthoptera Species File [2024-11-27] http://Orthoptera.SpeciesFile.org

Clark, D. P. (1965). On the sexual maturation, breeding and oviposition behaviour of the Australian plague locust, *Chorthoicetes terminifera* (Walk.). *Australian Journal of Zoology* 13:17–45.

Clark, L. R. (1950). On the abundance of the Australian plague locust *Chortoicetes terminifera* (Walker) in relation to the presence of trees. *Australian Journal of Agricultural Research* 1:64–75.

Collins, R. O., and Burns, J. M. (2007). *A history of Sub-Saharan Africa.* Cambridge University Press.

Cressman, K., and Elliott, C. (2014). *The FAO commission for controlling the desert locust in South-West Asia: A celebration of 50 years.* Food and Agriculture Organization of the United Nations.

Darwin, C. (1862). *On the various contrivances by which British and foreign orchids are fertilised by insects, and on the good effect of intercrossing.* John Murray.

de Carvalho, T. N., and Shaw, K. L. (2010). Nuptial gifts enhance sperm transfer in the Hawaiian swordtail cricket, *Laupala cerasina. Animal Behavior* 79:819–826.

Descamps, M. (1979). Eumastacoidea néotropicaux: Diagnoses, signalisations, notes biologiques. *Annales de la Société Entomologique de France* 15(1):117–155.

Desutter-Grandcolas, L., Jacquelin, L., Hugel, S., Boistel, R., Garrouste, R., Henrotay, M., Warren, B. H, Chintauan-Marquier, I. C., Nel, P., Grandcolas, P., and Nel, A. (2017). 3-D imaging reveals four extraordinary cases of convergent evolution of acoustic communication in crickets and allies (Insecta). *Scientific Reports* 7:7099. https://doi.org/10.1038/s41598-017-06840-6

Deveson, E. D. (2011). The search for a solution to Australian locust outbreaks: How developments in ecology and government responses influenced scientific research. *Historical Records of Australian Science* 22:1.

Devriese, H., Nguyen, E., and Husemann, M. (2023). An identification key to the genera and species of Afrotropical Tetrigini (genera *Paratettix, Leptacridium, Hedotettix, Rectitettix* nov. gen., and *Alienitettix* nov. gen.) with general remarks on the taxonomy of Tetrigini (Orthoptera, Tetrigidae). *Zootaxa* 5285(3):511–556.

Dey, L. S. (2023). Phylogenetic and phylogenomic analyses and distribution modelling of a challenging taxon – the band-winged grasshoppers. Chapter 6: Same same but different, pp. 229–286. PhD thesis, University of Hamburg. https://ediss.sub.uni-hamburg.de/handle/ediss/10252

Dey, L. S., Simões, M. V., Hawlitschek, O., Sergeev, M. G., Xu, S. Q., Lkhagvasuren, D., and Husemann, M. (2021). Analysis of geographic centrality and genetic diversity in the declining grasshopper species *Bryodemella tuberculata* (Orthoptera: Oedipodinae). *Biodiversity and Conservation* 30:2773–2796.

Diwakar, S., and Balakrishnan, R. (2006). Male and female stridulation in an Indian weta (Orthoptera: Anostostomatidae). *Bioacoustics* 16(1):75–85.

Diwakar, S., and Balakrishnan, R. (2007). The assemblage of acoustically communicating crickets of a tropical evergreen forest in southern India: Call diversity and diel calling patterns. *Bioacoustics* 16(2):113–135.

Dowle, E. J., Trewick, S. A., and Morgan-Richards, M. (2024). Fossil calibrated phylogenies of southern cave wētā show dispersal and extinction confound biogeographic signal. *Royal Society Open Science* 11(2). https://doi.org/10.1098/rsos.231118

Eades, D. C. (2001). Version 2 of the Orthoptera Species File Online. *Journal of Orthoptera Research* 10(2):153–163.

Eichler, W. (1928). Lebensraum und Lebensgeschichte der Dahlemer Palmenhausheuschrecke *Phlugiola dahlemica* nov. spec. (Orthopt. Tettigoniid.). *Deutsche Entomologische Zeitschrift* 73:498–570.

Engelhardt, E.K., Biber, M.F., Dolek, M., Fartmann, T., Hochkirch, A., Leidinger, J., Löffler, F., Pinkert, S., Poniatowski, D., Voith, J., Winterholler, M., Zeuss, D., Bowler, D.E., and Hof, C. (2022). Consistent signals of a warming climate in occupancy changes of three insect taxa over 40 years in central Europe. *Global Change Biology* 28:3998–4012. https://doi.org/10.1111/gcb.16200.

Euw, J.W., Fishelson, L., Parsons, J.A., Reichstein, T., and Rothschild, M. (1967). Cardenolides (heart poisons) in a grasshopper feeding on milkweeds. *Nature* 214:35–39.

Exploring Africa 2025. https://exploringafrica.matrix.msu.edu/. Visited 2025.

Fabre, J.H. (1922). Série 6. IX Le Dectique á front blanc–Les mouers. In *Souvenirs entomologiques: Études sur l'instinct et les moeurs des insectes*. Librairie Delagrave.

Fartmann, T., Jedicke, E., Stuhldreher, G., and Streitberger, M. (2021a). *Insektensterben in Mitteleuropa: Ursachen und Gegenmaßnahmen*. Eugen Ulmer.

Fartmann, T., Poniatowski, D., and Holtmann, L. (2021b). Habitat availability and climate warming drive changes in the distribution of grassland grasshoppers. *Agriculture, Ecosystems and Environment* 320:107565. https://doi.org/10.1016/j.agee.2021.107565

Fartmann, T., Brüggeshemke, J., Poniatowski, D., and Löffler, F. (2022). Summer drought affects abundance of grassland grasshoppers differentially along an elevation gradient. *Ecological Entomology* 47:778–790. https://doi.org/10.1111/een.13168

Ferreira, J., Desutter-Grandcolas, L., Nel, A., Josse, H., and Campos, L. de. (2024). First 3-D reconstruction of the male genitalia of a Cretaceous fossil cricket: Diving into the evolutionary history of the Oecanthidae family (Orthoptera: Grylloidea) with the incorporation of new fossils in its phylogeny and a total-evidence dating approach. *Systematic Entomology* (in press).

Field, L.H., ed. (2001). *The biology of wētās, king crickets and their allies*. Cabi Publishing.

Finck, J., and Ronacher, B. (2017). Components of reproductive isolation between the closely related grasshopper species C*horthippus biguttulus* and *C. mollis. Behavioral Ecology and Sociobiology* 71:70. https://doi.org/10.1007/s00265-017-2295-3

Fisher, B.F., and Hugel, S. (2022). Edible terrestrial arthropod traditions and uses on Madagascar. In *The new natural history of Madagascar*, S.M. Goodman, ed. Princeton University Press. Pp. 218–230.

Fishpool, L.D.C., and Popov, G.B. (1984). The grasshopper faunas of the savannas of Mali, Niger, Benin and Togo. *Bulletin de l'Institut Fondamental d'Afrique Noire (Série A)* (1981) 43:275–410.

Fitton, M., and Harman, K. (2007). The 'Linnaean' insect collection. Pp. 47–58 in Gardiner, B., and Morris, M., eds., *The Linnaean Collections. The Linnean Special Issue* 7:1–84.

Fletcher, M.T., Lowe, L.M., Kitching, W., and König, W.A. (2000). Chemistry of Leichhardt's Grasshopper, *Petasida ephippigera*, and its host plants, *Pityrodia jamesii, P. ternifolia*, and *P. pungens. Journal of Chemical Ecology* 26(10):2275–2290. https://doi.org/10.1023/A:1005518625764

Fontana, P., Tollis, P., Buzzetti, F.M., and Vigna Taglianti, A. (2001). The Orthopteroid insects of the Abruzzo National Park (Central Apennine, South Italy): A preliminary checklist. (Insecta: Blattodea, Mantodea, Orthoptera, Phasmatodea, Dermaptera). Poster abstract. *Metaleptea*, Special Meeting Issue, 45.

Fontana, P., Buzzetti, F.M., Cogo, A., and Odé, B. (2002). *Guida al riconoscimento e allo studio di cavallette, grilli, mantidi e insetti affini del Veneto. Guide Natura/1*, Museo Naturalistico Archeologico di Vicenza.

Fontana, P., García García, P.L., and Buzzetti, F.M. (2007). Listado preliminar de los Ortópteros de México. *Entomologia Mexicana* 6(2):1337–1342.

Fontana, P., Buzzetti, F.M., and Mariño-Pérez, R. (2008). *Chapulines, langostas, grillos y esperanzas de México: Guía fotográfica*. (Grasshoppers, locusts, crickets and katydids of Mexico: Photographic guide.) WBA Handbooks, 1.

Fontana, P., Buzzetti, F.M., Mariño-Pérez, R., Castellanos-Vargas, I., Monge-Rodríguez, S., and Cano-Santana, Z. (2017). *Ortópteros de Oaxaca*. (Orthopterans of Oaxaca). WBA Project, *WBA Monographs* 8.

Food and Agriculture Organization. 2024. Frequently asked questions about locusts. The Food and Agriculture Organization of the United Nations. https://www.fao.org/locusts/faqs/en/

Fournel, J., Micheneau, C., and Baider, C. (2015). A new critically endangered species of Angraecum (Orchidaceae) endemic to the island of Mauritius, Indian Ocean. *Phytotaxa* 222:211–220.

Gangwere, S.K., and Llorente, V. (1992). Distribution and habits of the Orthoptera (sens. lat.) of the Balearic Islands (Spain). *Eos* 68:51–87.

Gassó, E., Stork, L., Weber, A., Ameryan, M., and Wolstencroft, K. (2020). *Natuurkundige Commissie Archives Online*. Brill. https://doi.org/10.1163/97890043 36865

Gastón, J. (1969). *Sintesis historica de la langosta en la Argentina*. Secretaría de Estado de Agricultura y Ganadería.

Gerhardt, H.C., and Huber, F. (2002). *Acoustic communication in insects and anurans: Common problems and diverse solutions*. University of Chicago Press.

Glaw, F., and Glaw, T. (2023). Die Riesenblattschrecke Siliquofera grandis (Blanchard, 1853). [2023-05-22] https://freunde-zsm.de/2023/04/objekt-des-monats-april-2023-die-riesenblattschrecke-siliquofera-grandis-blanchard-1853/

– – –. (2024). Die Neuguinea-Riesenblattschrecke *Siliquofera grandis. Reptilia* 165:44–48.

Golding, F.D. (1948). The Acrididae (Orthoptera) of Nigeria. *Transactions of the Royal Entomological Society of London* 99:517–587.

Gottsberger, B., and Mayer, F. (2019). Dominance effects strengthen premating hybridization barriers between sympatric species of grasshoppers (Acrididae, Orthoptera). *Journal of Evolutionary Biology* 32(9): 921–930. https://doi.org/10.1111/jeb.13490

Gorochov, A. V. (2007). Notes on taxonomy of the subfamily Hexacentrinae with description of some taxa (Orthoptera: Tettigoniidae). *Zoosystematica Rossica* 16(2):209–214.

Greathead, D. J. (1962). A review of the insect enemies of Acridoidea (Orthoptera). *Transactions of the Royal Entomological Society of London* 114:437–517.

Greeff, M., Caspers, M., Kalkman, V., Willemse, L., Sunderland, B. D., Bánki, O., and Hogeweg, L. (2022). Sharing taxonomic expertise between natural history collections using image recognition. *Research Ideas and Outcomes* 8:e79187. https://doi.org/10.3897/rio.8.e79187

Grzywacz, B., Warchałowska-Śliwa, E., Kociński, M., Heller, K.-G., and Hemp, C. (2021). Diversification of the balloon bushcrickets (Orthoptera, Hexacentrinae, Aerotegmina) in the East African mountains. *Scientific Reports* 11:9878. https://doi.org/10.1038/S41598-021-89364-4

Günther, K. K. (1992). Revision der Familie Cylindrachetidae Giglio-Tos, 1914 (Orthoptera, Tridactyloidea). *Deutche Entomologische Zeitschrift* 39(4/5):233–291.

Gwynne, D. T. (1997). The evolution of edible 'sperm sacs' and other forms of courtship feeding in crickets, katydids and their kin (Orthoptera: Ensifera). In *The evolution of mating systems in insects and arachnids*, edited by Choe, J., and Crespie, B. J. Cambridge University Press.

– – – . (2001). Katydids and bush-crickets: Reproductive behavior and the evolution of the Tettigoniidae. Cornell University Press.

Harris, R., McQuillan, P., and Hughes, L. (2012). Patterns in body size and melanism along a latitudinal cline in the wingless grasshopper, *Phaulacridium vittatum*: Geographic variation in size and melanism. *Journal of Biogeography* 39(8):1450–1461.

Harris, R.M.B., McQuillan, P., and Hughes, L. (2013). Experimental manipulation of melanism demonstrates the plasticity of preferred temperature in an agricultural pest (*Phaulacridium vittatum*). *PLoS ONE* 8(11):e80243.

Hawlitschek, O., Morinière, J., Lehmann, G. U., et al. (2016). DNA barcoding of crickets, katydids and grasshoppers (Orthoptera) from Central Europe with focus on Austria, Germany and Switzerland. *Molecular Ecology Resources* 17:1037–1053. https://doi.org/10.1111/1755-0998.12638

Hawlitschek, O., Sadílek, D., Dey, L. S., Buchholz, K., Noori, S., Baez, I. L., Wehrt, T., Brozio, J., Trávníček, P., Seidel, M., and Husemann, M. (2023). New estimates of genome size in Orthoptera and their evolutionary implications. *PLoS ONE* 18:e0275551. https://doi.org/10.1371/journal.pone.0275551

Heads, S. W. (2009). New pygmy grasshoppers in Miocene amber from the Dominican Republic (Orthoptera: Tetrigidae). *Denisia* 26:69–74.

– – – . (2010). The first fossil spider cricket (Orthoptera: Gryllidae: Phalangopsinae): 20 million years of troglobiomorphosis or exaptation in the dark? *Zoological Journal of the Linnean Society* 158:56–65. https://doi.org/10.1111/j.1096-3642.2009.00587.x

Heads, S. W., and Wang, Y. (2013). First fossil record of *Melanoplus differentialis* (Orthoptera: Acrididae: Melanoplinae). *Entomological News* 123:33–37. https://doi.org/10.3157/021.123.0108

Hebard, M. (1923). Studies in the Dermaptera and Orthoptera of Colombia. Third paper. Orthopterous family Acrididae. *Transactions of the American Entomological Society* 49:165–313.

Hebert, P. D., Cywinska, A., Ball, S. L., and deWaard, J. R. (2003a). Biological identifications through DNA barcodes. *Proceedings of the Royal Society of London Series B: Biological Sciences* 270:313–321. https://doi.org/10.1098/rspb.2002.2218

Hebert, P.D.N., Ratnasingham, S., and de Waard, J. R. (2003b). Barcoding animal life: Cytochrome c oxidase subunit 1 divergences among closely related species. *Proceedings of the Royal Society of London Series B: Biological Sciences*. https://doi.org/10.1098/rsbl.2003.0025

Hegg, D., Morgan-Richards, M., and Trewick, S. A. (2022). High alpine sorcerers: Revision of the cave wētā genus *Pharmacus* Pictet and de Saussure (Orthoptera: Rhaphidophoridae: Macropathinae), with the description of six new species and three new sub-species. *European Journal of Taxonomy* 808:1–58. https://doi.org/10.5852/ejt.2022.808.1721

Heller, K.-G., and Hemp, C. (2018). Extremely divergent song types in the genus *Aerotegmina* Hemp (Orthoptera: Tettigoniidae: Hexacentrinae) and the description of a new species from the Eastern Arc Mountains of Tanzania (East Africa). *Bioacoustics* 28(3). https://doi.org/10.1080/09524622.2018.1443284

Heller, K.-G., Ostrowski, T. D., and Hemp, C. (2010). Singing and hearing in *Aerotegmina kilimandjarica* (Tettigoniidae: Hexacentrinae), a species with unusual low carrier frequency of the calling song. *Bioacoustics* 19(3):195–210.

Helversen, O. von, and Elsner, N. (1977). The stridulatory movements of acrid grasshoppers recorded with an opto-electronic device. *Journal of Comparative Physiology* 122:53–64.

Helversen, D. von, and Helversen, O. von (1975). Verhaltensgenetische Untersuchungen am akustischen Kommunikationssystem der Feldheuschrecken (Orthoptera, Acrididae). *Journal of Comparative Physiology* 104:273–323.

– – – . (1994). Forces driving coevolution of song and song recognition in grasshoppers. In Schildberger, K., and Elsner, N., eds., *Neural basis of behavioural adaptations. Fortschritte der Zoologie* 39:253–284.

Hemp, C. (2001). *Aerotegmina*, a new genus of African Listroscelidinae (Orthoptera: Tettigoniidae, Listroscelidinae, Hexacentrini). *Journal of Orthoptera Research* 10(1):125–132.

– – – . (2006). *Aerotegmina shengenae*, a new species of Listroscelidinae (Orthoptera: Tettigoniidae) from the Eastern Arc Mountains of East Africa. *Journal of Orthoptera Research* 15(1):99–103.

– – – . (2010). Cloud forests in East Africa as evolutionary motors for speciation processes of flightless *Saltatoria* species. In *Mountains in the mist: Science for conservation and management of tropical montane cloud forests*, edited by Bruijnzeel, L. A., Scatena, F. N., and Hamilton, L. S. University of Hawaii Press.

– – – . (2016). The Eastern Arc Mountains and coastal forests of East Africa – An archive to understand large-scale biogeographical patterns: *Pseudotomias*, a new genus of African Pseudophyllinae (Orthoptera: Tettigoniidae). *Zootaxa* 4126(4):480–490.

Hemp, C., Schultz, O., Hemp, A., and Wägele, W. (2007). New Lentulidae species from East Africa (Orthoptera: Saltatoria). *Journal of Orthoptera Research* 16:85–96.

Hemp, C., Heller, K.-G., Warchalowska-Sliwa, E., and Hemp, A. (2013). The genus *Aerotegmina* (Orthoptera, Tettigoniidae, Hexacentrinae): Morphological relations, phylogeographical patterns and the description of a new species. *Organisms, Diversity and Evolution* 13(4):521–530. https://doi.org/10.1007/s1317-013-0133-7.

Hemp, C., Scherer, C., Brandl, R., and Pinkert, S. (2020). The origin of the endemic African grasshopper family Lentulidae (Orthoptera: Acridoidea) and its climate-induced diversification. *Journal of Biogeography* 47:1805–1815.

Henneberry, T. J. (2008). Federal entomology: Beginnings and organizational entities in the United States Department of Agriculture, 1854–2006, with selected research highlights. *Agricultural Information Bulletins* 309846, United States Department of Agriculture, Economic Research Service.

Hill, J. G. (2018). The grasshopper fauna of southeastern grasslands: A preliminary investigation. In Hill, J. G., and Barone, J. A., eds., *Southeastern Grasslands: Biodiversity, Ecology, and Management.* University of Alabama Press.

Hochkirch, A., Deppermann, J., and Gröning, J. (2006). Visual communication behaviour as a mechanism behind reproductive interference in three pygmy grasshoppers (genus *Tetrix*, Tetrigidae, Orthoptera). *Journal of Insect Behavior* 19:559–571.

Hochkirch, A., Witzenberger K., Teerling A., and Niemeyer F. (2007). Translocation of an endangered insect species, the field cricket (*Gryllus campestris* Linnaeus, 1758) in northern Germany. *Biodiversity and Conservation* 16:3597–3607.

Hochkirch, A., Nieto, A., García Criado, M., Cálix, M., Braud, Y., Buzzetti, F. M., Chobanov, D., Odé, B., Presa Asensio, J. J., Willemse, L., Zuna-Kratky, T., Barranco Vega, P., Bushell, M., Clemente, M. E., Correas, J. R., Dusoulier, F., Ferreira, S., Fontana, P., García, M. D., Heller, K.-G., Iorgu, I. S., Ivkovic, S., Kati, V., Kleukers, R., Krištín, A., Lemonnier-Darcemont, M., Lemos, P., Massa, B., Monnerat, C., Papapavlou, K. P., Prunier, F., Pushkar, T., Roesti, C., Rutschmann, F., Sirin, D., Skejo, J., Szövényi, G., Tzirkalli, E., Vedenina, V., Barat Domenech, J., Barros, F., Cordero Tapia, P. J., Defaut, B., Fartmann, T., Gomboc, S., Gutiérrez-Rodríguez, J., Holuša, J., Illich, I., Karjalainen, S., Kočárek, P., Korsunovskaya, O., Liana, A., López, H., Morin, D., Olmo-Vidal, J. M., Puskás, G., Savitsky, V., Stalling, T., and Tumbrinck, J. (2016). *European Red List of Grasshoppers, Crickets and Bush-crickets*. Publications Office of the European Union.

Hochkirch, A., Casino, A., Penev, L., Allen, D., Tilley, L., Geirgiev, T., Gospodinov, K., and Barov, B. (2022). European Red List of Insect Taxonomists. Publications Office of the European Union. https://data.europa.eu/doi/10.2779/364246

Horsák, M., Chytrý, M., Hájková, P., Hájek, M., Danihelka, J., Horsáková, V., Ermakov, N., German, D. A., Kočí, M., Lustyk. P., Nekola, J. C., Preislerová, Z., and Valachovič, M. (2015). European glacial relict snails and plants: Environmental context of their modern refugial occurrence in southern Siberia. *Boreas* 44:638–657.

Houston, T. F. (2007). Observation of the biology and immature stages of the sandgroper *Cylindraustralia kochii* (Saussure), with notes on some congeners (Orthoptera: Cylindrachetidae). *Records of the Western Australian Museum* 23:219–234.

Hsu, P.-W., Hugel, S., Wetterer, J. K., Tseng, S.-P., Ooi, C.-S.M., Lee, C.-Y., and Yang, C.-C.S. (2020). Ant crickets (Orthoptera: Myrmecophilidae) associated with the invasive yellow crazy ant *Anoplolepis gracilipes* (Hymenoptera: Formicidae): Evidence for cryptic species and potential co-introduction with hosts. *Myrmecological News* 30:103–129. https://doi.org/10.25849/myrmecol.news_030:103

Hugel, S. (2022). Orthoptera, Caelifera and Ensifera, grasshoppers, katydids, crickets, *valala*, *angely*. In *The new natural history of Madagascar*, edited by Goodman, S. M. Princeton University Press.

Hugel, S., Micheneau, C., Fournel, J., Warren, B. H., Gauvin-Bialecki, A., Pailler, T., Chase, M. W., and Strasberg, D. (2010). Glomeremus species from the Mascarene Islands (Orthoptera, Gryllacrididae) with the description of the pollinator of an endemic orchid from the island of Réunion. *Zootaxa* 2545:58–68.

Hunter, D., and Cosenzo, E. L. (1990). The origin of plagues and recent outbreaks of the South American locust, *Schistocerca cancellata* (Orthoptera: Acrididae) in Argentina. *Bulletin of Entomological Research* 80:295–300.

Hurka, H., Friesen, N., Bernhardt, K.-G., Neuffer, B., Smirnov, S. V., Shmakov, A. I., and Blattner, F. G. (2019). The Eurasian steppe belt: Status quo, origin and evolutionary history. *Turczaninowia* 22(3):5-71.

Husemann, M., Dey, L.-S., Sadílek, D., Ueshima, N., Hawlitschek, O., Song, H., and Weissman, D. B. (2022). Evolution of chromosome number in grasshoppers (Orthoptera: Caelifera: Acrididae). *Organisms Diversity and Evolution* 22:649–657. https://doi.org/10.1007/s13127-022-00543-1

Ingrisch, S. (2002). Orthoptera from Bhutan, Nepal, and North India in the Natural History Museum Basel. *Entomologia Brasiliensia*, 24:123–159.

Ingrisch, S., and Köhler, G. (1998). *Die Heuschrecken Mitteleuropas*. Westarp Wissenschaften.

International Union for Conservation of Nature (2017). SSC Species Conservation Planning Sub-Committee (2017). *Guidelines for species conservation planning*. Version 1.0. International Union for Conservation of Nature.

Iorgu, I. S., Iorgu, E. I., Stalling, T., Puskás, G., Chobanov, D., Szövényi, G., Moscaliuc, L. A., Motoc, R., Tăuşan, I., and Fusu, L. (2021). Ant crickets and their secrets: *Myrmecophilus acervorum* is not always parthenogenetic (Insecta: Orthoptera: Myrmecophilidae). *Zoological Journal of the Linnean Society* 20:1–18.

Iorio, C., Scherini, R., Fontana, P., Buzzetti, F. M., Kleukers, R., Odé, B., and Massa, B. (2019). *Grasshoppers and crickets of Italy: A photographic field guide to all the species*. WBA Handbooks.

Jago, N. D. (1990). The genera of the Central and South American grasshopper family Proscopiidae (Orthoptera: Acridomorpha). *Eos: Revista española de Entomología* 65(1):249–307.

Jiménez-Mejias, P., Fernández-Mazuecos, M., Amat, M. E., and Vargas, P. (2015). Narrow endemics in European mountains: High genetic diversity within the monospecific genus *Pseudomisopates* (Plantaginaceae) despite isolation since the late Pleistocene. *Journal of Biogeography* 42(8):1455–1468. https://doi.org/10.1111/jbi.12507

Johns, P. M. (1997). The Gondwanaland wētā: Family Anostostomatidae (formerly in Stenopelmatidae, Henicidae or Mimnermidae): Nomenclatural problems, world checklist, new genera and species. *Journal of Orthoptera Research* 6:125–138. https://doi.org/10.2307/3503546

Jongema, Y. (2017). *World list of edible insects*. Wageningen University.

Joyce, R. J. V. (1952). The ecology of grasshoppers in East Central Sudan. *Anti-Locust Bulletin* 11:1–103.

Karsch, F. (1890). Neue westafrikanische, durch Herrn Premierleutnant Morgen von Kribi eingesendete Orthopteren. *Entomologische Nachrichten* 17/18:257–276.

Kasalo, N., Fisher, N. J., Creek, E., and Connors, M. (2023). *Tepperotettix reliqua* (Orthoptera: Tetrigidae), a lonely Papuan relict in Australia. *Australian Zoologist* 43(1):67–78.

Kasalo, N., Skejo, J., and Husemann, M. (2023). DNA barcoding of pygmy hoppers: The first comprehensive overview of the BOLD systems' data shows promise for species identification. *Diversity* 15(6):696.

Kataoka, H., Koita, N., Kondo, N. I., Ito, H. C., Nakajima, M., Momose, K., . . . and Teraoka, H. (2022). Metabarcoding of feces and intestinal contents to determine carnivorous diets in red-crowned cranes in eastern Hokkaido, Japan. *Journal of Veterinary Medical Science* 84(3):358–367.

Kearney, M. R. (2018). The matchstick grasshopper genus *Warramaba* (Morabidae: Morabinae): A description of four new species and a photographic guide to the group. *Zootaxa* 4482(2):201–244. https://doi.org/10.11646/zootaxa.4482.2.1

Kekeunou, S., and Tamesse, J. L. (2016). Consumption of the variegated grasshopper in Africa: Importance and threats. *Journal of Insects as Food and Feed* 2(3):213–222.

Khattar, N. (1972). A description of the adult and the nymphal stages of *Schizodactylus monstrosus* (Drury) (Orthoptera). *Journal of Natural History* 6:589–600.

Khatua, S., Dhara, S., Sahu, R., and Ghorai, S. K., (2020). Burrowing adaptations and anthropogenic threats of a dune cricket, *Schizodactylus monstrosus* (Drury): A case study from Champa riverbank, West Bengal, India. *International Journal of Experimental Research and Review* 23:43–51.

Kleukers R., Ode, B., and Willemse, L. (2018). First record of the Atlantic beach-cricket *Pseudomogoplistes vicentae* on the Spanish mainland. *Articulata* 33:131–134.

Knowles, L. L. (2000). Tests of Pleistocene speciation in montane grasshoppers from the sky islands of western North America (genus *Melanoplus*). *Evolution* 54:1337–1348.

Knowles, L. L., and Carstens, B. C. (2007a.) Delimiting species without monophyletic gene trees. *Systematic Biology* 56:887–895.

– – – . (2007b). Estimating a geographically explicit model of population divergence. *Evolution* 61:477493.

Knowles, L. L., Huang, J. P., Wood, P. L., Mariño-Pérez, R., Hill, J. G., and Sanabria-Urbán, S. (2023). Phylogenetic tests of diversification models: Did repeated colonization or in situ divergence build-up an endemic Mexican fauna of grasshoppers? (In review).

König, S., Krauss, J., Keller, A., Bofinger, L., and Steffan-Dewenter, I. (2022). Phylogenetic relatedness of food plants reveals highest insect herbivore specialisation at intermediate temperatures along a broad climatic gradient. *Global Change Biology* 28:4027–4040.

Koot, E.M., Morgan-Richards, M., and Trewick, S.A. (2022). Climate change and alpine adapted insects: Modelling environmental envelopes of a grasshopper radiation. *Royal Society Open Science* 9:211596.

Korsunovskaya, O., Berezin, M., Heller, K.G., Tkacheva, E., Kompantseva, T., and Zhantiev R. (2020). Biology, sounds and vibratory signals of hooded katydids (Orthoptera: Tettigoniidae: Phyllophorinae). *Zootaxa* 4852:309–322.

Krenn, H.W., Fournel, J., Bauder, J.A., and Hugel, S. (2016). Mouthparts and nectar feeding of the flower visiting cricket *Glomeremus orchidophilus* (Gryllacrididae). *Arthropod Struct Dev* 45:221–229.

Kuřavová, K., Sipos, J., Wahab, R.A., Kahar, R.S., and Kočarek, P. (2017). Feeding patterns in tropical groundhoppers (Tetrigidae): A case of phylogenetic dietary conservatism in a basal group of Caelifera. *Zoological Journal of the Linnean Society* 179(2):291–302.

Le Gall, M., Overson, R., and Cease, A.J. (2019). A global review on locusts (Orthoptera: Acrididae) and their interactions with livestock grazing practices. *Frontiers in Ecology and Evolution* 7:1–24.

Lecoq, M. (1978). Le problème sauteriaux en Afrique soudano-sahélienne. *Agronomie Tropicale* 33:241–258. https://www.researchgate.net/publication/235409847

– – – . (1988). Les criquets du Sahel [Sahelian grasshoppers]. Collection Acridologie Opérationnelle, n°1. Comité Inter-Etats de Lutte contre la Sécheresse dans le Sahel, Département de Formation en Protection des végétaux (Niamey). http://locust.cirad.fr/ouvrages_pratiques/pdf/DFPV1.pdf

– – – . (1992). Une structure de population originale chez *Poekilocerus bufonius hieroglyphicus* (Klug, 1832) dans le Tamesna nigérien, en saison sèche (Orth., Pyrgomorphidae). *Bulletin de la Société entomologique de France* 97:55–60.

– – – . (2022). *Schistocerca gregaria* (desert locust). Crop Protection Compendium. CAB International. https://doi.org/10.1079/cabicompendium.49833

– – – .(2022). *Oedaleus senegalensis* (Senegalese grasshopper). Crop Protection Compendium. CAB International. https://doi.org/10.1079/cabicompendium.37101

– – – . (2023). *Locusta migratoria* (migratory locust). Crop Protection Compendium. CAB International.

Lecoq, M., and Bazelet, C.S. (2019). Red Locust *Nomadacris septemfasciata* (Serville, 1838) (Acrididae). In Lecoq, M., and Zhang, L., eds., *Encyclopedia of pest Orthoptera of the world*. China Agricultural University Press.

Lecoq, M., and Cease, A.J. (2022). What have we learned after millennia of locust invasions? *Agronomy* 12:1–10.

Lecoq, M., and Zhang, L., eds. (2019). *Encyclopedia of pest Orthoptera of the world*. China Agricultural University Press.

Lee, C., and McPherson, S. (2023). *Nature's tricks: How animals and plants use disguises and deception*. Don Hanson Charitable Foundation.

Lee, J., Famoso, N.A., and Lin, A. (2024). Microtomography of an enigmatic fossil egg clutch from the Oligocene John Day Formation, Oregon, USA, reveals an exquisitely preserved 29-million-year-old fossil grasshopper oothecal. Park Stewardship Forum, University of California.

Leguével de Lacombe, B.F., and de Froberville, E. (1840). *Voyage à Madagascar et aux îles Comores (1823 à 1830)*. Desessart.

Lehmann, G.U.C. (2012). Weighing costs and benefits of mating in bushcrickets (Insecta: Orthoptera: Tettigoniidae), with an emphasis on nuptial gifts, protandry and mate density. *Frontiers in Zoology* 9:19.

Lehmann, G.U.C., and Lehmann, A.W. (2000a). Spermatophore characteristics in bushcrickets vary with parasitism and remating interval. *Behavioral Ecology and Sociobiology* 47:393–399.

– – – . (2000b). Female bushcrickets mated with parasitized males show rapid remating and reduced fecundity (Orthoptera: Phaneropteridae: *Poecilimon mariannae*). *Naturwissenschaften* 87:404–407.

– – – . (2006). Potential lifetime reproductive success of male bushcrickets parasitized by a phonotactic fly. *Animal Behaviour* 71:1103–1110.

– – – . (2007). Sex differences in "time out" from reproductive activity and sexual selection in male bushcrickets (Orthoptera: Zaprochilinae: *Kawanaphila mirla*). *Journal of Insect Behavior* 20:215–227.

– – – .(2008). Bushcricket song as a clue for spermatophore size? *Behavioral Ecology and Sociobiology* 62:569–578.

– – – .(2016). Material benefit of mating: The bushcricket spermatophylax as a fast uptake nuptial gift. *Animal Behaviour* 112:267–271.

Lehmann, G.U.C., Lehmann, K., Neumann, B., Lehmann, A.W., Scheler, C., and Jungblut, P.R. (2018). Protein analysis of the spermatophore reveals diverse compositions in both the ampulla and the spermatophylax in a bushcricket. *Physiological Entomology* 43:1–9.

Lewis, S.M., Vahed, K., Koene, J.M., Engqvist, L., Bussiere, L.F., Perry, J.C., Gwynne, D., and Lehmann, G.U.C. (2014). Emerging issues in the evolution of animal nuptial gifts. *Biology Letters* 10:20140336.

Ley, W. (1951). *Dragons in amber: Further adventures of a romantic naturalist*. Sidgwick and Jackson.

Li, R., Ying, X., Deng, W., Rong, W., and Li, X. (2021). Mitochondrial genomes of eight Scelimeninae species (Orthoptera) and their phylogenetic implications within Tetrigoidea. *PeerJ* 9:e10523.

Liana, A. (1980). Matériaux pour la connaissance des Proscopiidae (Orthoptera). *Mitteilungen aus dem Hamburgischen Zoologischen Museum und Institut* 77:229–260.

Linnaeus, C. (1758). Systema naturæ per regna tria naturæ, secundum classes, ordines, genera, species, cum characteribus, differentiis, synonymis, locis. Tomus I. Editio decima, reformata. Pp. [1–4], 1–824. Holmiæ. (Salvius).

Llorente del Moral, V., and Presa Asensio, J. J. (1997). *Los pamphagidae de la península ibérica: (Insecta, Orthoptera, Caelífera)*. Universidad de Murcia.

Lockwood, J. A. (2004). *Locust: The devastating rise and mysterious disappearance of the insect that shaped the American frontier*. Basic Books.

Lockwood, J. A., and DeBrey, L. D. (1990). A solution for the sudden and unexplained extinction of the Rocky Mountain locust, *Melanoplus spretus* (Walsh). *Environmental Entomology* 19:1194–1205.

Lockwood, J. A., Thompson, C. D., DeBrey, L. D., Love, C. M., Nunamaker, R .A., and Pfadt, R. E. (1991). Preserved grasshopper fauna of Knife Point Glacier, Fremont County, Wyoming, U.S.A. *Arctic and Alpine Research* 23:108–114.

Lockwood, J. A., Guzzo, A. M., and Carlisle, A. H. (2020). Librettos, sopranos, and science: Communicating ecology through opera. *Bulletin of the Ecological Society of America* July: 1–7 (e01730). https://doi.org/10.1002/bes2.1730

Löffler, F., Poniatowski, D., and Fartmann, T. (2019). Orthoptera community shifts in response to land-use and climate change: Lessons from a long-term study across different grassland habitats. *Biological Conservation* 236:315–323. https://doi.org/10.1016/j.biocon.2019.05.058

Loeffler-Henry, K., Kang, C., Dawson, J. W., and Sherratt, T. N. (2023). Mimicry in motion: A grasshopper species that looks like, and moves like, a sympatric butterfly. *bioRxiv* 2023.06.04.543626. https://doi.org/10.1101/2023.06.04.543626

Loureiro, J., Correia, P., Nel, A., and de Jesus, A. (2010). *Lusitaneura covensis* n. gen., n. sp., first Caloneurodea from the Carboniferous of Portugal (Insecta: Pterygota: Panorthoptera). *Annales de la Société entomologique de France* 46:242–246. https://doi.org/10.1080/00379271.2010.10697664

Louveaux, A., Amédégnato, C., Poulain, S., and Desutter-Grandcolas, L. Orthoptères Acridomorpha de l'Afrique du Nord-ouest. Version [24 February 2024]. http://acrinwafrica.mnhn.fr/

MacArthur, R. H., and Wilson, E. O. (1967). *The theory of island biogeography*. Princeton University Press.

Magor, J. I., Lecoq, M., and Hunter, D. M. (2008). Preventive control and desert locust plagues. *Crop Protection* 27:1527–1533.

Mantle, B., LaSalle, J., and Fisher, N. (2012). Whole-drawer imaging for digital management and curation of a large entomological collection. *ZooKeys* 209:147–163. https://doi.org/10.3897/zookeys.209.3169

Mařan, J. (1958). Eine neue Art der Gattung Isophya Br. W. aus der Tschechoslowakei, Orthoptera: Tettigoniidae. *Acta Entomologica Musei Nationalis Pragae* 32(523):5–24.

Mariño-Pérez, R., and Song, H. 2023. Keys to Pyrgomorphidae genera of the world. *Miscellaneous Publications of the Museum of Zoology, University of Michigan* 212:1–185.

Marshall, D. C., and Hill, K.B.R. (2009). Versatile aggressive mimicry of cicadas by an Australian predatory katydid. *PLoS ONE* 4(1):e4185. https://doi.org/10.1371/journal.pone.0004185

Matenaar, D., Bröder, L., Bazelet, C. S., and Hochkirch, A. (2014). Persisting in a windy habitat: Population ecology and behavioral adaptations of two endemic grasshopper species in the Cape region (South Africa). *Journal of Insect Conservation* 18:447–456.

Matenaar D., Bröder, L., and Hochkirch, A. (2016). A preliminary phylogeny of the South African Lentulidae. *Hereditas* 153:1–8.

Matenaar, D., Fingerle, M., Heym, E., Wirtz, S., and Hochkirch, A. (2018). Phylogeography of the endemic grasshopper genus *Betiscoides* (Lentulidae) in the South African Cape Floristic Region. *Molecular Phylogenetics and Evolution* 118:318–329.

Mason, S. C.,Jr., Betancourt, I. S., and Gelhaus, J. K. (2020). Importance of building a digital species index (spin-dex) for entomology collections: A case study, results and recommendations. *Biodiversity Data Journal* 8: e58310. https://doi.org/10.3897/BDJ.8.e58310

Massa, B., ed. (1995). Arthropoda di Lampedusa, Linosa e Pantelleria. *Il Naturalista Siciliano* 19.

Massa, B. (2013). Pamphagidae (Orthoptera: Caelifera) of North Africa: Key to genera and the annotated check-list of species. *Zootaxa* 3700 (3):435–475.

– – –. (2016). On some interesting African Katydids (Orthoptera Tettigoniidae). *Entomologia* 4:303.

Massa, B., and Fontana, P. (2020). Endemism in Italian Orthoptera. *Biodiversity Journal* 11(2):405–434.

Massa, B., Fontana, P., Buzzetti, F.M., Kleukers, R., and Odè B. (2012). Orthoptera. *Fauna d'Italia* XLVIII. Calderini.

Massa, B., Cusimano, C. A., Fontana, P., and Brizio, C. (2022). New unexpected species of *Acheta* (Orthoptera, Gryllidae) from the Italian volcanic island of Pantelleria. *Diversity* 14:802. https://doi.org/10.3390/d14100802

Matt, S., Flook, P.K., and Rowell, C.H.F. (2008). A partial molecular phylogeny of the Eumastacoidea s. lat. (Orthoptera, Caelifera). *Journal of Orthoptera Research* 17(1):43–55.

McCartney, J., Potter, M. A., Robertson, A. W., Telscher, K., Lehmann, G.U.C., Lehmann, A. W., Helversen, D. von, Reinhold, K., Achmann, R., and Heller, K.-G. (2008). Understanding nuptial gift size in bush-crickets: An analysis of the genus *Poecilimon* (Tettigoniidae; Orthoptera). *Journal of Orthoptera Research* 17:231–242.

Medina, H., Cease, A., and Trumper, E. (2017). The resurgence of the South American locust (*Schistocerca cancellata*). *Metaleptea* 37:17–21.

Mendelson, T.C., and Shaw, K. L. (2002). Genetic and behavioral components of the cryptic species boundary between *Laupala cerasina* and

L. kohalensis (Orthoptera: Gryllidae). *Genetica* 116:301–310.

Mendelson, T. C., and Shaw, K. L. (2005). Rapid speciation in an arthropod. *Nature* 433:375–376.

Mestre, J., and Chiffaud J. (2023). *Les acridiens d'Afrique occidentale et nord-cenbtrale,* [4] + 770 p., 360 Figs, 479 cartes.

Mey, E. (1994). In memoriam Wolfdietrich Eichler (1912–1994). *Rudolstädter naturhistorische Schriften* 6:107–108.

Meza-Joya, F. L., Morgan-Richards, M., Koot, E. M., and Trewick, S. A. (2023). Global warming leads to habitat loss and genetic erosion of alpine biodiversity. *Journal of Biogeography* 50:961–975.

Meza-Joya, F. L., Morgan-Richards, M., and Trewick, S. A. (2024). Phenotypic and genetic divergence in a cold-adapted grasshopper may lead to lineage-specific responses to rapid climate change. *Diversity and Distributions* 30:e13848.

Michelsen, A. (1998). Biophysics of sound localization in insects. In Hoy, R. R., Popper, A. N., and Fay, R. R., eds., *Comparative hearing: Insects*. Springer.

Micheneau, C., Fournel, J., Warren, B. H., Hugel, S., Gauvin-Bialecki, A., Pailler, T., Strasberg, D., and Chase, M. W. (2010). Orthoptera, a new order of pollinator. *Annals of Botany* 105:355–364.

Modder, W.W.D. (1983). Diurnal variation in feeding and gut activity in nymphs of the African pest grasshopper. *Zonocerus variegatus. Insect Science and Its Application* 5:527–531.

Morgan-Richards, M., Marshall, C. J., Biggs, P. J., and Trewick, S. A. (2023). Insect freeze-tolerance downunder: The microbial connection. *Insects* 14:89.

Mugleston, J. D., Naegle, M., Song, H., and Whiting, M. F. (2018). A comprehensive phylogeny of Tettigoniidae (Orthoptera: Ensifera) reveals extensive ecomorph convergence and widespread taxonomic incongruence. *Insect Systematics and Diversity* 2:1–25. https://doi.org/10.1093/isd/ixy010

Mullen, S. P., Mendelson, T. C., Schal, C.and Shaw, K. L. (2007). Rapid evolution of cuticular hydrocarbons in a species radiation of acoustically diverse Hawaiian crickets (Gryllidae: Trigonidiinae: *Laupala*). *Evolution* 61:223–231.

Nickle, D. A. (2002). New species of katydids (Orthoptera: Tettigoniidae) of the Neotropical genera *Arachnoscelis* (Listroscelidinae) and *Phlugiola* (Meconematinae), with taxonomic notes. *Journal of Orthoptera Research* 11:125–133.

Nocke, H. (1972). Physiological aspects of sound communication in crickets (*Gryllus campestris* L.). *Journal of Comparative Physiology* 80:141–162.

Nuhlíčková, S., Svetlík, J., Kaňuch, P., Krištín, A., and Jarčuška, B. (2023). Movement patterns of the endemic flightless bush-cricket, *Isophya beybienkoi. Journal of Insect Conservation* 28:141–150. https://doi.org/10.1007/s10841-023-00529-0

Ogan, S., Paulus, C., Froehlich, C., Renker, C., Kolwelter, C., Schendzielorz, M., Danielczak, A., Müller, K., Eulering, H., and Hochkirch, A. (2022). Re-surveys reveal biotic homogenization of Orthoptera assemblages as a consequence of environmental change. *Diversity and Distribution* 28:1795–1809.

Oliver, T. H., and Morecroft, M. D. (2014). Interactions between climate change and land use change on biodiversity: Attribution problems, risks, and opportunities. *WIREs Climate Change* 5:317–335.

Ortego, J., and Knowles, L. L. (2022). Geographic isolation versus dispersal: Relictual alpine grasshoppers support a model of interglacial diversification with limited hybridization. *Molecular Ecology* 31(1):296–312.

Ortis, G., Marini, L., Cavaletto G., and Mazzon, L. (2023). Increasing temperatures affect multiyear life cycle of the outbreak bush-cricket *Barbitistes vicetinus* (Orthoptera, Tettigoniidae). *Insect Science* 30:530–538. https://doi.org/10.1111/1744-7917.13094

Otte D. (1970). A comparative study of communicative behavior in grasshoppers. *Miscellaneous Publications of the Museum of Zoology, University of Michigan* 141:1–167.

– – – . (1981). *The North American grasshoppers*. Vol. 1. Acrididae: Gomphocerinae y Acridinae. Harvard University Press.

– – – . (1984). *The North American grasshoppers*. Vol. 2. Acrididae: Oedipodinae. Harvard University Press.

– – – . (1994). *The crickets of Hawaii: Origin, systematics and evolution*. Orthopterists' Society.

– – – . (2002). Studies of *Melanoplus* 1. Review of the Viridipes Group (Acrididae: Melanoplinae). *Journal of Orthoptera Research* 11:31–118.

– – – . (2012). Eighty new species of *Melanoplus* from the United States. *Transactions of the American Entomological Society* 138:73–167.

– – – . (2024). Keys to southern African Lentulidae (Orthoptera: Acridoidea) and revision of *Eremidium*. *Transactions of the American Entomological Society* 150:22–158.

Paranjape, S. Y., and Bhalerao, A. M. (1994). Distribution and etho-ecology of grouse locusts of certain localities in Maharashtra with a note on the status of tetrigid taxonomy. *Records of the Zoological Survey of India*, 351–366.

Pfeiffer, I. (1861). *Reise nach Madagaskar: Nebst einer Biographie der Verfasserin, nach ihren eigenen Aufzeichnungen,* Volume 1. Gerold.

Pocco, M. E., Scattolini, M. C., Lange, C. E., and Cigliano, M. M. (2014). Taxonomic delimitation in color polymorphic species of the South American grasshopper genus *Diponthus* Stål (Orthoptera, Romaleidae, Romaleini). *Insect Systematics and Evolution* 45(4):303–350.

Pocco, M. E., Minutolo, C., Dinghi, P. A., et al. (2015). Species delimitation in the Andean grasshopper genus *Orotettix* Ronderos and Carbonell (Orthoptera: Melanoplinae): An integrative approach combining morphological, molecular and biogeographical data. *Zoological Journal of the Linnean Society* 174:733–759. https://doi.org/10.1111/zoj.12251

Pocco, M. E., Guzmán, N., Plischuk, S., Confalonieri, V., Lange, C. E., and Cigliano, M. M. (2018). Diversification patterns of the grasshopper genus *Zoniopoda* Stål (Romaleidae, Acridoidea, Orthoptera) in open vegetation biomes of South America. *Systematic Entomology* 43(2):290–307. https://doi.org/10.1111/syen.12277

Pocco, M. E., Lange, C. E., and Cigliano, M. M. (2023). Relationships and taxonomy of the genus *Diponthus* Stål (Orthoptera: Acridoidea: Romaleidae). *Zootaxa* 5336(1):033–081.

Poniatowski, D., and Fartmann, T. (2011a). Weather-driven changes in population density determine wing dimorphism in a bush-cricket species. *Agriculture, Ecosystems and Environment* 145:5–9. https://doi.org/10.1016/j.agee.2010.10.006

Poniatowski, D., and Fartmann, T. (2011b). Dispersal capability in a habitat specialist bush-cricket: The role of population density and habitat moisture. *Ecological Entomology* 36:717–723. https://doi.org/10.1111/j.1365-2311.2011.01320.x

Poniatowski, D., Heinze, S., and Fartmann, T. (2012). The role of macropters during range expansion of a wing-dimorphic insect species. *Evolutionary Ecology* 26:759–770. https://doi.org/10.1007/s10682-011-9534-2

Poniatowski, D., Münsch, T., Helbing, F., and Fartmann, T. (2018). Arealveränderungen mitteleuropäischer Heuschrecken als Folge des Klimawandels. *Natur und Landschaft* 93(12):553–561.

Poniatowski, D., Beckmann, C., Löffler, F., Münsch, T., Helbing, F., Samways, M. J., and Fartmann, T. (2020). Relative impacts of land-use and climate change on grasshopper range shifts have changed over time. *Global Ecology and Biogeography* 29:2190–2202. https://doi.org/110.1111/geb.13188

Popov, G. B. (1985). Ecological studies on oviposition by swarms of the desert locust (*Schistocerca gregaria* Forskål) in eastern Africa. *Anti-Locust Bull.* 31:1–72.

– – – . (1988). Sahelian grasshoppers. *ODNRI Bulletin* 5. http://gala.gre.ac.uk/11048

– – – . (1989). *Nymphs of the Sahelian grasshoppers: An illustrated guide.* Overseas Development Natural Resources Institute. https://doi.org/10.1017/S0007485300046046 (abstract)

Prange, H. D., and Pinshow, B. (1994). Thermoregulation of an unusual grasshopper in a desert environment: The importance of food source and body size. *Journal of Thermal Biology* 19(1):75–78. https://doi.org/10.1016/0306-4565(94)90011-6

Ragge, D. R., and Reynolds, W. J. (1998) . *The songs of the grasshoppers and crickets of Western Europe.* Harley Books.

Rentz, D. (1996). *Grasshopper Country.* University of New South Wales Press.

Rentz, D.C.F. (1991). Orthoptera. In Naumann, I. D., Carne, P. B., Lawrence, J. F., Nielson, E. S., Spradbery, J. P., Taylor, R. W., Whitten, M. J., and Littlejohn, M. J., eds. *The Insects of Australia Volume I*, 2nd ed. Melbourne University Press.

– – – . (2010). *A guide to the katydids of Australia.* CSIRO Publishing.

Rentz, D.C.F., and Monteith, G. B. (2022). A review of the Australian endemic genus *Cooloola* Rentz, 1980, with description of seven new species and summary of biology (Orthoptera: Anostostomatidae: Anostostomatinae: Cooloolini). *Australian Entomologist* 49(3):209–253.

Rentz, D.C.F., Lewis, R. C., Su, Y. N., and Upton, M. S. (2003). *A Guide to Australian Grasshoppers and Locusts.* Natural History Publications (Borneo).

Richards, A. M. (1964). Insects of Campbell Island. Orthoptera: Rhaphidophoridae of Auckland and Campbell Islands. *Pacific Insects Monographs*: 216–225.

Ries, M. W., Adriaansen, C., Aldobai, S., Berry, K., Bal, A. B., Catenaccio, M. C., Cigliano, M. M., Cullen, D. A., Deveson, E. D., Diongue, A., Foquet, B., Hadrich, J., Hunter, D., Johnson, D. L., Karnatz, J. P., Lange, C. E., Lawton, D., Lazar, M., Latchininsky, A. V., Lecoq, M., Gall, M. L., Lockwood, J., Manneh, B., Overson, R., Peterson, B. F., Piou, C., Poot-Pech, M. A., Robinson, B. E., Rogers, S. M., Song, H., Springate, S., Therville, C., Trumper, E., Waters, C., Woller, D. A., Youngblood, J. P., Zhang, L., and Cease, A. (2024). Global perspectives and transdisciplinary opportunities for locust and grasshopper pest management and research. *Journal of Orthoptera Research* 33:169–216.

Roberts, H. R. (1941). A comparative study of the subfamilies of the Acrididae (Orthoptera) primarily on the basis of their phallic structures. *Proceedings of the Academy of Natural Sciences, Philadelphia* 93:201–246.

Römer, H. (1998). The sensory ecology of acoustic communication in insects. In Hoy, R. R., Popper, A. N., and Fay, R. R., eds., *Comparative hearing: Insects.* Springer.

Rowell, C.H.F. (2013). *The grasshoppers (Caelifera) of Costa Rica and Panama.* The Orthopterists' Society.

Rowell, C.H.F., Hemp, C., and Harvey, A. W. (2015). Jago's Grasshoppers of East and North East Africa, Vol. 1. Blurb.

Rühr, P. T., Edel, C., Frenzel, M., and Blanke, A. (2024). A bite force database of 654 insect species. *Scientific Data* 11(58):1–7. https://doi.org/10.1038/s41597-023-02731-w

Rust J. (1998). Biostratinomie von Insekten aus der Fur-Formation von Dänemark (Moler, oberes Paleozän/unteres Eozän). *Paläontologische Zeitschrift* 72:41–58.

Sahagún, B. (1577). *Florentine Codex.* Mexico. Disponible en Códice Florentino Digital/Digital Florentine Codex, editado por Kim N. Richter y Alicia Maria Houtrouw, "Libro 11: Cosas terrenales," fol. ir, Getty Research Institute, 2023. https://florentinecodex.getty.edu/es/book/11/folio/ir Consultado el 22 mayo 2024

Sanabria-Urbán, S., Mariño-Pérez, R., and Song, H. (2019). Corn grasshopper *Sphenarium purpurascens* Charpentier (Orthoptera: Pyrgomorphidae). In *Encyclope-*

dia of Pest Orthoptera of the World. China Agricultural University Press.

Schirmel, J., Gerlach, R., and Buhk, C. (2019). Disentangling the role of management, vegetation structure and plant quality for Orthoptera in lowland meadows. *Insect Science* 26:366–378.

Schirmel, J., Entling, M. H., and Eckerter, P. (2020). Eichenschrecken als Hauptbeute des neozoischen Stahlblauen Grillenjägers *Isodontia mexicana* (Hymenoptera: Sphecidae) in der südpfälzischen Agrarlandschaft. *Articulata* 35:149–160.

Schubnel, T., Legendre, F., Roques, P., Garrouste, R., Cornette, R., Perreau, M., Perreau, N., Desutter-Grandcolas, L., and Nel, A. (2021). Sound vs. light: Wing-based communication in Carboniferous insects. *Communications Biology* 4:794. https://doi.org/10.1038/s42003-021-02281-0

Schultz O., Hemp, C., Hemp, A., and Wägele, W. (2007). Molecular phylogeny of the endemic East African flightless grasshoppers *Altiusambilla* Jago, *Usambilla* (Sjöstedt) and *Rhainopomma* Jago (Orthoptera: Acridoidea: Lentulidae). *Systematic Entomology* 32:712–719.

Sevastianov, N., Neretina, T., and Vedenina, V. (2023). Evolution of calling songs in the grasshopper subfamily Gomphocerinae (Orthoptera, Acrididae). *Zoologica Scripta* 52:154–175.

Shaw, K. L. (2000). Further acoustic diversity in Hawaiian forests: Two new species of Hawaiian cricket (Orthoptera: Gryllidae: Trigonidiinae: *Laupala*). *Zoological Journal of the Linnean Society* 129:73–91.

Shin, S., Baker, A. J., Enk, J., McKenna, D., Foquet, B., Vandergast, A. G., Weissman, D. B., and Song, H. (2024). Orthoptera-specific target enrichment (OR-TE) probes resolve relationships over broad phylogenetic scales. *Scientific Reports* 14:21377. https://doi.org/10.1038/s41598-024-72622-6

Showler, A. T., and Lecoq, M. (2021). Incidence and ramifications of armed conflict in countries with major desert locust breeding areas. *Agronomy* 11:114.

Simon, S., Letsch, H., Bank, S., Buckley, T. R., Donath, A., Liu, S., Machida, R., Meusemann, K., Misof, B., Podsiadlowski, L., Zhou, X., Wipfler, B., and Bradler, S. (2019). Old World and New World Phasmatodea: Phylogenomics resolve the evolutionary history of stick and leaf insects. *Frontiers in Ecology and Evolution* 7. https://doi.org/10.3389/fevo.2019.00345

Sivyer, L., Morgan-Richards, M., Koot, E., and Trewick, S. A. (2018). Anthropogenic cause of range shifts and gene flow between two grasshopper species revealed by environmental modelling, geometric morphometrics and population genetics. *Insect Conservation and Diversity* 11(5):415–434.

Skejo, J., Pushkar, T. I., Kasalo, N., Pavlović, M., Deranja, M., Adžić, K., et al. (2022). Spiky pygmy devils: Revision of the genus *Discotettix* (Orthoptera: Tetrigidae) and synonymy of Discotettiginae with Scelimeninae. *Zootaxa* 5217(1):1–64.

Skejo, J., Kasalo, N., Fontana, P., Ivković, S., Škorput, J., Buzzetti, F. M., Gomboc, S., Scherini, R., Ćato, S., Rebrina, F., Heller, K.-G., Adžić, K., Deranja, M., and Tvrtković, N., (2023). *Dinarippiger* gen. nov. (Tettigoniidae: Bradyporinae: Ephippigerini), a new saddle bush-cricket genus for *Ephippiger discoidalis* Fieber, 1853 from the Dinaric Alps. *Zootaxa* 5271 (1):049–090. https://doi.org/10.11646/zootaxa.5271.1.2

Sjöstedt, Y. (1909). 17. Orthoptera. 6. Locustodea: 125–148, 7. Acridiodea: 149–200. In Sjöstedt, Y., ed., *Wissenschaftliche Ergebnisse der Schwedischen Zoologischen Expedition nach dem Kilimanjaro, dem Meru und den umgebenden Massaissteppen Deutsch-Ostafrikas 1905–1906*.

Song, H., Amédégnato, C., Cigliano, M. M., Desutter-Grandcolas, L., Heads, S., Huang, Y., Otte, D., and Whiting, M. F. (2015). 300 million years of diversification: Elucidating the patterns of orthopteran evolution based on comprehensive taxon and gene sampling. *Cladistics* 31(6):1–31.

Song, H., Foquet, B., Mariño-Pérez, R., and Woller, D. A. (2017). Phylogeny of locusts and grasshoppers reveals complex evolution of density-dependent phenotypic plasticity. *Scientific Reports* 7 :1–13. https://doi.org/10.1038/s41598-017-07105-y

Song, H., Mariño-Pérez, R., Woller, D. A., and Cigliano, M. M. (2018). Evolution, diversification, and biogeography of grasshoppers (Orthoptera: Acrididae). *Insect Systematics and Diversity* 2(4):3, 1–25. https://doi.org/10.1093/isd/ixy008

Spira, T. P. (2011). *Wildflowers and plant communites of the southern Appalachian Mountains and Peidmont: A naturalist's guide to the Carolinas, Virgina, and Tennessee and Georgia*. University of North Carolina Press.

Stalling, T. (2024). *Myrmecophilus jordanicus*, a new species of ant cricket from Jordan and Israel, with notes on the synonymy of *Myrmecophilus nigricornis* (Orthoptera: Myrmecophilidae). *Israel Journal of Entomology* 53:25–32.

Stalling, T., and Birrer, S. (2013). Identification of the ant-loving crickets, *Myrmecophilus* Berthold, 1827 (Orthoptera: Myrmecophilidae), in Central Europe and the northern Mediterranean Basin. *Articulata* 28(1/2):1–11.

Stamps, G. F., and Shaw, K. L. (2019). Male use of chemical signals in sex discrimination of *Laupala pruna*. *Animal Behavior* 156:111–120.

Steedman, A. (1990). *Locust handbook*. 3rd ed. Natural Resources Institute.

Stoll, C. (1787). *Représentation exactement colorée d'après nature des Spectres ou Phasmes, des Mantes, des Sauterelles, des Grillons, des Criquets et des Blattes. Qui se trouvent dans les quatre parties du monde. l'Europe, l'Asie, l'Afrique et l'Amérique; Rassemblées et Décrites par Caspar Stoll*. Pages 1–56, plates 1–18, figures 1–68.

– – – . (1813). *Représentation exactement colorée d'après nature des Spectres ou Phasmes, des Mantes, des Sauterelles, des Grillons, des Criquets et des Blattes. Qui se trouvent dans les quatre parties du monde. l'Europe, l'Asie, l'Afrique et l'Amérique; Rassemblées et Décrites par Caspar Stoll.* Pages 57–74, plates 19–25, figures 69–100.

Sultana, R., and Wagan, M. S. (2015). Grasshoppers and locusts of Pakistan. Higher Education Commission-Pakistan.

Sword, G. A., Lecoq, M., and Simpson, S. J. (2010). Phase polyphenism and preventative locust management. *Journal of Insect Physiology* 56:949–957.

Tan, M. K. (2012). *Orthoptera in the Bukit Timah and Central Catchment Nature Reserves (Part 1): Suborder Caelifera.* Raffles Museum of Biodiversity Research, National University Singapore, Singapore. Uploaded 4 May 2012.

– – – . (2017). *Orthoptera in the Bukit Timah and Central Catchment Nature Reserves (Part 2): Suborder Ensifera.* 2nd ed. Lee Kong Chian Natural History Museum, National University of Singapore, Singapore. Uploaded 16 June 2017.

Tan, M. K., and Robillard, T. (2012). Two new cricket species (Orthoptera: Gryllidae and Mogoplistidae) from the mangrove areas of Singapore. *Raffles Bulletin of Zoology* 60(2):411–420.

Tan, M. K., and Robillard, T. (2021). Population divergence in the acoustic properties of crickets during the COVID-19 pandemic. *Ecology (The Scientific Naturalist)* 102(7):e03323. https://doi.org/10.1002/ecy.3323

Tan, M. K., Artchawakom, T., Wahab, R. A., Lee, C.-Y., Belabut, D. M., and Tan, H.T.W. (2017). Overlooked flower visiting Orthoptera in Southeast Asia. *Journal of Orthoptera Research* 26(2)143–153. https://doi.org/10.3897/jor.26.15021

Tan, M. K., Leem, C.J.M., and Tan, H.T.W. (2017). High floral resource density leads to neural constraint in the generalist, floriphilic katydid, *Phaneroptera brevis* (Orthoptera: Phaneropterinae). *Ecological Entomology* 42(5):535–544. https://doi.org/10.1111/een.12414

Tan, M. K., Montealegre-Z., F., Wahab, R. A., Lee, C.-Y., Belabut, D. M., Japir, R., and Chung, A.Y.C. (2019). Ultrasonic songs and stridulum anatomy of *Asiophlugis* crystal predatory katydids (Tettigonioidea: Meconematinae: Phlugidini). *Bioacoustics* 29(6):619–637. https://doi.org/10.1080/09524622.2019.1637783

Tatamic, N. J., Umbers, K.D.L., and Song, H. (2013). Molecular phylogeny of the *Kosciuscola* grasshoppers endemic to the Australian alpine and montane regions. *Invertebrate Systematics* 27:307–316. https:/ /doi.org/10.1071 /IS 12072

Taylor-Smith, B. L., Morgan-Richards, M., and Trewick, S. A. (2013). New Zealand ground wētā (Anostostomatidae: *Hemiandrus*): Descriptions of two species with notes on their biology. *New Zealand Journal Zoology* 40:314–329.

Taylor-Smith, B. L., Trewick, S. A., and Morgan-Richards, M. (2016). Three new ground wētā and a redescription of *Hemiandrus maculifrons. New Zealand of Zoology* 43:363–383.

Therville, C., Anderies, J. M., Lecoq, M., and Cease, A. (2021). Locusts and people: Integrating the social sciences in sustainable locust management. *Agronomy* 11:951.

Think Africa 2025. https://thinkafrica.net/. Visited 2025.

Thorn, S., König, S., Fischer-Leipold, O., Gombert, J., Griese, J., and Thein, J. (2022). Temperature preferences drive additive biotic homogenization of Orthoptera assemblages. *Biology Letters* 18:20220055. https://doi.org/10.1098/rsbl.2022.0055

Tomar, M., and Diwakar, S. (2020). Investigating host plant association, calling activity, and sexual dimorphism in Indian *Gryllacropsis* sp. (Orthoptera: Anostostomatidae). *Ecology and Evolution* 10(21):11850–11860.

Tonzo, V., and Ortego, J. (2021). Glacial connectivity and current population fragmentation in sky islands explain the contemporary distribution of genomic variation in two narrow-endemic montane grasshoppers from a biodiversity hotspot. *Diversity and Distributions* 27(9):1619–1633. https://doi.org/10.1111/ddi.13306

Trewick, S. A. (2021). A new species of large *Hemiandrus* ground wētā (Orthoptera: Anostostomatidae) from North Island, New Zealand. *Zootaxa* 4942(2).

Trewick, S. A., and Morgan-Richards, M. (2019). *Wild Life New Zealand.* Hand-in-Hand Press. https://sites.massey.ac.nz/wildlifenz/

Trewick, S. A., and Morgan-Richards, M. (2000). Artificial wētā roosts: A technique for ecological study and population monitoring of tree wētā (*Hemideina*) and other invertebrates. *New Zealand Journal of Ecology* 24: 201–208.

Trewick, S. A., Taylor-Smith, B. L., and Morgan-Richards, M. (2020). Ecology and systematics of the wine wētā and allied species, with description of four new *Hemiandrus* species. *New Zealand Journal of Zoology* 48:47–80.

Trewick, S., Hegg, D., Morgan-Richards, M., Murray, T., Watts, C., Johns, P., and Michel, P. (2022). *Conservation status of Orthoptera (wētā, crickets and grasshoppers) in Aotearoa New Zealand.* New Zealand Threat Classification Series 39. Department of Conservation, Wellington.

Trewick, S. A., Koot, E. M., and Morgan-Richards, M. (2023). Māwhitiwhiti Aotearoa: Phylogeny and synonymy of the silent alpine grasshopper radiation of New Zealand (Orthoptera: Acrididae). *Zootaxa* 5383:225–241.

Trewick, S. A., Taylor-Smith, B. L., and Morgan-Richards, M. (2024). Wētā Aotearoa: Polyphyly of the New Zealand Anostostomatidae (Insecta: Orthoptera). *Insects* 15:787.

Trumper, E. V., Cease, A. J., Cigliano, M. M., Bazán, F. C., Lange, C. E., Medina, H. E., Overson, R. P., Therville, C., Pocco, M. E., Piou, C., Zagaglia, G., and Hunter, D. (2022). A review of the biology, ecology, and management of the South American locust, *Schisto-cerca cancellata* (Serville, 1838), and future prospects. *Agronomy* 12:1–20.

Tumbrinck, J. (2014). Taxonomic revision of the Cladonotinae (Orthoptera: Tetrigidae) from the islands of South-East Asia and from Australia, with general remarks to the classification and morphology of the Tetrigidae and descriptions of new genera and species from New Guinea and New Caledonia. In Telnov, D., ed., *Biodiversity, biogeography and nature conservation in Wallacea and New Guinea*, Vol. 2. Entomological Society of Latvia.

Turnell, B. R., and Shaw, K. L. (2015). Polyandry and postcopulatory sexual selection in a wild population. *Molecular Ecology* 24:6278–6288.

Umbers, K.D.L., Tatarnic, N. J., Holwell, G. I., et al. (2013). Bright turquoise as an intraspecific signal in the chameleon grasshopper (*Kosciuscola tristis*). *Behavioral Ecology and Sociobiology* 67:439–447. https://doi.org/10.1007/s00265-012-1464-7

Umbers, K.D.L., Slatyer, R. A., Tatamic, N. J., Muschett, G. R., Wang, S., and Song, H. (2022). Phylogenetics of the skyhoppers (*Kosciuscola*) of the Australian Alps: Evolutionary and conservation implications. *Pacific Conservation Biology* 28:298–299.

Uvarov, B. (1977). *Grasshoppers and Locusts*, Vol. 1. Cambridge University Press.

Uvarov, B. P. (1921. A revision of the genus *Locusta*, L.(= *Pachytylus*, Fieb.), with a new theory as to the periodicity and migrations of locusts. *Bulletin of Entomological Research* 12:135–163.

– – – . (1936). New Orthoptera from Cyprus. *Annals and Magazine of Natural History* 18:505–515.

– – – . (1949). On the insect fauna of Cyprus. Results of the expedition of 1939 by Harald, Hakan, and Lindberg, P. H. IV. Tettigoniidae and Acrididae. *Societas Scientiarum Fennica Commentationes Biologicae* 10:1–6.

Uvarov, B. P., and Volkonsky, M. A. (1939). Notes on a desert grasshopper with digging habits, *Eremogryllus hammadae* Krauss, 1902 (Orthoptera, Acrididae). *Proceedings of the Royal Entomological Society of London A* 14:19–23.

Vahed, K. (1998). The function of nuptial feeding in insects: A review of empirical studies. *Biological Reviews* 73(1):43–78.

– – – . (2019). The life cycle of the Atlantic beach-cricket, *Pseudomogoplistes vicentae* Gorochov, 1996. *Journal of Insect Conservation* 24:473–485.

Vahed, K., and Bourgaize, T. (2020). Overwintered adult male scaly crickets, *Pseudomogoplistes vicentae* Gorochov, found in Guernsey in March/April. *British Journal of Entomology and Natural History* 33:35–40.

Van Itterbeeck, J., Rakotomalala Andrianavalona, I. N., Rajemison, F.I., Rakotondrasoa, J. F., Ralantoarinaivo, V. R., et al. (2019). Diversity and use of edible grasshoppers, locusts, crickets, and katydids (Orthoptera)

in Madagascar. *Foods* 8:666. https://doi.org/10.3390/foods8120666

Vedenina, V. Y. (2015). Courtship song analysis in two hybrid zones between sibling species of the *Chorthippus albomarginatus* group (Orthoptera, Gomphocerinae). *Entomological Review* 95:166–180.

Vedenina, V. Y., and Helversen, O. von (2003). Complex courtship in a bimodal grasshopper hybrid zone. *Behavioural Ecology and Sociobiology* 54:44–54.

Vedenina, V., and Mugue, N. (2011). Speciation in gomphocerine grasshoppers: Molecular phylogeny versus bioacoustics and courtship behavior. *Journal of Orthoptera Research* 20:109–125. https://www.jstor.org/stable/23034228

Vedenina, V. Y., Panyutin, A. K., and Helversen, O. von (2007). The unusual inheritance pattern of the courtship songs in closely related grasshopper species of the *Chorthippus albomarginatus*-group (Orthoptera: Gomphocerinae). *Journal of Evolutionary Biology* 20:260–277.

Vedenina, V., Fähsing, S., Sradnick, J., Klöpfel, A., and Elsner, N. (2013). A narrow hybrid zone between the grasshoppers *Stenobothrus clavatus* and *S. rubicundus* (Orthoptera: Gomphocerinae): Female preferences for courtship songs. *Biological Journal of the Linnean Society* 108:834–843. https://doi.org/10.1111/bij.12005

Voigt, C. C., Lehmann, G.U.C., Michener, R. H., and Joachimski, M. M. (2006). Nuptial feeding is reflected in tissue nitrogen isotope ratios of female katydids. *Functional Ecology* 20:656–661.

Voigt, C. C., Kretzschmar, A. S., Speakman, J. R., and Lehmann, G.U.C. (2008). Female bushcrickets fuel their metabolism with nuptial gifts. *Biology Letters* 4:476–478.

Waloff, Z. (1960). The fluctuating distribution of the desert locust in relation to the strategy of control. Report of the 7th Commonwealth Entomological Conference, London, July 1960.

– – – . (1976). Some temporal characteristics of desert locust plagues. *Anti-Locust Memoir* 13. Anti-Locust Research Center.

Whitman, D. W. (1990). Grasshopper chemical communication. In *Biology of Grasshoppers*, edited by Chapman, R. F., and Joern, A. John Wiley and Sons.

Willemse, F. (1966). List of new taxa of Orthoptera described by C. Willemse. *Publicaties van het Natuurhistorisch Genootschap in Limburg* 16:31–42.

– – – . (1971). The genus *Oropodisma* Uvarov, 1942, with the description of two new species (Orthoptera, Acrididae, Catantopinae). *Publicaties van het Natuurhistorisch Genootschap in Limburg* 20:19–25.

Willemse, F., and Willemse, L. (2008). An annotated checklist of the Orthoptera-Saltatoria from Greece including an updated bibliography. *Articulata–Beiheft* 13:1–91.

Willemse, L., Kleukers, R., and Odé, B. (2018). *The grasshoppers of Greece*. Naturalis Biodiversity Center.

Wim, C., Prakash, A., Müller, A., and Lazutkaite, E. (2023). Insecticide use against desert locust in the Horn of Africa 2019–2021 reveals a pressing need for change. *Agronomy* 13:819.

Wipfler, B., Letsch, H., Frandsen, P. B., Kaplie, P., Mayer, C., Bartel, D., Buckley, T. R., Donath, A., Edgerly-Rooks, J. S., Fujita, M., Liu, S., Machida, R., Mashimo, Y., Misof, B., Niehuis, O., Peters, R. S., Petersen, M., Podsiadlowski, L., Schütte, K., Shimizu, S., Uchifune, T., Wilbrandt, J., Yan, E., Zhou, X., and Simon, S. 2019. Evolutionary history of Polyneoptera and its implications for our understanding of early winged insects. *Proc. Nat. Acad. Sci. U.S.A.* 116: 3024–3029. https://doi.org/10.1073/pnas.1817794116

Witzenberger, K. A., and Hochkirch. A. (2008). Genetic consequences of animal translocations: A case study using the field cricket, *Gryllus campestris L. Biological Conservation* 141:3059–3068.

Woller, D. A. (2017). Xerophillic flightless grasshoppers (Orthoptera: Acrididae: Melanoplinae: *Melanoplus*: The Puer Group) of the southeastern U.S.A.: An evolutionary history. Dissertation submitted to Texas A & M University.

Woller D. A., Fontana, P., Mariño-Pérez, R., and Song, H. (2014). Studies in Mexican grasshoppers: *Liladownsia fraile*, a new genus and species of Dactylotini (Acrididae: Melanoplinae) and an updated molecular phylogeny of Melanoplinae. *Zootaxa* 3793(4):475–495. https://doi.org/10.11646/zootaxa.3793.4.6

Wright, A. M., Bapst, D. W., Barido-Sottani, J., and Warnock, R.C.M. (2022). Integrating fossil observations into phylogenetics using the fossilized birth–death model. *Annual Review of Ecology, Evolution, and Systematics* 53:251–273. https://doi.org/10.1146/annurev-ecolsys-102220-030855

Wright, D. E. (1987). Analysis of the development of major plagues of the Australian plague locust *Chortoicetes terminifera* (Walker) using a simulation model. *Australian Journal of Ecology* 12:423–437.

Würmli, M. (1973). Ergebnisse der Bhutan-Expedition 1972 des Naturhistorischen Museums in Basel. Orthoptera: Gryllacridoidea. *Verh. Naturf. Ges. Basel* 83:337–347.

Yadav, S., Stow, A. J., and Dudaniec, R. Y. (2019). Detection of environmental and morphological adaptation despite high landscape genetic connectivity in a pest grasshopper (*Phaulacridium vittatum*). *Molecular Ecology* 28(14):3395–3412.

Yadav, S., Stow, A., and Dudaniec, R. Y. (2020). Elevational partitioning in species distribution, abundance and body size of Australian alpine grasshoppers (*Kosciuscola*). *Austral Ecology*. https://doi.org/10.1111/aec.12876

Yetchom-Fondjo, J. A., Kekeunou S., Kenne, M., Missoup, A. D., and Xu, S. Q. (2020). Diversity, abundance and distribution of grasshopper species (Orthoptera: Acrididea) in three different types of vegetation with different levels of anthropogenic disturbances in the Littoral Region of Cameroon. *Journal of Insect Biodiversity* 014(1):016–033.

Zha, L., Yu, F., Boonmee, S., Eungwanichayapant, P. D., and Wen, T. (2017). The subfamily Cladonotinae (Orthoptera: Tetrigidae) from China with description of a new monotypic genus. *Journal of Natural History* 51(25–26):1479–1489.

Zhang, L., Lecoq, M., Latchininsky, A., and Hunter, D. (2019). Locust and grasshopper management. *Annual Review of Entomology* 64:15–34.

INDEX

Page numbers in italics indicate figures.

Abracris flavolineata, 208
Acalypha diversifolia, 223
Acanthacris: *A. ruficornis*, 178, 179; *A. ruficornis citrina*, 178
Acheta, 164; *A. domesticus*, 27; *A. pantescus*, 164, 165
Acinipe, 159
Acorypha clara, 179
Acrida bicolor, 177, 178, 184
Acrididae, 26, 27, 41, 53, 86, 98, 121, 129, 173, 176, 198, 205, 206, 209, 211
Acridinae, 49, 87, 208
Acridoderes strenuus, 184
Acridoidea, 183, 212
acridoid grasshoppers, 281
Acripeza reticulata, 113, 114
Acrostira: *A. bellamyi*, 249; *A. euphorbiae*, 48
Acrotylus: *A. longipes*, 46; *A. patruelis*, 31, 46, 47
Adimantus ornatissimus, 209
Adriatic marbled bush cricket, 253
Aemodogryllinae, 38, 139
Aeropedellus: *A. clavicornis*, 130; *A. variegatus*, 129, 130, 152
Aerotegmina, 190, 191; *A. kilimandjarica*, 190, 191, 191, 192, 193, 193; *A. megaloptera*, 193; *A. shengenae*, 193; *A. taitensis*, 193; *A. vociferator*, 193
African gaudy grasshoppers, 55
African Pneumoridae, 87
Afromastax zebra zebra, 182, 183, 184
Aganacris velutina, 59
Agnapha, 114
Aiolopus thalassinus, 257
Akamasacris, 284
Alectoria superba, 114
Alpine bush crickets, 156
Alpine dark bush cricket, 53
Alpine grasshoppers, 258
alpine ground wētā, 125
alpine scree wētā, 125, 126
Amblycoryphini, 59
Amedegnatiana, 158
Ammodramus savannarum, 51
Amorphopus, 41
Amphinotus nymphula, 251
Amusurgus caerulus, 140
Amytta: *A. kilimandjarica*, 193; *A. meruensis*, 193; *A. merumontana*, 193; *A. olindo*, 193
Anabrus simplex, 86
Anacridium melanorhodon, 80
Anadolua, 152; *A. schwarzi*, 151
Anchotatus-Anchocoema, 218

Anderus, 127; *A. brucei*, 128; *A. maculifrons*, 128
Angraecum, 66, 69; *A. cadetii*, 66, 67, 68, 69; *A. jeannineanum*, 67; *A. sesquipedale*, 66
Anisoura nicobarica, 127, 127
Anonconotus mercantouri, 156
Anoplolepis gracilipes, 60
Anoplophilinae, 38
Anostostomatidae, 26, 27, 88, 90, 95, 114, 124, 133, 202, 205
ant crickets, 27, 60
Anterastes, 152
Anthophiloptera dryas, 48
Apiales, 49
Apioscelis, 218
Apoboleus degener, 180, 181, 184
Arachnitus, 202
Arcyptera microptera, 154
Argiope bruennichi, 52
Arphia pseudonietana, 31
Arphia simplex, 31
Asclepiadaceae, 175
Asiophlugis temasek, 141, 143
Asteraceae, 209, 215
Asterales, 49
Astroma, 218
Atlantic beach cricket, 166–67
Atractomorpha acutipennis, 184
Auckland tree wētā, 125
Australian plague locust, 80, 80
Austroicetes pusilla, 46
Aztecacris laevis, 203

Bactrophorinae, 212
Baetica ustulata, 158
Balkan predatory bush cricket, 150
band-winged grasshoppers, 30, 55
Barbitistes: *B. constrictus*, 82; *B ocskayi*, 82; *B. serricauda*, 82; *B. vicetinus*, 82–83
Batrachideinae, 41, 42, 43, 144
bee-eaters, 52
bee flies, 52
beetles, 60
Bei-Bienko's plump bush cricket, 170–71
Betiscoides, 195
Betiscoides sp., 196; *B. muris*, 196
bladder grasshoppers, 87
Blatta acervorum, 61
Blattodea, 17, 19, 20
blister beetles, 52
bog bush cricket, 254, 255
Bolidorhynchus-Microcoema, 218

Bombay locust, 80
Bombyliidae, 52
Brachycaulopsis jovelensis, 202
Brachytrupes, 87; *B. megacephalus*, 161; *B. membrana-ceus*, 97; *B. membranaceus colosseus*, 281, *282*
Bradyporinae, 158
Breviphisis, *193*
Bryodema luctuosum, *29*, *31*
Bryodemella tuberculata, *28*, *31*, 263–64, *264*
Bryodemini, 132
Bryophyta, *49*
Bulgarian stone grasshopper, *150*
bulldog raspy cricket, 108
Bunkeya sp., *27*; *B. congoensis*, 184
Burmecaelidae, 22
bush crickets, 26, *27*, *58*, 59, 62, 65, 94, *156*
butterflies, 60
Bycanistes brevis, 190

Caelifera, 21, 22, 23, 26, *49*, 86, 88, 119, 180, 185, 218, 281
Calamacris, 205
Califera, 205
Calliptamus italicus, 80, 132
Calotropis procera, 175
camel crickets, *27*, 36, 40, 124, 137
Camelotettix curvinotus, *145*
Carabidae, 52
Carliola carinata, 268, *269*
Carphoproscopia, 218
Caryophyllales, *49*
Catantopinae, 116
Catantops stramineus, 184
cave crickets, 26, 36, 48
cave wētā, 124
Cebidae, 52
Celes variabilis, 154
Central American locust, 80
Cephalocoema sp., *219*
Ceuthophilinae, *38*
Championica montana, 56
Chapadamastax diamantina, 222
chapulines, 286
Chauliogryllacris acaropenates, 108
Chirista compta, 184
Chlorobalius leucoviridis, *110*, 114
Chlorophyta, *49*
Chloroplus cactocaetes, 48
Chorotypidae, *27*
Chorthippus, *31*, 102, 129, 130, 132, 159; *C. albomargina-tus*, *99*, 101, 103, *104*, *105*; *C. biguttulus*, 102, 106; *C. biroi*, 161; *C. brunneus*, 102; *C. corsicus*, 161; *C. dorsatus*, *99*, 101; *C. jacobsi*, 102; *C. karelini*, 103, *104*; *C. mollis*, 102, 106; *C. oreophilus*, *130*; *C. oschei*, 103, *104*, *105*; *C. pascuorum*, 161; *C. pullus*, 264
Chorthopodisma cobellii, 158
Chortoicetes terminifera, 46, 80, *80*
Chortophaga viridifasciata, *29*, *31*
Chromacris, 202, 212, 215; *C. speciosa*, 215

Chrotogonus senegalensis, 184, 185
Ciconia ciconia, 51
Circotettix: *C. carlinianus*, *29*, *31*; *C. latifasciata*, *29*; *C. undulatus*, *31*
Cladonofus, *41*
Cladonotinae, *41*, 43, *145*
Cladoramus, *41*
Clonopsis gallica, *19*
clown hoppers, 221
Cocconotus, 202
cockroaches, 20, *61*
Colemania, 185
Comicus, 135
common kestrel, 52
common predatory bush cricket, 60
Conocephalinae, 202
Conocephalus ebneri, 154
Conophyma, 130, *130*
Conophyminae, 130
Conozoa sulcifrons, *31*
Conzoa carinata, *31*
Cooloola, *110*, 114
Cooloola monsters, *110*, 114, 119
Cooloolidae, 119
Cophopodisma pyrenea, 158
Copiocerinae, 209
Copiphora, 202; *C. gorgonensis*, *92*
corn grasshopper, 283–84
Corynorhynchus, 220
Coryphosima stenoptera, 183, 184
Cota, *41*
Crau Plain grasshopper, *252*
crested tooth-grinder, 111
crickets, 94, 96
Crustacea, 166
crystal predatory katydid, *141*, 143
Cylindrachetidae, 26, *27*
cylindrachetids, 119
Cylindraustralia, 119; *C. kochii*, 119, 120, *120*; *C. tindalei*, 119, 120
Cylindroryctes spegazzini, 119
Cylindrotettix dorsalis, 206, *207*
Cyphocerastis, 183
Cyphoderis monstrosa, *92*
Cyphoderris, 90, 95, 97; *C. monstrosa*, 90, *91*
Cyrtacanthacridinae, 210

Daguerreacris tandiliae, 221
Decticus, 158; *D. albifrons*, 62
Deinacrida: *D. connectens*, 125, 126, *126*; *D. heteracantha*, *125*, 126
Deltonotus, *41*
Dericorys albidula, 175
Dericorythidae, 130
Dermaptera, *17*, *18*
desert locust, *73*, 74, *79*, 80, *80*, 173, 175
desert long-horned grasshoppers, *27*
Detritus, *49*

Devylderia, 195
Diacamma sp., 59
Diaphanogryllacris, 134, *134*
Dichopetala, 202
Dichroplus, 210; *D. maculipennis*, 210
Dictyophorus, 185; *D. spumans*, *187*
Dictyoptera, *17*
Diestramima, 137, *139*; *D. tsongkhapa*, 137;
    D. asynamora, *27*
Digentia, 183
Dinarippiger, 158
Dinocras cepalotes, 19
Dinotettix, *41*
Diotarus, *41*
Diponthus argentinus, 215
Diraneura, 209
Discotettigini, 144, *145*
Discotettix, *41*
Discotettix kirscheyi, *145*
Dociostaurus, 245; *D. maroccanus*, 47, 80, 132;
    D. minutus, 161
Dolichopodainae, *38*
Dolichopodinae, 36

Ecphantus quadrilobus, 108, *109*, 111
Elcanidae, 21, 22, 23
Embia thyrrenica, *18*
Embioptera, *17*, *18*, 20
Eneopterinae, 86
Ensifera, 21, 22, 23, 26, 48, *49*, *58*, 86, 88, *88*, 90, *92*, 94,
    95, 119, 124, 281
Entomophaga grylli, 53
Ephippiger, 158; *E. camillae*, 158; *E. carlottae*, 158;
    E. ruffoi, 158
Epicauta, 52
Epigrypa, 218
Epipodisma pedemontana, 158
Episactidae, 221
Epsigrypa, 218
Eremogryllinae, 175
Eremogryllus hammadae, 175, *176*
Euchorthippus sardous, 161
Eugaster, *156*, 158
Eugryllacris, 134
Eukinolabia, *17*
Eumastacidae, *27*, 221
Eumastacinae, 222
Eumastacoidea, 132, 183
Eumastax, 223
Eunapiodes, 159
Eupholidoptera, 152, 161; *E. francesia*, *162*
Euphorbiaceae, 223
European Gomphocerinae, 28
European hop hornbeam, 82
Euryparyphes, 159
Euschmidtia congana, 184
Eyprepocnemis plorans, 184

Fabales, *49*
Falco tinnunculus, 52
field cricket, 250
field grasshoppers, *27*, 88
Fijitettigini, *41*
Fijitettix, *41*
flesh flies, 52
flightless grasshopper, 267, *268*, *269*
forbhoppers, *27*
Forficula apennina, *18*
Fraxinus ornus, 82

Gammarotettiginae, *38*
Gastrimargus africanus, 178, 184
gaudy grasshoppers, *27*
Geomantis larvoides, *18*
giant hooded katydid, 108, 274, *275*
giant wētā, *125*, 126
gladiators, 16
Glomeremus: *G. orchidophilus*, 68, 69;
    G. orchiophilus, 66, 67; *G. paraorchicophilus*, 67
Glyphanus obtusus, 278, *279*
Gomera stick grasshopper, *249*
Gomphocerinae, 48, *49*, 86, 87, 98, *99*, 101, 129, 130,
    205, 209
Gomphocerippus rufus, *99*, 101
Gomphocerus sibiricus, *51*, 98, *99*, 129
Gomphomastax, 132
Gondwanan wasp katydids, *110*, 114
Goniaea spp., 111
grasshoppers, 48, 73
grasshopper sparrow, 51
green bush cricket, 54
greenhouse camel cricket, 36, *38*
green mountain grasshopper, *51*
green tree ant, 114
green-winged grasshopper, *257*
grigs, 94
ground beetles, 52
ground wētā, 127–28
Gryllacrididae, 26, *27*, 66, 69, 93, 95, 205
Gryllacris, 134
Gryllacropsis, 133; *G. magniceps*, 133
Gryllacrydidae, *92*
Gryllidae, *27*, 86, 87, *89*, 90, *92*, 93, *97*, 205
Gryllidea, *89*, *97*
Gryllinae, 87
Grylloblatta, *18*
Grylloblattodea, *17*, *18*, 20
Grylloidea, *49*, 86, *92*, 93, 94, 95
Gryllotalpa sp., *88*, *97*
Gryllotalpidae, *27*, 48, 87, 90, *97*, 119, 205, 281
Gryllotalpoidea, 94, 97
Gryllus: *G. bimaculatus*, *92*; *G. campestris*, 250
Guyanese cricket, 87
Gymnidium, 195

Haglidae, 21
Hagloidea, 88, *92*, 94
halgania grasshoppers, 111
Halmenus, 210
Hawaiian crickets, 32, 35
hawk moth, 66
Heideina, 125
Heliastus, 209; *H. subroseus*, *31*
Helicomastax, 223
Helicopacris modesta, 212, *213*
Hemiandrus, 127; *H. bilobatus*, *128*; *H. focalis*, 125;
    H. jacinda, 127–28, *128*; *H. pallitarsus*, 128
Hemideina: *H. crassidens*, *124*, 125; *H. maori*, 125, *126*;
    H. thoracica, 125
Hemierianthus, 183
Heteromallus spina, 36, 38, *39*
Heteropternis thoracica, 184
Hetrodinae, 158
Hexacentrus unicolor/japonicus, *193*
Hintzia, 183
Histrioacrida roseipennis, 111
Holoarcus sp., *145*
Holochlora biloba, 89
Holopercna, 183; *H. gerstaeckeri*, 184
Homeomastax dereixi, *222*
Homoeogryllus reticulatus, *89*
horse-headed grasshoppers, 218
hoverflies, 60
Hyalopterix rufipennis, 209
Hyalopteryx rufipennis, *208*
Hybusa, 218
Hybusinae, 218
Hymenotes, *41*

Ichthiacris, 205
Ichthyotettix, 205
Indian wētā, 133
invertebrates, *49*
Isophya, 152, 154; *I. beybienkoi*, 170, *171*; *I. gulae*, 154;
    I. rectipennis, 152
Isoplectron: *I. armatum*, 40; *I. parallum*, 36, *37*
Isoptera, 20
Italian locust, 80, 132
Italohippus, 159
Italopodisma, *157*, 158, *258*

Jerusalem cricket, *27*

Karruia sp., *196*
katydids, *58*, *59*, 62, 94, 95
Kilimanjaro balloon bush cricket, *192*, 193
Kisella, 159
Kosciuscola, 116, 118; *K. tristis*, 116, *117*, 118;
    K. usitatus, 118
Kraussaria angulifera, 178

Lactista, 209; *L. azteca*, *29*; *L. aztecus*, *31*
Lamiales, *49*

Laniidae, 52
Lanius collurio, 54
Laupala, 32, 35; *L. cerasina*, *33*; *L. kona*, *33*; *L. pruna*, *34*
Lebinthus luae, *141*, 143
Leichhardt's grasshopper, 114
Lentulid, *195*
Lentulidae, *27*, 196
Leptophyes, 158; *L. axeli*, *162*
Leptysminae, 206
Lerneca fuscipennis, 87
Lichenomorphus, 202
Liladownsia, *204*
Listroscelidinae, 202
Lithidiopsis sp., *197*
Locusta: *L. migratoria*, 74, *75*, 76, 80, *80*, 132, 280, 281;
    L. migratoria migratoria, 47
Locustana pardalina, 80
Locustopsidae, 21, 23
locusts, 48, 72, 73
long-horned grasshoppers, 62
Longiphisis, *193*
long-legged sandhopper, 108
Lophotettiginae, 43
Loveridgacris, 185
lubber grasshoppers, *27*, 212, 215

Macropathinae, 36, *38*
Maeacris aptera, 209
maize cricket, 135
Malenamastax, 222
Malvaceae, 223
sword-tailed cricket, *141*
mantises, 20
Mantodea, *17*, *18*, 20
Mantophasmatodea, 16, *17*, *18*
Maotoweta virescene, 38, *39*, 40
Marellia remipes, 206
Marelliinae, 206
Maripa, 222
Markia, 202
Marsabitacris citronota, 188
Masynteinae, 221
Masyntes, 221
matchstick grasshopper, 108, *109*
Mato Grosso locust, 80
Maura rubroornata, 185, *186*
Meconema, 48, 53; *M. meridionale*, *52*
Meconematinae, 143
Megalopyrga monochroma, 188
Melanoplinae, 23, *49*, 198, 205, 209
Melanopline grasshoppers, 209
Melanoplini, *204*
Melanoplus, 198, 200, 205, 284, 286; *M. bivitattus*, 198;
    M. deceptus, *201*; *M. femurrubrum*, 198;
    M. frigidus, 152, 159; *M. indigens*, *201*;
    M. magdalenae, *201*; *M. sanguinipes*, 198;
    M. spretus, 73, 265–66
Melastomataceae, 223

Meloidae, 52
Mercantour Alpine bush cricket, *156*
Meropidae, 52
Mesasippus, 132; *M. ammophilus*, 130;
    M. kozhevnikovi iliensis, 130, *131*
Metarhizium acridum, 73
Metrioptera, 158; *M. brachyptera*, *254*, 255, 256, *256*
Metrodorinae, *41*, 43, 44
Microcentrum spp., 286
Micropodisma salamandra, 158
migratory locust, 74, *75*, 76, *76*, 80, *80*, 81, 132
Mimetica, 58; *M. incisa*, 58
Minutophasma richtersveldense, *18*
Minyacris nana, 108
Mioscirtus wagneri, 29, *29*, 30, *31*
Miotopus richardsae, 36, *37*
Miramella, 159; *M. alpina*, *51*
Mogoplistidae, *27*, 205
mole crickets, *27*, 48, 87, 94, 95, 96, 119, 281
monkey grasshoppers, 132
monkey hoppers, 221
Montana medvedevi, 154
Morabidae, *27*
Morabinae, 108
Morgenia, 180
Moritala sp., 108, *109*
Mormon cricket, 86
Moroccan locust, 80, 132
Morphacris fasciata, *31*, 183, 184
Morseinae, 221
Motuweta: *M. isolata*, 126; *M. riparia*, 126
mountain katydid, *113*, 114
Mount Cook flea, 40
mud crickets, 26
Mylabris, 52
Myrmecophilidae, *27*, 60, 205
Myrmecophilinae, 60
Myrmecophilus, *27*, 60; *M. acervorum*, 60, *61*;
    M. albicinctus, 60; *M. quadrispinus*, 60

Nadigella, 159
Neobarretia, 202; *N. spinosa*, *203*
Neoconocephalus spp., 286; *N. triops*, *27*
Neonetus n. sp1, 36, *37*
Neonetus n. sp2, 38, *39*
Neonetus n. sp3, 38, *39*
Neonetus variegatus, 38, *39*
Neoscapteriscus borelli, *27*
Neotropical monkey grasshoppers, 132
Nepheliphila raptor, *193*
Nesonotus vulneratus, 88, *89*
Nesotettix, *41*
Nesotettix cheesmanae, 42
Netrosoma, 205
New Zealand alpine grasshoppers, 121
Nilgiri tahr, 267
Nilgiritragus hylocrius, 267
Nocaracris, 152; *N. bulgaricus*, *150*

Nodutus, 220
Nomadacris septemfasciata, 80, *80*, 280
northernmost monkey grasshopper, *132*
Notoplectron campbellense, 36

oak bush cricket, *52*, 53
Oaxaca, 205
Occidentosphena uvarovi, 185
Ochrilidia nuragica, 161, *163*
Odontopodisma schmidtii, 158
Oecanthidae, 95, 202, 205
Oecophylla smaragdina, 114
Oedaleus: *O. nigeriensis*, 178; *O. senegalensis*, 178, *179*
Oedipoda: *O. aurea*, 29, *31*; *O. caerulescens*, 46
Oedipodinae, 23, 46, 49, 87, 129, 205
Oedischiidae, 21
Ommatolampidinae, 206
Omocestus minutus, 99, 101
Onconotus servillei, 151, 154
Opaonella tenuis, 206
Ormia ochracea, 53
ornate bright bush cricket, *53*
Ornebius: *O. lupus*, 140; *O. tampines*, 140
Ornithacris cavroisi, 178, 179
Oropodisma, 152, 258, 260, 261–62; *O. chelmosi*, *261*,
    262; *O. erymanthosi*, *261*, 262; *O. kyllinii*, *261*, 262;
    O. taygetosi, *261*, 262
Orthacridinae, 205
Orthoptera, 16, *17*, 21, 22, 23, 24, 26, 40, *43*, 48, 50, 52,
    53, 66, 86, 93, 108, 115, 124, 159, 175, 205
Ostrya carpinifolia, 82
Oxya hyla, 180, *181*, 184, 281
Oxycatantops congoensis, 184
Oxyinae, 116

Pachyrhamma, 40; *P. edwardsii*, 38, *39*; *P. longipes*, 38, *39*
painted grasshopper, 189
Palmenhaus cricket, 168–69
Pamphagidae, 152, 159, *174*, *252*
Pamphaginae, 159
Pamphagus, *157*, 159; *P. sardeus*, *157*
Paracinema tricolora, 281
Paracinipe, 159
Paraeumigus, 159
Paragryllus sp., *88*
Paramastacides ramachendrai, 267
Paramastacinae, 221
Paramastax, 221; *P. rosenbergi*, *223*
Paramphibotettix sanguinolentus, *145*
Paranocarodes straubei, 154
Parapetasia, 183, 185; *P. femorata*, 183, *187*; *P. rammei*, 183
Paraphymateus roffeyi, 188
Paraselina, *41*
Paratettix, *41*, 42
Paratettix sp., *145*; *P. nigrescens*, 42
Parepisactinae, 221, 222
Parepisactus, 222; *P. norcentralis*, *225*
Parnassiana, 152

Parorthacris somalica, 188
Parudenus falklandicus, 36
Parvotettis domesticus, 36
Patanga succincta, 80
Paulinia acuminata, 206, *206*
Pauliniinae, 206
peacock katydid, 57
Pedies, 205
Pentacentrus sp., *89*
Pepsis, 59
Periplaneta americana, 19
Peripodisma, 152, 158, 258
Perixerus, 205
Peruvian locust, 80
Petasida ephippigera, 114
Phaedrotettix, 205
Phalangopsidae, *88, 89*, 205
Phaneroptera brevis, 142, 143
Phaneropteridae, *88, 89*
Phaneropterinae, *49*, 202
Pharmacus montanus, 40
Phasmatodea, *17*, 20, 55
Phasmida, *19*, 20
Phasmodes, 108
Phasmodinae, 108
Phaulacridium, 116, 118; *P. vittatum*, 114, *115*, 116
Phlugidia, 169
Phlugiola, 169; *P. dahlemica*, 168–69; *P. redtenbacheri*, 169
Pholidoptera, 152; *P. aptera, 53*; *P. brevipes*, 154
Pholidopterini, 152
Phonochorion, 154; *P. uvarovi*, 152, *153*
Phricta spinosa, 108, 111
Phryganistria, 20
Phrynotettix, 212
Phylotettix rhombeus, 42
Phymateus, 185; *P. saxosus, 282*; *P. viridipes*, 185
Physemophorus sokotranus, 188
Phyteumas, 185
Phytomastax, 132, *132*
Piscacris, 205
plague locusts, 46
Platycleis, 158; *P. concii*, 161; *P. kibris*, 161; *P. monticola*, 161; *P. ragusai*, 161
Plecoptera, 16, *17, 19*
Pleioplectron hudsoni, 36, *37*
Pleioplectron simplex, 36, *37*, 40
Pneumoridae, 26
Poales, 48, *49*
Podisma, 158, 258; *P. amedegnatoae*, 158; *P. cantabricae*, 158; *P. carpetana*, 158, 260; *P. carpetana carpetana, 259*; *P. dechambrei*, 158; *P. emiliae*, 158; *P. goidanichi*, 158; *P. magdalenae*, 158; *P. pedestris*, 158; *P. ruffoi*, 158
Podismini, 258
Poecilimon, 152, 154, 158; *P. ampliatus*, 152; *P. bosphoricus*, 154; *P. gerlindae, 63*; *P. gracilis, 63, 64*; *P. intermedius*, 154; *P. ornatus, 53*
Poecilocloeus, 209, *211*

Poekilocerus, 185; *P. buronius hieroglyphicus*, 175
Polyneoptera, 16, 20
possums, 126
Potua, 41
praying mantises, 20
prickly gorse, 126
Prionolopha serrata, 212
Prionotropis rhodanica, 252
Proctolabinae, 209
Prolaupala, 32
Prophalangopsidae, 90, 93, 94, 95, 97
Prosarthria, 220
Proscopia, 218
Proscopiidae, 52
Proscopiinae, 218
Proscopiini, 218, 220
Proscopildae, *27*
Prosphena, 205
Proteaceae, *195*
Pseudoamigus, 159
Pseudomastacinae, 221, 222
Pseudomastax, 221, 222; *P. personata, 224*
Pseudomogoplistes, 166; *P. vicentae*, 166
Pseudopodisma fieberi, 158
Pseudoproscopia, 220
Pseudoprumna baldensis, 158
Psiloscirtus bolivianus, 208
Psophus stridulus, 31, 264
Psorodonotus, 152, 154
Pterochroza ocellata, 57
Pteropera, 183
Pterotiltus, 183, 184; *P. hollisi*, 184
pygmy grasshoppers, *27*, 40, 41, 42, 144, *145*, 146
pygmy mole crickets, 26, *27*, 205
Pyrgacridae, *27*
Pyrgomorpha: *P. cognata*, 185; *P. vignaudii*, 177–78, 185
pyrgomorphid, *282*
Pyrgomorphidae, *27*, 185, 202, 205
Pyrgomorphinae, 205
Pyrgomorphoidea, 183
Pyrgomorphula serbica, 152, *153*, 154
Pyrgotettix, 205

Quiva sp., *97*

Ramburiella turcomana, 278, *279*
Rammepodisma, 152
Raniliella testudo, 111
Ranunculales, *49*
raspy crickets, 66, 67, 143
rats, 126
rattle grasshopper, 264
Rattus spp., 126
red-backed shrike, *54*
red locust, 80, *80*, 81
red velvet mites, 52
Restionaceae, *195*
Rhacocleis maculipedes, 161

316

Rhammatocerus schistocercoides, 76, 80, 209, *209*
Rhaphidophora, 137, *139*; *R. angulata*, 137; *R. taiwana*, *139*
Rhaphidophoridae, 26, *27*, 36, 40, 48, 95, 124, 137, 205
Rhaphidophorinae, *38*, 137, *139*
Rhopalosomatidae, 52
Rhynchotettix, *41*
Rhytidochrotinae, 206
Richnoderma, 212
Ripipterygidae, 22, 26, *27*, 119
Ripipteryx tricolor, *204*
rock/ice-crawlers, 20
Rocky Mountain locust, 24, 73, 265–66
Roeseliana roeselii, *255*, 256, *256*
Roesel's bush cricket, *255*
Romalea: *R. eques*, 215; *R. microptera*, 215; *R. microtera*, *27*
Romaleidae, *27*, 202, 212
Romaleinae, 215
Rosaceae, 223
Rosales, *49*
rove beetles, 52
Royitettix, *41*
Rubus spp., 223
Russalpiina, 121

Saga: *S. campbelli*, 152, 278; *S. ephippigera*, *150*, 152;
        S. hellenica, 278; *S. natoliae*, 152, 278, *279*;
        S. pedo, 60, 152, 154
Saginae, 152
sandgropers, 26, 119, 120
sand hoppers, 166
Sarahan grasshoppers, 175
Sarcophagidae, 52–53
Sardoplatycleis galvagnii, 158, 161
Sathrophylliopsis longepilosa, 56
Scara, *41*
Scelimena celebica, *233*
Scelimeninae, *41*, 43, *145*
Scelimenini, 144, *145*
Schistocerca sp., 47, 284, 286, *287*; *S. americana*, *27*;
        S. cancellata, 80, *80*, 210; *S. gregaria*, 74, 80, *80*, 173,
        173; *S. interritta*, 80; *S. piceifrons*, 76, 80
Schizodactylidae, 26, *27*, 93, 95
schizodactylids, 135
Schizodactyloidea, 135
Schizodactylus, 135; *S. hesperus*, 135; *S. inexpectatus*, 135;
        S. minor, 135; *S. monstrosus*, 135, *136*;
        S. sindhenesis, 135
Scintharista notabilis, *31*
Senegalese grasshoppers, 178
Serbian stick grasshopper, 152, *153*
Serpusia, 183; *S. opacula*, 184
short-backed saddle bush cricket, *161*
short-horned grasshoppers, 26, 53, 121
shrikes, 52
Siberian grasshopper, *51*
Sida spp., 223
Sigaus, 121; *S. australis*, 122, *122*, 123; *S. campestris*, 123;
        S. childi, 122; *S. minutus*, *121*, 122;

S. nitidus, 122, *122*, 123; *S. nivalis*, 122, *122*, 123;
        S. piliferus, *122*, 123; *S. robustus*, 122; *S. villosus*, *121*,
        122, 123
Siliquofera grandis, 108, 274, 275
silverfishes, 60
Singapuriola separata, *140*
sky-island grasshoppers, *201*
Solanaceae, 209, 212
South American locust, 77, 80, *80*
southern barbed-wire bush cricket, *151*
Spathosternum pygmaeum, 184
speckled buzzing grasshopper, 28, 263–64
speckled Sardinian bush cricket, *161*
spectacular crested katydid, 114
Sphecidae, 52–53
sphecid wasps, 52–53
Sphenacris, 205
Sphenarium, 205, *285*, 286; *S. histrio*, 284;
        S. mexicanum, 284; *S. purpurascens*, 283, *284*
Sphenotettix, 205
Sphingidae, 66
Sphingonotini, 132
Sphingonotus, 209; *S. nebulosus*, *29*, *31*; *S. pilosus*, *29*, *31*
spine-kneed grasshopper, 180, *181*
spiny rain-forest katydid, 108
splay-footed crickets, 26, 135, *136*
spotted predatory katydid, *110*, 114
Staleochlora ronderosi, 212, *216*
Staphyliniidae, 52
Stenobothrus, 159; *S. clavatus*, 102, 103, *104*, 106;
        S. cotticus, 152; *S. eurasius*, *87*; *S. fischeri*, *100*, 101;
        S. rubicundus, 102, 103, *104*, 106
Stenocrobylus festivus, *182*, 183, 184
Stenopelmatidae, 26, *27*, 93, 202, 205
Stenopelmatus, *203*; *S. piceiventris*, *27*
Stenopola puncticeps, 206, *207*
stick grasshoppers, 52, 218, 220
stick insects, 20
Stilpnochlora, 202
stinging ant, *59*
Stiphra, 220
stone grasshoppers, *157*, 278, *279*
stone wētā, 125
Supersonus aequoreus, 91
Svistella chekjawa, 140, *141*
swordtail crickets, 32
Systella, 59
Systolederus, *41*

Tachinidae, 52
tachinid flies, 52
Tachycines, 137, *139*
Tachycines asynamorus, 36, *38*, *139*
Taeniopoda, 286
Talitridae, 166
Talitropsis: *T. sedelloti*, 40; *T. sedilotti*, 36, 37
Tanaoceridae, *27*, 202
Tanaocerus: *T. koebelei*, *27*; *T. rugosus*, 205

Taphronota, 185; *T. calliparea*, 183, 184, 185, *187*
Temnomastacinae, 221, 222
Temnomastacini, 221
Terminalia, 133
termites, 20
Tessellana lagrecai, 161
Tetanorhynchini, 218
Tetanorhynchus, 220; *T.* cf. *carbonelli*, *220*
Tetrigidae, *27*, 40, 41, 42, *43*, 44, 144, 146, *251*
Tetriginae, *41*, *43*, *49*, *145*
Tetrigoidea, 22, 26, 183
Tetrix, 42; *T. japonica*, 146; *T. subulata*, *27*; *T. tuerki*, 264
Tettigonia, 158; *T. cantans*, 158; *T. hispanica*, 158;
    T. longispina, 158; *T. silana*, 158; *T. viridissima*, *54*
Tettigoniidae, 26, *27*, 82, 90, *92*, 93, 133, 202, 205
Tettigoniidea, *88*, *89*, *97*
Tettigoniinae, *49*
Tettigonioidea, *94*, *97*
Tettigonoidea, *92*
Thericleidae, *27*
tiny grasshopper, 108
Titanacris, 202, 212
Titanoptera, 21, *24*
Trachyrhachis kiowa, 29, 30
Trachyrhachys kiowa, *29*, *31*
tree locust, 80
tree wētā, 124–26
Trichosurus vulpecula, 126
Tridactylidae, 26, *27*, 48, 119, *204*
Tridactyloidea, 22, 119, 183
Tridacytlidae, 205
Trigonidiidae, 205
Trigonidiinae, 32
Trigonidium, 32
Trigonidlidae, *27*
Trilophidia: *T. annulata*, *29*, *31*; *T. conturbata*, 183
Trimerotopis latifasciata, *31*
Trimerotropis, 209; *T. cyaneipennis*, *31*; *T. sparsa*, *29*
Tripetalocera, *41*
Tripetalocerina, *41*
Tripetalocerinae, 43
Tristiridae, *27*
Troglophilinae, 36, *38*
Trombidiidae, 52
Tropidacris, 202, 212, 215; *T. collaris*, 212, *214*, 215;
    T. cristata, 212
Tropidischia xanthostoma, *38*
Tropidischiinae, *38*
true crickets, *27*
Truncotettix, *41*
Truxalis sp., *174*
Trypophyllum, *41*
Tuarega insignis, *174*
tusked wētā, 126–27, *127*
Typophyllum spurioculis, 91

Ulex europaeus, 126
Ulmus, 82
Urnisiella rubropunctata, 108
Uromenus, 158; *U. brevicollis*, *161*

Velarifictorus, 272
Veria colorata, *110*, 114
Vilerna rugulosa, 208
Vittisphena somalica, 188

Warramaba ngadju, 108
Warramaba virgo, 108
wartbiter, 62
wasp spider, 52
web-spinners, 20
Wellington tree wētā, 125
wētā, 124, *124*
white stork, 51–52
wingless grasshopper, *115*, 116

Xanthopan: *X. morgani praedicta*, 66; *X. praedicta*, 66
Xenephias socotranus, 188
Xeniinae, 218
Xenonomia, 17
Xerophyllini, *41*
Xiphipyrgus tunstalli, 188
Xistrella, *41*
Xya japonica, *27*
Xyronotidae, 202
Xyronotus, 205; *X. aztecus*, *204*, 205; *X. cohni*, 205;
    X. hubbelli, 205

Zaprochilinae, 108
Zeuneriana: *Z. amplipennis*, 154; *Z. marmorata*, *253*
Zoniopoda tarsata, 215, *217*
Zonocerus, 185; *Z. elegans*, 185, *187*; *Z. variegatus*, *27*,
    184, 185, 188, 189, *189*
Zoraptera, 16, *17*, *18*
Zorotypus asymmetricus, *18*
Zubowski's grasshopper, 87
Zygophlaeoba sinuatocollis, 267

Published by Princeton University Press in 2026
41 William Street, Princeton, NJ 08540, USA
99 Banbury Road, Oxford, OX2 6JX, UK
press.princeton.edu

GPSR Authorized Representative: Easy Access System Europe – Mustamäe tee 50, 10621 Tallinn, Estonia, gpsr.requests@easproject.com

ISBN 978-0-691-28105-6
Ebook ISBN: 978-0-691-28106-3
Library of Congress Control Number: 2025937667
British Library Cataloging-in-Publication Data is available

Graphic design by pooldesign, Zurich
Cover image: blickwinkel/G. Fischer
Cover design: Wanda España
Cover images: *(Front)*: The spiny devil katydid *Panacanthus cuspidatus* is a truly bizarre looking representative of the order Orthoptera. blickwinkel / G. Fischer. *(Spine)*: *Phytomastax* cf. *artemisiana* (Eumastacidae) from Bartogay, Kazakhstan. Photo: N. Sevastianov. *(Back cover)*: *(Left)*: *Pseudophyllus hercules* (Tettigoniidae) is a large leaf-mimicking katydid from Borneo. Photo: Chien. C. Lee. *(Middle)*: Gregarious nymphs of *Tropidacris collaris* (Romaleidae), Formosa, Argentina. Photo: Martina Pocco. *(Right)*: *Holoarcus truncatus* (Tetrigidae) is a deadleaf mimicking pygmy hopper from New Guinea. Photo: Chien. C. Lee.

10 9 8 7 6 5 4 3 2 1

Printed in the Czech Republic